PLANT MEMBRANES

CELL BIOLOGY: A SERIES OF MONOGRAPHS

Series Editor, E. Edward Bittar

Volume 1: PHOTOSYNTHESIS
Christine H. Foyer

Volume 2: CELL AND TISSUE REGENERATION
Margery G. Ord and Lloyd A. Stocken

Volume 3: PLANT MEMBRANES: ENDO- AND PLASMA MEMBRANES OF PLANT CELLS
David G. Robinson

PLANT MEMBRANES

Endo- and Plasma Membranes of Plant Cells

DAVID G. ROBINSON
Göttingen University
Göttingen, West Germany

A WILEY-INTERSCIENCE PUBLICATION
JOHN WILEY & SONS
New York • Chichester • Brisbane • Toronto • Singapore

Library of Congress Cataloging in Publication Data:

Robinson, David G.
Plant membranes.

(Cell biology; v. 3)
"A Wiley-Interscience publication."
Includes bibliographies and index.
1. Plant cell membranes. 2. Plant cells and tissues.
I. Title. II. Series: Cell biology (New York, N.Y.); v. 3.

QK725.R726 1984 581.87′5 84-7539
ISBN 0-471-86210-X

Printed in the United States of America

10 9 8 7 6 5 4 3 2 1

To
Gisela,
Eva,
Alexander,
and Andreas

Mikroskope und Fernrohre verwirren
nur den reinen Menschensinn.

J. W. GOETHE

SERIES PREFACE

The aim of the Cell Biology Series is to focus attention upon basic problems and show that cell biology as a discipline is gradually maturing. In its largest aim, each monograph seeks to be readable and informative, scholarly, and the work of a single mind. In general, the topics chosen deal with major contemporary issues. Together they represent a rather large domain whose importance has grown enormously in the course of the last generation. The introduction of new techniques has no doubt ushered in a small revolution in cell biology. However, we still know very little about the cell as an ordered structure. As will become abundantly clear to the reader, real progress is not just a matter of progress of technique but also a matter of close interaction between advances in different fields of study, as well as genesis of new approaches and generalized concepts.

E. EDWARD BITTAR

Madison, Wisconsin
January 1984

PREFACE

Cytologists have to accept a lot on faith, and their belief is strengthened when different methods lead to the same, or more or less the same, result. They must also be adept in translating two-dimensional images into three-dimensional structures. Moreover, the static micrographs which they obtain should be interpreted in terms of the dynamic events occurring in the cell at the moment of fixation. This is by no means easy and can often result in overinterpretation, but this may often be useful in the long run by focusing attention on the problem and provoking or stimulating research. One interprets one's data in the light of results already published, and, for plant cytologists who are working on endo- or plasma membranes, this means being aware of the developments in this area in animal cells in addition to keeping abreast with the situation in plants. It is with this necessity in mind that I have written this monograph. In so doing I have attempted, whenever possible, to present results from the two systems side by side rather than give a "biological synthesis." Instead, I have tried a synthesis of another sort, namely, that of structure and function. A dual approach of this type is quite prevalent among animal cell biologists, but there are few plant cell biologists who feel equally at home with the techniques of ultrastructure research as they do with biochemical methods. I hope their numbers grow and, at the risk of being supercilious, I hope too that this monograph might contribute in a small way toward this.

Both the title and the subject matter of this monograph are selective. The reader will search in vain for a section pertaining to plastids or mitochondria. There are two reasons for this: one is the relative lack of experience of the author with these two organelles and, second, they are of such importance that a monograph could easily be devoted to each. As far as the content of this monograph is concerned the accent has been on higher plants. Work on algae has only been included where examples serve to illustrate a point in a particularly vivid manner. This has been done, not because algae are unimportant but because there is such a variation in their cell structure and, when compared to higher plants, there has been relatively little work done on organelle isolation from them. Hence, it is much easier to present a unified structure–function approach when one restricts oneself to the higher plant.

The preparation of this monograph would have been impossible without the help of numerous colleagues throughout the world. Some have sent micrographs, some preprints so that the finished work might be well illustrated as well as up to date. Others (in particular Professors Hoch, Matile, and Mühlethaler) have cast a critical eye over certain chapters, suggesting deletions and providing improvements. I owe them all a debt of gratitude. Last but not least a special word of recognition and thanks is due my technician, Mrs. Heike Freundt, whose help in a secretarial capacity has been of immeasurable value.

DAVID G. ROBINSON

Göttingen, West Germany
November 1984

CONTENTS

PART ONE: The Structure and Isolation of Plant Cell Membranes

PART THREE: Biogenesis and Turnover of Plant Cell Membranes

Part 1

The Structure and Isolation of Plant Cell Membranes

1

INTRODUCTION

Every cytologist knows that the electron micrographs which he or she obtains are nothing more than artifacts. They are essentially complicated density maps of heavy-metal concentration which, if one is lucky, might faithfully represent the contours of organelles. From *in vitro* tests we know which substances or groups react with heavy metals (see, e.g., Hayat, 1970) but the situation changes when we go to the cellular environment. Perhaps the best example for this is the origin of the black–white–black staining profile for a biomembrane (see Robertson, 1981). Glutaraldehyde cross-links proteins and osmium tetroxide reacts heavily with fats, particularly of the unsaturated kind. These "facts" are found in every textbook on electron microscopy, but they are insufficient to explain the final image seen in a thin section in the electron microscope. Apparently, the extreme hydropholic environment of the lipid bilayer of a biomembrane causes a displacement of the somewhat polar primary reaction product OsO_3 which, instead of being deposited at the reaction site (i.e., the hydrocarbon chains of the lipids) becomes localized in the hydrophilic regions of the bilayer (i.e., at the head groups) (Stoeckenius, 1962). OsO_4 not only stains membranes in an unexpected manner but also induces extreme morphological changes in bacteria. Thus, mesosomes, which are present as extensive invaginations of the plasma membrane (PM) in sections from osmium-fixed material, are not seen in cryosections prepared from unfixed specimens (see Dubochet et al., 1983). This latter technique is in its infancy and there could be many surprises in store for us, but the similarity in results from freeze-fracturing and $KMnO_4$ fixations make me feel somewhat confident in the belief that that which we now recognize as a cell organelle will remain so.

From the structural viewpoint plant cells are almost as easy to handle as animal cells. Fixation, embedding, and sectioning schedules are often similar, the only difference being in the osmolarity of the fixative employed. In plants vacuoles and cell walls are present which lead to the development of turgor, a characteristic not seen in animal cells. Well-fixed plant tissue reflects this turgescent state showing a PM pressing against the cell wall and revealing intact vacuolar or tonoplast (TP) membranes. The "watery nature" of plant cells, however, makes their preparation for freeze-fracture electron microscopy much more of a problem than for animal cells. Only by prefixing and impregnating with a cryoprotectant such as glycerol, or better still with the help of the recently developed spray-freezing techniques, can one guarantee good replicas of large, "water-rich" cells which are free from ice crystal damage. Unfortunately, since higher plant cells, with the exception of gametes, are always present as a tissue, the spray-freeze techniques which require a cell suspension are less applicable. Conventional freezing techniques—that is, plunging specimens directly into liquid Freon without cryoprotection—usually can give good results for the PM, but the quality of the internal or endomembranes in replicas of higher plant material leaves much to be desired. In contrast, unicellular algae, which are less vacuolate or may even possess a contractile vacuole, have a much greater cytoplasm–water ratio and are therefore much more suited for freeze-fracture electron microscopy. Thus, since this text deals mainly with the structure and function of the endomembranes of higher plant cells, there are very few freeze-fracture micrographs included and those which are present are almost exclusively from algae.

The presence of a vacuole also makes life unpleasant for the cell biologist wishing to isolate organelles from plant tissues. When a cell wall is present, which necessitates greater shear forces in homogenization, the breakage of the vacuole is inevitable and this creates problems that do not exist with animal cells. For a start the homogenate is much more dilute and then there is the presence of a plethora of harmful substances which were contained in the vacuole. Dilution can be overcome by ultrafiltration, but there is no patent recipe for combatting the degradative effect of the vacuolar contents. Each plant cell or tissue must be considered anew and the presence of phenolics and lipolytic and proteolytic enzymes determined and corrected for (Anderson, 1968; Galliard, 1974; Loomis, 1974). The preparation of protoplasts followed by their gentle rupture (see Chapter 6.), which avoids vacuolar damage, is clearly a method of choice, but this has not yet been extensively used and carries with it the implicit assumption that the organelles in a protoplast are functionally the same as they were before plasmolyzing the cells and removing their walls enzymatically. Once one has obtained a homogenate with reasonably intact, undegraded membranous organelles from plant cells, the methods employed to separate the various organelles from one another are the same as those for animal tissues. Separations based on sedimentation coefficients or density in sucrose (see Figure 1.1) or other gradient media (Price, 1974), separations based on surface properties (e.g., phase partition) (Albertson, 1974), and free-flow electrophoresis (Hannig and Heidrich, 1974) are available and, with the exception of the

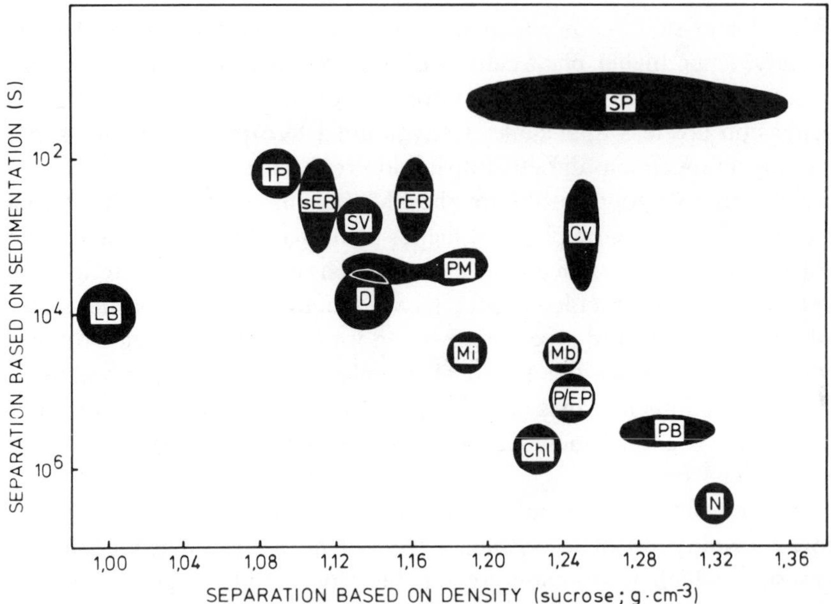

Figure 1.1. Diagram for the major organelles occurring in plant cells. The abcissa is the equilibrium density in sucrose and the ordinate is the sedimentation coefficient expressed logarithmically. Chl—chloroplast; CV—coated vesicle; D—dictyosome; LB—lipid body; Mb—Microbody; Mi—mitochondrion; N—nucleus; PB—protein body; P/EP—pro/etioplast; PM—plasma membrane (dumbell-shaped distribution due to the differing equilibrium densities of this membrane in mono- and dicotyledenous plants); rER—rough ER; sER—smooth ER; SP—soluble proteins; SV—secretory vesicles; TP—tonoplast.

latter technique, are in frequent use by plant cell biologists. Some organelles might be different from those present in animal cells (e.g., vacuole, protein–lipid bodies), but as far as the endo and plasma membranes of plant cells are concerned, there exists no separation method which has not been previously introduced on animal tissues.

REFERENCES

Albertsson, P. A. (1974). In *Methods in Enzymology* (S. Fleischer and L. Packer, eds.), Vol. XXXI, pp. 761–769. Academic Press, New York.

Anderson, J. W. (1968). *Phytochem.* **7,** 1973–1988.

Dubochet, J., McDowall, A. W., Menge, B., Schmid, E. N., and Lickfeld, K. G. (1983). *J. Bacteriol.* **155,** 381–390.

Galliard, T. (1974). In *Methods in Enzymology* (S. Fleischer and L. Packer, eds.), Vol. XXXI, pp. 520–528. Academic Press, New York.

Hannig, K. and Heidrich, H. G. (1974). In *Methods in Enzymology* (S. Fleischer and L. Packer, eds.), Vol. XXI, pp. 746–761. Academic Press, New York.

Hayat, M. A. (1970). *Principles and Techniques of Electron Microscopy*. Vol. 1, Biological Applications. Van Nostrand Reinhold Company, New York.

Loomis, W. D. (1974). In *Methods in Enzymology* (S. Fleischer and L. Packer, eds.), Vol. XXXI, pp. 528–544. Academic Press, New York.

Price, C. A. (1974). In *Methods in Enzymology* (S. Fleischer and L. Packer, eds.), Vol. XXXI, pp. 501–519. Academic Press, New York.

Robertson, J. D. (1981). *J. Cell. Biol.* **91,** 1895–2045.

Stoeckenius, W. (1962). *J. Cell. Biol.* **12,** 221–229.

2

ENDOPLASMIC RETICULUM

On a protein basis the endoplasmic reticulum (ER) is the most abundant membrane type in the cell. Directly or indirectly, all other single-membrane organelles appear to arise from this membrane system. Because of its ability to temporarily link up with polysomes, the ER is not only the site of storage or secretory protein synthesis (see Chapters 10 and 11) but is also a major site of membrane protein and lipid synthesis (see Chapter 15). The structure and functions of this most important endomembrane in plant cells has been reviewed on several occasions (most recently by Chrispeels, 1980; Chrispeels and Jones, 1980/81) and a chapter dealing with its isolation is also now available (Lord, 1983).

2.1. STRUCTURE AND MORPHOLOGY

It is almost impossible to estimate accurately the amount of ER in a plant cell partly because the transition from the cisternal or platelike to tubular forms cannot be adequately predicted from single electron micrographs of thin sections. Estimates have been given (6–12 $\mu m^2 \cdot \mu m^{-3}$ and 1 $\mu m^2 \cdot \mu m^{-3}$) for tapetal and root tip cells, respectively (Gunning and Steer, 1975) which, although perhaps not quite correct, do at least signify the considerble amounts of this membrane present in the cell.

It is easier to appreciate the relationship between tubular and cisternal ER when thick (μm) sections from zinc–iodine–osmium (ZIO) fixed material are observed with a high-voltage electron microscope (Figure 2.1a,b). Cisternal "plaques," each with a fenestrated periphery of tubular elements, are connected with one another through these tubular extensions. By having connections between neighboring plaques above, below, and laterally, a complex network can be constructed. Normally the cisternal plaques are arranged randomly, but there are cases where regular, even semicrystalline arrays are present.

During the differentiation of sieve cells, the degradation of the nucleus is accompanied by a stacking of ER cisternae which also lose their ribosomes. Quite often the stacks are oriented perpendicular to the PM [see Figure 2.2a and Sjolund and Shih (1983)] but by curving may also occasionally run parallel to it. With further maturation complexer, anastomosing, forms of ER are then developed which eventually cover almost the whole of the PM [see Evert (1983) and Thorsch and Esau (1981) for literature]. In some cases these ER complexes attain huge proportions

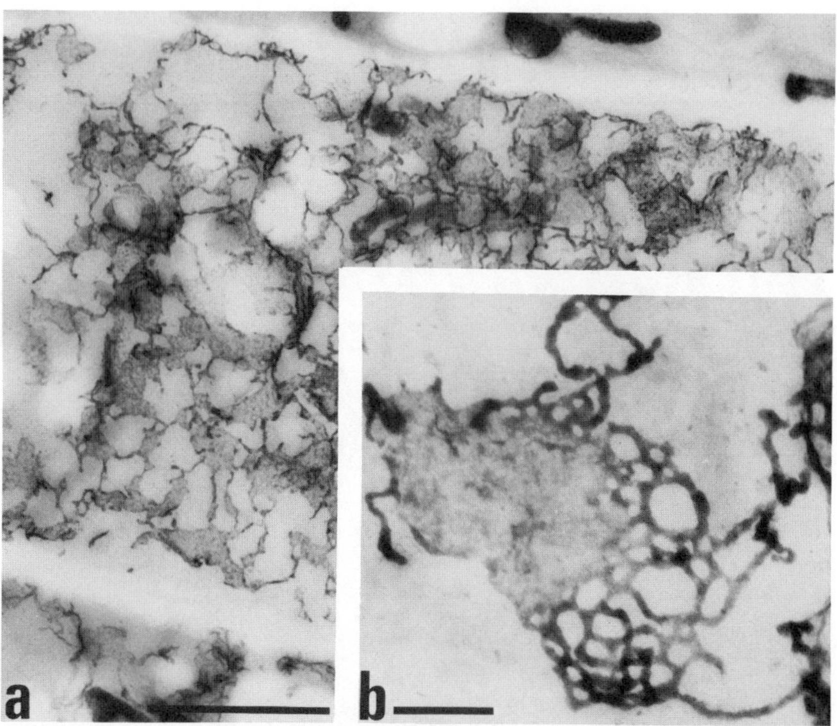

Figure 2.1. (a) 1-μm section from a maize root tip cell fixed according to the ZIO procedure and observed at 1 MeV. Bar = 3 μm. (b) Higher magnification of 2.1(a) revealing clearly the cisternal and tubular portions of the ER. Bar = 0.5 μm. Unpublished micrographs of Hawes and Juniper.

and adopt a semicrystalline nature not unlike that of the prolamellar body of the etioplast [cf. Gunning (1965) with Behnke (1981)]. Since identical ER aggregates have also been found in leaf parenchyma cells of virus-infected *Helleborus niger* plants (see Figure 2.2b), their formation cannot be regarded as being restricted to sieve cells. Virus infections also give rise to a number of other types of ER aggregations, one of the most frequently occurring forms of which is known as the "pinwheel" type (e.g., Hooper and Wiese, 1972; Lawson and Hearon, 1971).

Although the proliferation of ER can also be experimentally induced by the addition of metabolic inhibitors (see Chapter 17), it must not be assumed that their presence is restricted to pathological conditions. Stacks or whorls of ER appear characteristic of resting buds (Shih and

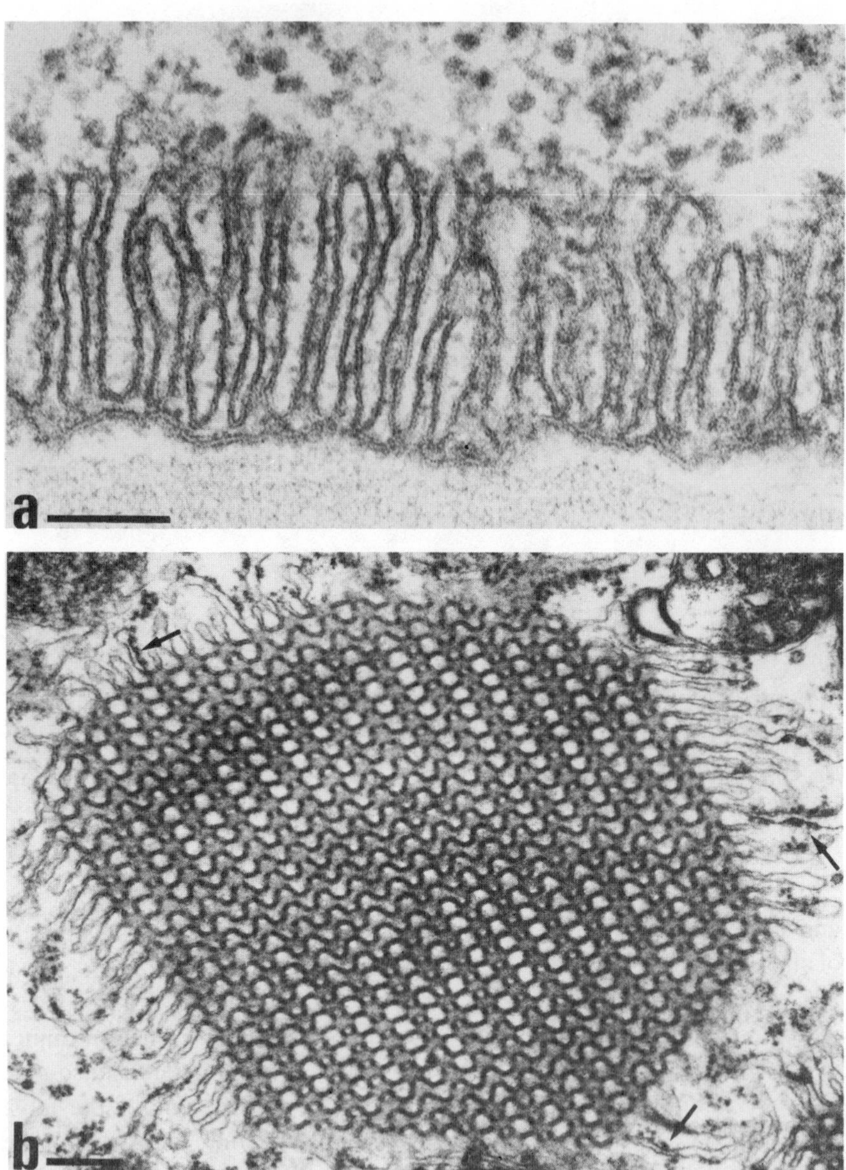

Figure 2.2. (a) Stacked sER perpendicular to the PM in developing sieve elements in callus cultures of *Straptanthus tortuosus*. Bar = 0.1 μm. (b) ER aggregate in virus-infected leaf parenchyma cells of *Helleborus niger*. Note the attachment of ribosomes at the peripheral outgrowths (arrows). Bar = 0.25 μm. Published (a) micrograph of Sjolund and Shigh (1983); unpublished (b) micrograph of Lesemann.

Rappaport, 1971) or seeds (Bouvier-Durand et al., 1981) and signify a relationship between the physiological status of a cell and the morphological appearance of its organelles.

ER is usually described as being either rough (see Figure 2.3) or smooth, depending on the presence on absence of polysomes. Work on animal cells indicates that in general terms the membranes of rER and sER are similar (e.g., Amar-Costesec et al., 1974) and is visually strengthened by the numerous continuities present between the two membrane types (Kreibich, Ulrich, and Sabatini, 1978). However, specific ribosomal-binding proteins, called *ribophorins* (Kreibich, Czacho-Graham, Grebenau, Mok, Rodriguez-Boulan, and Sabatini, 1978), are now recognized which are only localized in the rER. The fact that polysomes tend to lie in a spiral configuration when they become attached to the surface of the rER implies that a certain minimal surface area must be present, a requirement that probably cannot be fulfilled by sER, which usually has a tubular configuration.

Our knowledge of secretory protein synthesis and segregation in animal cells (see Chapter 10) makes it clear that those polysomes which are seen attached to the ER do not remain permanently there. After the polypeptide has been translated and inserted into the ER lumen, the polysome is released from the membrane surface to be replaced by another. The populations of permanently bound and free ribosomes one often finds described in the older literature are really a thing of the past.

2.2. MARKERS FOR ISOLATION

2.2.1. Redox Reactions

Nonphosphorylating electron transport systems are present in the microsomal fractions of both plant and animal cells. All involve the transfer of electrons from NAD(P)H via a flavoprotein reductase to *b*- or *c*-type cytochromes (Lehninger, 1975). In comparison to the situation in plants, the function of these electron transfer reactions in animal cells are known and well documented; for example cytochrome *b* transfers electrons to a stearyl CoA desaturase, and cytochrome P-450 (a CO-binding *b*-type pigment) transfers electrons to a large variety of dealkylation and hydrox-

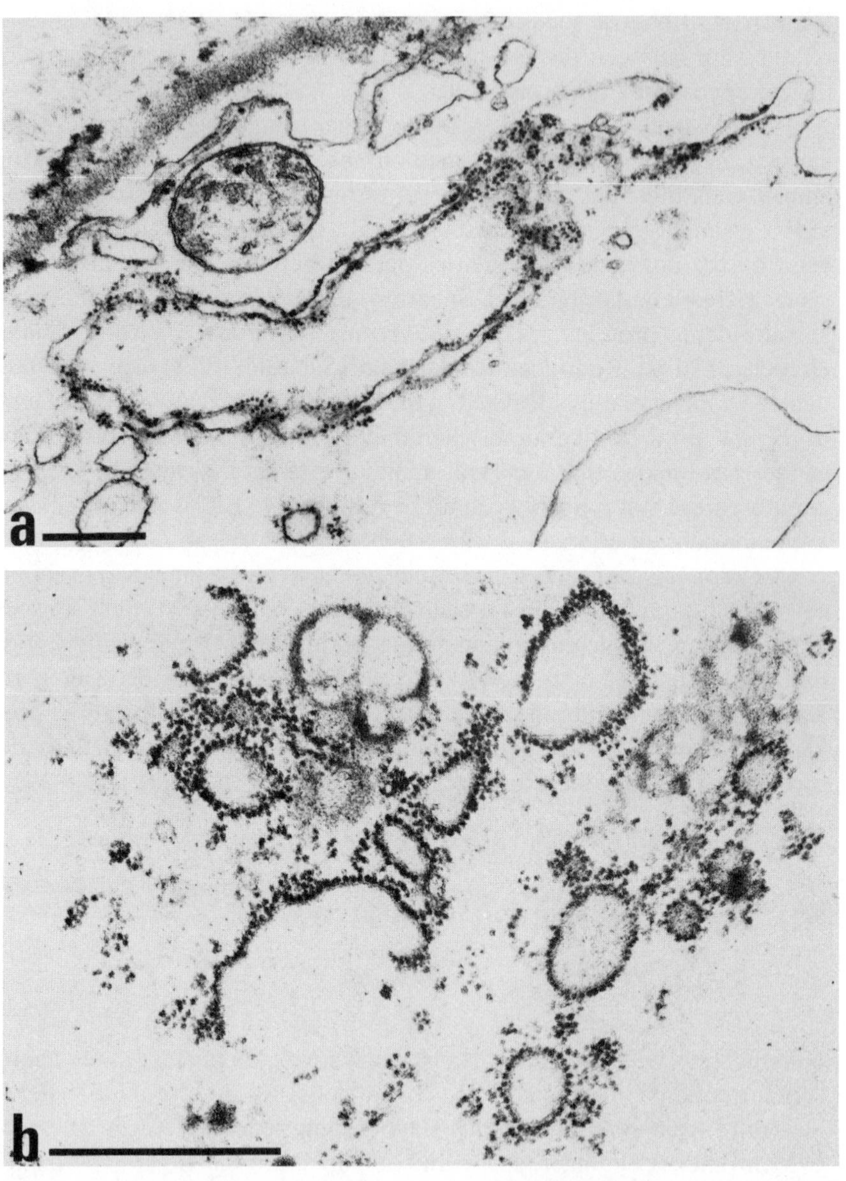

Figure 2.3. rER. (a) *In situ*. Portion of an *Cucurbita* endosperm cell. (b) *In vitro*. Isolated from anion root tips. Bar = 0.5 μm. Unpublished (a) micrograph of Hafemann and Robinson; published (b) micrograph of Philipp et al. (1976).

ylation reactions (Estabrook et al., 1972; Hagihara et al., 1975; Lemberg and Barrett, 1973).

An antimycin A-insensitive cytochrome having difference spectrum peaks at 559, 527, and 427 nm has been noted on several occasions for the microsomal fraction from plant cells (Lord et al., 1973; Philipp et al., 1976; Rungie and Wiskich, 1972). These features are characteristic of cytochrome *b*, but the responsible pigment has occasionally received a different name by other authors (see Franke et al., 1975). It is unclear whether cytochrome P-450 is present in plants. Based on the dithionite versus NADH difference spectra, Lord et al. (1973) have claimed its presence in castor bean ER fractions and Markham et al. (1972) have done the same for microsomal fractions of bean and maize cotyledons. The participation of cytochrome P-450 in the hydroxylation of cinnamic acid by Sorghum microsomes has also been reported (Potts et al., 1974). Philipp et al. (1976), however, have not been able to detect this cytochrome in onion tissue.

The flavoprotein enzymes NADH-cytochrome b_5 reductase and NADH-cytochrome *c* and NADPH–cytochrome *c* reductase (CCR) are generally accepted markers for the ER in animal cells. The latter are antimycin and rotenone insensitive, thus differentiating them from the mitochondrial form of this enzyme (De Pierre and Ernster, 1977). Numerous papers attest to the ER localization of CCR activity plant cells [for references see Franke et al. (1975), Nagahashi and Beevers (1978), Philipp et al. (1976), and Quail (1979)], but it is noteworthy that there are, in addition to the membrane-bound form, two soluble (4S and 8S) species which apparently represent nitrate reductase (Wallace and Johnson, 1978). The specific activities of NADH–CCR in plant cells are similar, if a little lower, to those from animal cells (Franke, 1974; Kasper, 1974), and the activity of NADPH–CCR is usually only about 10% of that of NADH–CCR (Bowles and Kauss, 1976; Hendriks, 1977; Lord et al., 1973; Ray, 1977). According to Philipp et al. (1976), optimal activity is obtained at 30° and at pH 8 in the presence of 0.1% Triton X-100. Pupillo and de Luca (1982), however, maintain that this concentration of Triton is inhibitory for NADH–CCR activity in microsomal fractions from maize or pumpkin.

As far as other redox enzymes are concerned, there are conflicting claims as to the presence of NADPH oxidase activity in the plant microsomal membranes (cf. Philipp et al. (1976) with Pupillo and de Luca

(1982)]. However, according to the latter authors this activity is not entirely restricted to the ER. Its use therefore as an ER marker must be taken with caution.

There is now considerable evidence in support of the contention that the cytochromes and flavin enzymes mentioned above are not exclusively localized in the ER in animal cells. Significant (10–30%) proportions of the total microsomal activities of NADH–CCR and NADPH–CCR are measurable in the GA (Jarasch et al., 1979; Howell et al., 1978; Ito and Palade, 1978) and PM (Jarasch et al., 1979). In addition, antibodies raised against ER NADPH–CCR also react with elements of GA fractions (Ito and Palade, 1978). Similarly, the cytochromes b_5 and P-450 have also been detected both spectrophotometrically and immunologically in GA and PM fractions from hepatocytes. According to Jarasch et al. (1979), PM pigments are much more quickly degraded than those in the ER, hence the lower values for these other membranes. Critical work of this nature has not yet been carried out with plant cells, but there are indications that a similar situation exists. Pupillo and de Luca (1982) claim that it is possible to differentiate between ER and PM NADH-CCR activities by the addition of exogenous quinones (e.g., duroquinone or ubiquinone) to the assay mixture. The NADH-CCR activity on the PM is then stimulated, whereas that of the ER apparently is not. Quinone enhancement of NADPH-CCR activity is however observed for both membrane types.

2.2.2. Glycosylation Reactions

The attachment of a $GlcNac_2Man_9Glc_3$ oligosaccharide is an Asn residue which occurs via a lipid intermediate constitutes the "core glycosylation" of a nascent polypeptide cotranslationally synthesized at the ER (see Section 10.1). Hence lipid-linked mannosyl- and *N*-acetylglucosamine transfer reactions can be utilized as markers for the identification of the ER. Such reactions have been characterized (Ericson and Delmer, 1977) and localized in the ER in the case of developing seeds (see Section 11.3) but are not in general use as ER markers in tissues not geared up for storage protein synthesis. The reason for this is probably the small amounts of *N*-glycosidic-Asn linkages found in plant glycoproteins in general (Selvendran and O'Neill, 1982).

2.2.3. Phospholipid Biosynthesis

Phospholipid biosynthesis is a feature of the microsomal fraction (see Morré, 1975; Robinson and Kristen, 1982, for reviews). In the well-studied germinating castor bean it is quite clear that the ER is the principal site of the reactions involved (Bowden and Lord, 1975; Lord, 1975, 1976; Lord et al., 1973; Moore et al., 1973; Moore, 1976). In growing, nonstorage tissue the ER has also been shown to be a major site of phospholipid biosynthesis (Hoch and Hartmann, 1981; Montague and Ray, 1977; Morré et al., 1970). This is based upon the localization of the enzyme cytidine diphosphatecholine–diglyceride phosphorylcholine transferase (CGT) which catalyzes the last reaction in the pathway leading to phosphatidylcholine. The reaction prior to this catalyzed by cytidine triphosphate–phosphorylcholine-cytidyl transferase (CCT) is however found in both mitochondrial and ER fractions (Hoch and Hartmann, 1981; Morré et al., 1970). According to Montague and Ray (1977), both enzymes are present in the GA isolated from pea stem tissue and represent about 25% of the total cellular activity. However, these authors, as well as Morré et al. (1970) who worked with onion stems, have emphasized that on a protein basis the GA shows the higher specific activity for these two enzymes.

It thus depends on the tissue type (castor bean endosperm is reputed to have few dictyosomes) whether one can reliably use enzymes of phospholipid biosynthesis as markers for the ER in plant cells.

2.2.4. Other Enzymes

Glucose-6-phosphatase is considered a typical marker enzyme for the ER in animal cells, especially in liver (Bergeron et al., 1973; Leskes et al., 1971). Claims for its presence in the ER *in situ* (Roland, 1969) and in isolated fractions (Lai et al., 1971) from plants cells have been made but have also not been confirmed (Philipp et al., 1976). According to Morré et al. (1974), this enzyme is found in the supernatant and may represent instead a nonspecific acid phosphatase (Lau and Lygre, 1973; Philipp et al., 1976).

2.3. SEPARATION

ER is usually separated from other membranes by a single isopycnic centrifugation in a sucrose density gradient. Under conditions where Mg^{2+} ions are kept at a minimum (presence of EDTA in the homogenization medium), the ER is in the smooth form and bands at a density of 1.10–1.12 $g \cdot cm^{-3}$. The separation is not sharp, there being an overlap with TP

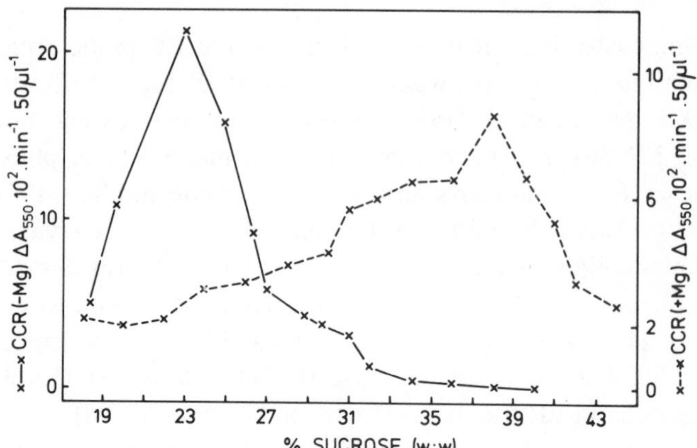

Figure 2.4. "Mg^{2+} shifting" of ER-based CCR activity in carrot root homogenates centrifuged isopycnically on linear sucrose gradients. Published results (redrawn) of Wienecke et al. (1982).

vesicles (see Chapter 6) and elements of the GA (see Chapter 4) at the light and heavy regions, respectively. Apart from taking fractions containing only the peak values for CCR activity, further attempts at purifying ER membranes from plant cells have not been undertaken.

The inclusion of m*M* (usually 3–5) amounts of $MgCl_2$ in the homogenizing and gradient media allows ER in the rough form to be isolated (see Figure 2.3b and Ray, 1979). The presence of ribosomes can be recognized optically in thin sections as well as by measurement of RNA (e.g., Lord et al., 1973; Ray et al., 1976). Where Mg^{2+} ions are low or absent, RNA usually remains at the top of isopycnic sucrose gradients. In contrast the coincident peaks of CCR activity and RNA in high Mg^{2+} gradients lie in the range 1.15–1.18 $g \cdot cm^{-3}$—that is, between regions containing GA and mitochondria, respectively (Figure 2.4). Since this part of the gradient is also occupied by PM vesicles, both in low and high Mg^{2+} conditions (see Chapter 7), a better separation of ER is not obtained as a result of this so-called Mg^{2+} shifting.

The amount of Mg^{2+} required to achieve a high proportion of retained ribosomes is critical, but also variable. Too much and unspecific membrane aggregates are formed which are centrifuged out of the gradient; too little and a high proportion of sER is obtained. Only through experimentation can these extremes be determined. Even then it is often not possible to avoid the presence of sER in high Mg^{2+} gradients relating to the existence of sER *in situ*.

REFERENCES

Amar-Costesec, A., Beaufay, H., Wibo, M., Thines-Sempoux, D., Feytmans, E., Robbi, M., and Berthet, J. (1974). *J. Cell. Biol.* **61,** 201–212.

Behnke, H.-D. (1981). *Nord. J. Bot.* **1,** 381–400.

Bergeron, J. J. M., Ehrenreich, J. H., Siekevitz, P., and Palade, G. E. (1973). *J. Cell. Biol.* **59,** 73–88.

Bouvier-Durand, M., Dereuddre, J., and Come, D. (1981). *Planta* **151,** 6–14.

Bowden, L., and Lord, J. M. (1975). *FEBS Lett* **49,** 369–371.

Bowles, D. J. and Kauss, H. (1976). *Biochem. Biophys. Acta* **443,** 360–374.

Chrispeels, M. J. (1980). In *The Biochemistry of Plants* (N. E. Tolbert, ed.), Vol. 1, pp. 389–412. Academic Press, New York.

Chrispeels, M. J. and Jones, R. L. (1980/81). *Israel J. Botany* **29,** 225–245.

de Pierre, J. W. and Ernster, L. (1977). *Ann. Rev. Biochem.* **46,** 201–262.

Ericson, M. C. and Delmer, D. P. (1977). *Plant Physiol.* **59,** 341–347.

Estabrook, R. W., Baron, J., Franklin, J., Mason, M., Waterman, M., and Peterson, L. (1972). In *The Molecular Basis of Electron Transport* (J. Schultz and B. F. Cameron, eds.), pp. 197–230. Academic Press, New York.

Evert, R. F. (1983). In *Contemporary Problems in Plant Anatomy* (R. A. White and W. C. Dickison, eds.), pp. 28–47. Academic Press, New York.

Franke, W. W. (1974). *Int. Rev. Cytol. Suppl.* **4,** 71–236.

Franke, W. W., Jarasch, E.-D., Hertz, W., Scheer, U., and Zerban, H. (1975). *Prog. Botany* **37,** 1–21.

Gunning, B. E. S. (1965). *Protoplasma* **60,** 111–130.

Gunning, B. E. S. and Steer, M. W. (1975). *Ultrastructure and the Biology of Plant Cells*. Edward Arnold, London.

Hagihara, B., Sato, N., and Yamanaka, T. (1975). In *The Enzymes* **11** (P. D. Boyer, ed.), pp. 549–593. Academic Press, New York.

Hendriks, T. (1977). *Plant Sci. Lett.* **9,** 351–363.

Hoch, K. and Hartmann, E. (1981). *Plant Sci. Lett.* **21,** 389–396.

Hooper, G. R. and Wiese, M. V. (1972). *Virology* **47,** 664–672.

Howell, K. E., Ito, A., and Palade, G. E. (1978). *J. Cell. Biol.* **59,** 581–589.

Ito, A. and Palade, G. E. (1978). *J. Cell. Biol.* **79,** 590–597.

Jarasch, E.-D., Kartenbeck, J., Bruder, G., Fink, A., Morré, D. J., and Franke, W. W. (1979). *J. Cell. Biol.* **80,** 37–52.

Kasper, C. B. (1974). In *The Cell Nucleus 1* (H. Busch, ed.), pp. 349–384. Academic Press, New York.

Kreibich, G., Czacho-Graham, M., Grebenau, R., Mok, W., Rodriguez-Boulan, E., and Sabatini, D. D. (1978). *J. Supramol. Struct.* **8,** 279–302.

Kreibich, G., Ulrich, B., and Sabatini, D. D. (1978). *J. Cell Biol.* **77,** 464–487.

Lai, Y. F., Thompson, J. E., and Barell, R. W. (1971). *Phytochem.* **10,** 41–49.

Lau, W. P. and Lygre, D. G. (1973). *Biochim. Biophys. Acta* **309,** 318–327.

Lawson, R. H. and Hearon, S. S. (1971). *Virology* **44,** 454–456.

Lehninger, A. L. (1975). *Biochemistry.* Worth Publishers, New York.

Lemberg, R. and Barrett, J. (1973). *Cytochromes.* Academic Press, New York.

Leskes, A., Siekevitz, P., and Palade, G. E. (1971). *J. Cell Biol.* **49,** 264–287.

Lord, J. M. (1975). *Biochem. J.* **151,** 451–453.

Lord, J. M. (1976). *Plant Physiol.* **57,** 218–223.

Lord, J. M. (1983). In *Isolation of Membranes and Organelles from Plant Cells* (J. L. Hall and A. L. Moore, eds.), pp. 119–134. Academic Press, New York.

Lord, J. M., Kagawa, T., Moore, T. S., and Beevers, H. (1973). *J. Cell. Biol.* **57,** 659–667.

Markham, A., Hartman, G. C., and Parke, D. C. (1972). *Biochem. J.* **130,** 90.

Montague, M. J. and Ray, P. M. (1977). *Plant Physiol.* **59,** 225–230.

Moore, T. S. (1976). *Plant Physiol.* **57,** 383–386.

Moore, T. S., Lord, J. M., Kagawa, T., and Beevers, H. (1973). *Plant Physiol.* **52,** 50–53.

Morré, D. J. (1975). *Ann. Rev. Plant Physiol.* **26,** 441–481.

Morré, D. J., Lembi, C. A., and van der Woude, W. J. (1974). In *Biochemische Cytologie der Pflanzenzelle* (G. Jacobi, ed.), pp. 147–172. Thieme Verlag, Stuttgart.

Morré, D. J., Nyquist, S., and Rivera, E. (1970). *Plant Physiol.* **45,** 800–804.

Nagahashi, J. and Beevers, L. (1978). *Plant Physiol.* **61,** 451–459.

Philipp, E.-I., Franke, W. W., Keenan, T. W., Stadler, J., and Jarasch, E.-D. (1976). *J. Cell. Biol.* **68,** 11–29.

Potts, J. R. M., Weklych, R., and Conn, E. E. (1974). *J. Biol. Chem.* **249,** 5019–5026.

Pupillo, P. and de Luca, L. (1982). In *Plasmalemma and Tonoplast: Their Functions in Plant Cell* (D. Marme, E. Marre, and R. Hertel, eds.), pp. 321–328. Elsevier Biomedical Press, Amsterdam.

Quail, P. H. (1979). *Ann. Rev. Plant Physiol.* **30,** 425–484.

Ray, P. M. (1977). *Plant Physiol.* **59,** 594–599.

Ray, P. M. (1979). In *Plant Organelles. Methodological Surveys (B): Biochemistry 9* (E. Reid, ed.), pp. 135–146. Horwood Publishers, Chichester.

Ray, P. M., Eisinger, W. R., and Robinson, D. G. (1976). *Ber. Dtsch. Bot. Ges.* **89,** 121–146.

Robinson, D. G. and Kristen, U. (1982). *Int. Rev. Cytol.* **77,** 89–127.

Roland, J.-C. (1969). *Comp. Rendus. Acad. Sci. (Paris)* **268,** 2052–2055.

Rungie, J. M. and Wiskich, J. T. (1972). *Austr. J. Biol. Sci.* **25,** 103–113.

Selvendran, R. R. and O'Neill, M. A. (1982). *Encyl. Plant Physiol.* **13A,** 515–583.

Shih, C. Y. and Rappaport, L. (1971). *Plant Physiol.* **48,** 31–35.

Sjolund, R. D. and Shih, C. Y. (1983). *J. Ultr. Res.* **82,** 111–121.

Thorsch, J. and Esau, K. (1981). *J. Ultr. Res.* **74,** 183–194.

Wallace, W. and Johnson, C. B. (1978). *Plant Physiol.* **61,** 748–752.

3

NUCLEAR ENVELOPE

With the possible exception of chloroplasts, which are not universally present in plant cells, the nucleus is the largest membrane-bound organelle. Because of its size and its contents (chromatin/choromosomes) it has long been recognized cytologically. Although the existence of a nuclear envelope (NE) had been deduced from a number of light microscopic observations stretching back a hundred years or more (see Franke et al., 1981, for a historical review), it only became clear that this was a membranous structure with the advent of electron nicroscopy.

There are numerous review articles dealing with one or more features of the NE (Franke, 1974a,b; Franke and Scheer, 1974; Franke et al., 1974, 1981; Harris, 1978; Maul, 1977). Additionally, review chapters on methods pertaining to nuclear and NE isolation are also available (Dunham and Bryant, 1983; Franke, 1974c; Mascarenhas et al., 1974).

3.1. STRUCTURE AND MORPHOLOGY

3.1.1. General Aspects

In section the NE appears to consist of two 5–7 nm thick membranes, and it is quite often implied that this is a "double-membrane" structure. True the inner and outer faces of the NE have different properties, but the existence of pores (see Figure 3.1) that provide continuities between them together with a number of features in common with the ER (see below) make it clear that the NE is in fact a complicated one-membrane system. One usually thinks of the nucleus as being roughly spherical in interphase and this is generally true, but there are numerous cases, particularly in plants, where the nucleus is lobed (e.g., Jordan and Chapman, 1973) or is characterized by fingerlike protrusions (Bell, 1972; Jordan, 1974; Quatrano, 1972). These are examples of true evaginations of the NE and differ therefore from the highly labyrinthine projections around the primary nucleus of *Ace tabularia,* which are without direct connection to the NE (Woodcock and Miller, 1973).

3.1.2. Similarities to the ER

The NE may be considered a specialized, locally differentiated portion of the ER. Apart from similar biochemical properties (see below) there

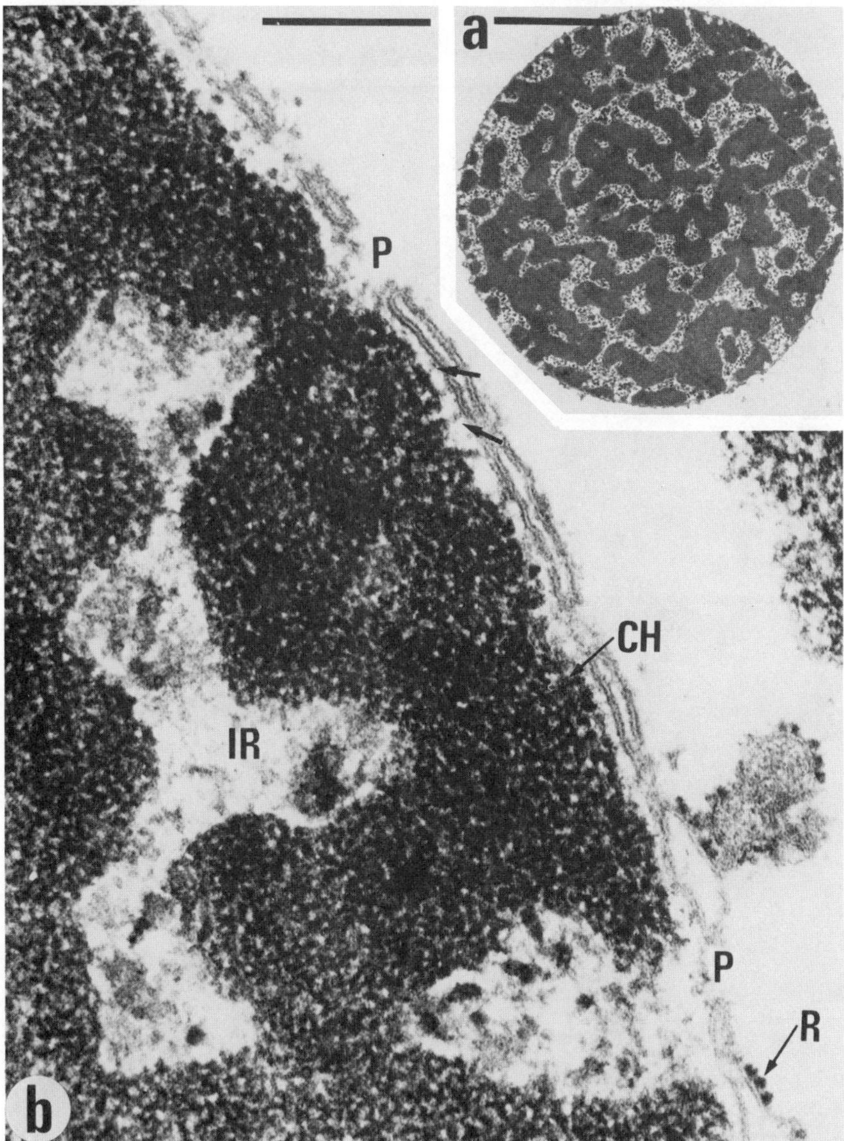

Figure 3.1. (a) Nucleus isolated from an onion root tip. Bar = 2 μm. (b) High magnification view of the peripheral region of a nucleus isolated from an onion root tip cells. The cisternal nature of the NE is clearly seen, so too the pores (P). Condensed chromation (CH) and interchromatic regions (IR) are present. Ribosomes (R) can be recognized on the outer surface of the NE. Fine connections between the inner surface of the NE and the chromatin are indicated by arrows. Bar = 0.2 μm. Published micrographs of Philipp et al., 1976.

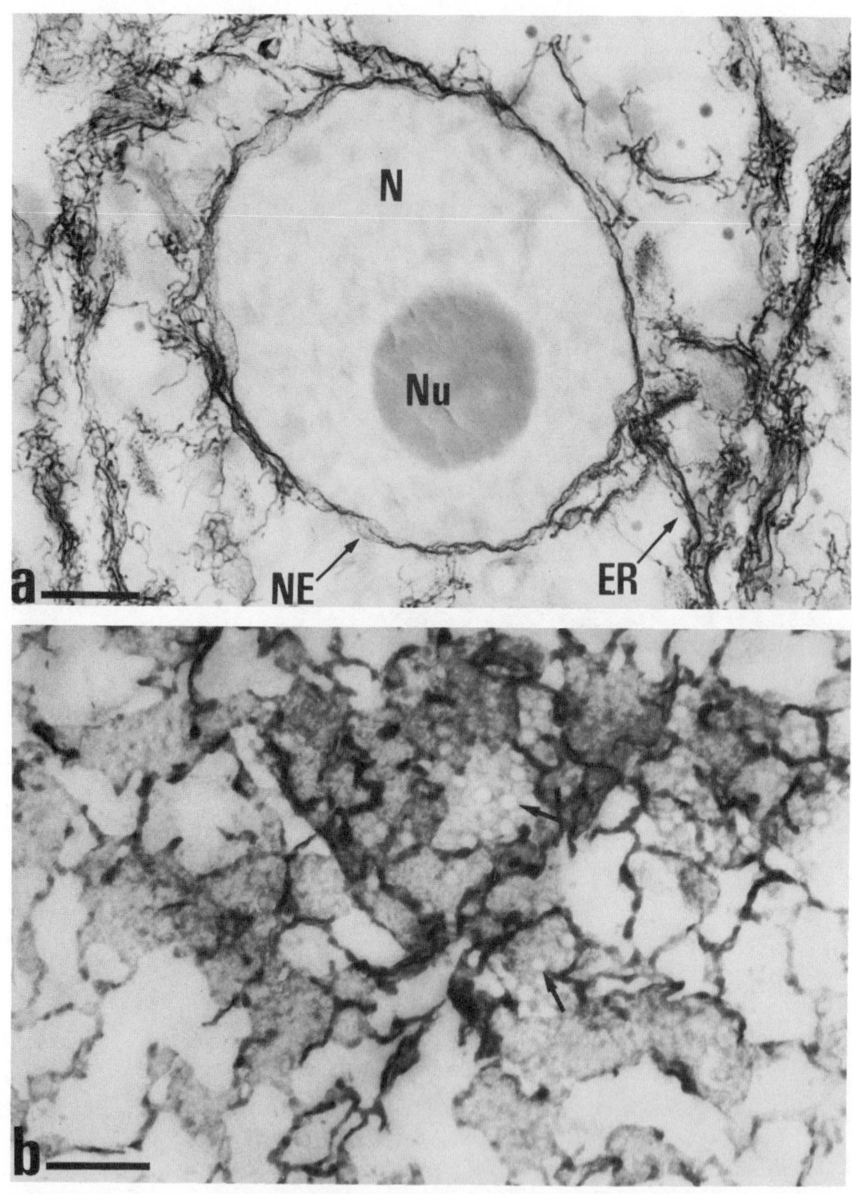

Figure 3.2. (a) 1-μm section through a ZIO-fixed maize root cap cell showing the close relationship between the NE and ER. (Nu = nucleolus; N = nucleus). Section viewed at 1 MeV. Bar = 3 μm. (b) Formation of the NE-diving telophase in a dividing maize root tip cell. Pores are indicated by arrows. 1-μm section of ZIO-fixed material viewed at 1 MeV. Bar = 0.5 μm. Unpublished micrographs of C. Hawes and B. L. Juniper.

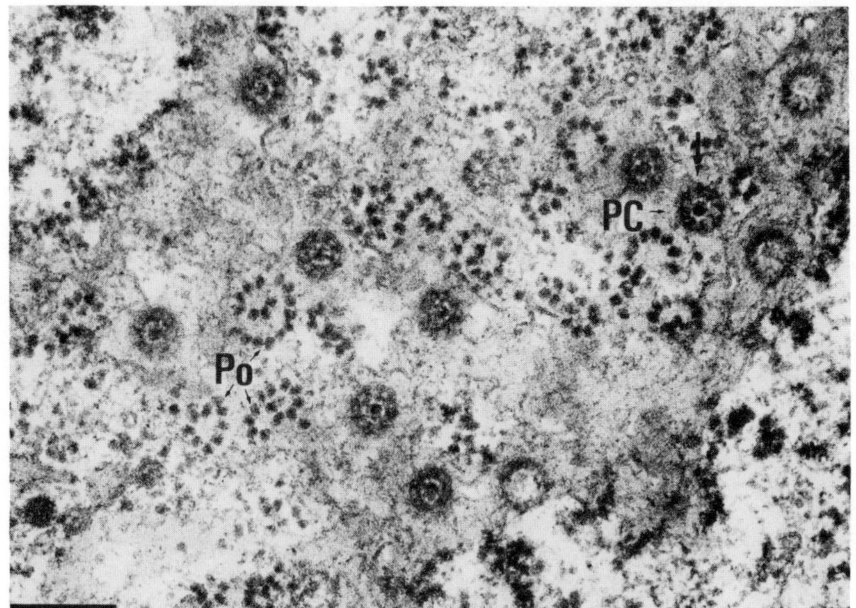

Figure 3.3. Tangential section through the NE of cultured (*Haplopappus gracilis* cells Polysomes (Po) and pore complexes (PC) are shown. The arrow indicates one PC in which the 9 annular and one central particle are well seen. Bar = 0.2 μm. Unpublished micrograph of W. Herth.

are a number of structural features which also accrue to this. They are as follows (not all, of course, occurring in the same organism):

1. Numerous examples of direct ER–NE connections have been recorded, particularly in the algal literature (e.g., Falk and Kleinig, 1968; Gibbs, 1962; see also Franke, 1974a,b, for references). Close associations, if not connections, between ER and the NE are also readily appreciable in thick sections viewed at high voltages (Figure 3.2a).
2. The NE can assume some of the roles of the ER with respect to other organelles; for example,
   a. in being able to bud-off vesicles destined for fusion with the GA (Falk and Kleinig, 1968; Leedale, 1969).

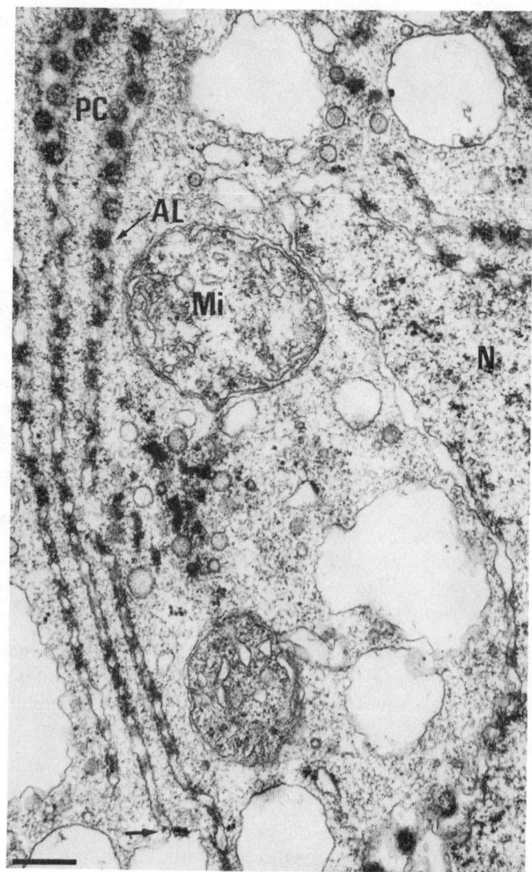

Figure 3.4. Annulate lamellae (AL) in developing pollen grains of *Canna generalis*. A transition between the AL and ER is indicated by the arrow. N = nucleus; Mi = mitochondria; PC = pore complex. Bar = 0.25 μm. Published micrograph of Scheer and Franke (1972).

b. in carrying ribosomes (see Figures 3.1 and 3.3) and in being able to synthesize and temporarily store (glyco)proteins in the NE lumen (Bouck, 1971; Krutrachue and Evert, 1978).

3. During mitosis NE fragments become indistinguishable from ER and the latter appear to contribute to the reformation of the NE during telophase (e.g., Esau and Gill, 1969; Hepler, 1980; see also Figure 3.2b).

4. Pore complexes (see below) are not restricted to the NE but are also found in the ER. Since the ER that bears such structures is usually stacked, it is customary to speak of "annulate lamellae, AL" (see Figure 3.4). They tend to occur in reproductive cells, both of animal (Kessel, 1968; Maul, 1977; Wischnitzer, 1970) and plant (Franke et al., 1974; Scheer and Franke, 1972) origin. The pore complexes in AL differ from those in the NE in being more symmetrically structured (there is no nucleoplasm!) and often much more densely packed.

3.1.3. The Inner Surface of the NE

There are numerous examples in the literature of the attachment of chromatin to the inner portion of the NE (see Franke, 1974a; Franke and Scheer, 1974, for references). Probably the most vivid cases are seen at the beginning of cell division when chromatin is present in the condensed form as chromosomes. In this respect, attachments are clearly recognizable when the dispersal of the NE does not occur (e.g., in dinoflagellate and hypermastigote flagellates) (Kubai, 1975) or is delayed (e.g., in meiotic prophase when the connection is characterized by the presence of a synaptinemal complex) (Moses, 1956). In interphase cells chromatin abuts on the inner surface of the NE in such a way that it would be surprising if interactions of some sort or another did not take place (see Franke et al., 1981). Thus, tenuous, attachments are in fact still visible in isolated nuclei where shrinkage has caused a withdrawal from the NE (arrows, Figure 3.1).

The existence of a layer 15–80 nm thick of nonchromatin, proteinaceous material immediately within the NE has been recorded on several occasions for animal cells (e.g., Fawcett, 1966; Harris, 1978) but, as far as I know, its existence has not yet been confirmed in plant cells. The filament meshwork nature of this lamina can be recognized after dissolution of the membrane with detergent (Dwyer and Blobel, 1976; Gerace et al., 1978). This material consists characteristically of three major (MW 60–80 kd) and three minor (MW 200, 160, and 125 kd) polypeptides (Krohne, Franke, and Scheer 1978). Although similar-sized polypeptides are present as nonhistone proteins in whole nuclei or chromatin preparations (Berezney, 1979), immunocytochemistry has confirmed the location of the lamina triplet polypeptides at the periphery of the nuclear matrix (Ely et al., 1978; Krohne, Franke, Ely, d'Arcy, and Jost 1978).

3.1.4. Pore Complexes

Pores (P in Figures 3.1–3.4) are characteristic of the NE and AC. In any one given species the membrane–membrane pore diameter is constant but values as a whole vary between 60–90 nm. However, the exclusion limit for transport through the pores, as determined from microinjection experiments with various-sized particles or viruses, is much smaller (in the range of 18–20 nm; see Bonner, 1978, for references).

The pores are not just simply holes, as might be thought when looking at ZIO-fixed material in the high-voltage electron microscope (Figure 3.2b), but have a complicated symmetrical arrangement of particles at both external (cytoplasmic) and internal (nucleoplasmic) openings. Since the architecture of these pore complexes has been the subject of numerous papers, I shall restrict my description to a minimum, referring the reader to the appropriate sections in the various review articles cited at the beginning of this chapter.

The components of the pore complex are as follows:

1. Two annuli, one at each orifice, and each consisting of eight symmetrically placed 10–25-nm particles (see large arrow, Figure 3.3).
2. Eight rod-shaped structures projecting radially into the lumen.
3. A centrally located particle.
4. Filaments of 4–8 nm diameter attached to the internal annulus.

These features are combined in the drawing of a nuclear pore given in Figure 3.5.

As mentioned above, the membrane of the NE can be dissolved away by detergent and/or high salt treatment. In addition to the lamina proteins, the pore complex also remains intact and is connected to others via filamentous structures which are presumably part of the lamina. Thus, the components of the pore complex are not membranous and are part of a nuclear cytoskeleton.

Just as there is great variability in pore diameter there are also considerable differences in pore frequency (from 1 to 60 pores·μm^{-2} NE surface area). As stated by Franke et al. (1981), "Although correlations of pore numbers and pore frequencies with certain nuclear activities e.g.

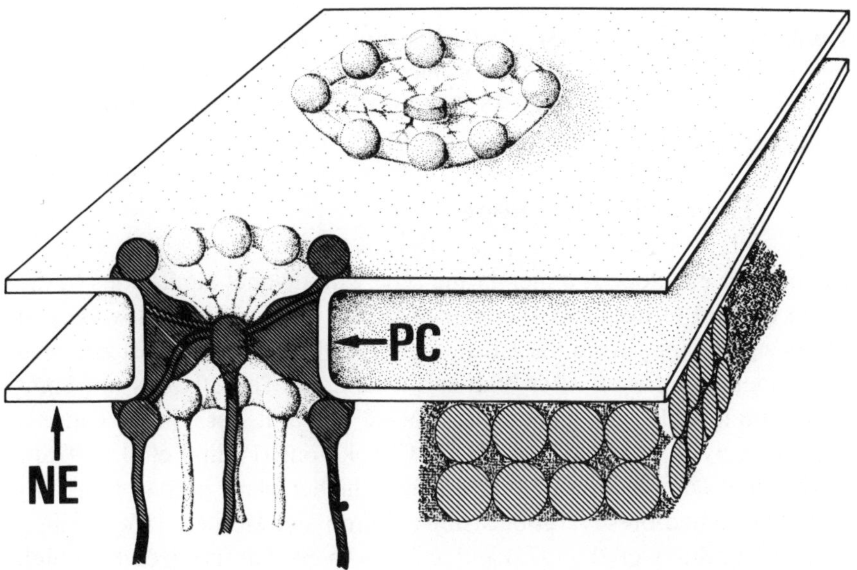

Figure 3.5. Schematic representation of the nuclear pore complex (PC). Chromatin and nonhistone triplet polypeptide material are represented by hatched circles and dots, respectively, at the inner surface of the NE. Published drawing of Franke et al., 1981.

transcription are sometimes suggestive, the functional associations of pore morphology and number cannot be resolved at the moment."

3.2. MARKERS FOR ISOLATION

3.2.1. The Nucleus

Since the nucleus is usually isolated before separation of the nuclear membrane, the majority of the markers deal with the content rather than the envelope of the nucleus. Due to its size and high nucleic acid content, the nucleus can easily be recognized in the phase contrast microscope or after suitable staining reactions (e.g., Feulgen or methyl green) in the light microscope. Marker enzymes for the nucleoplasm such as DNA and RNA polymerases (Quail, 1979) which involve the *in vitro* incorporation of radioactivity from nucleoside triphosphates into TCA-insoluble material have been frequently employed but are not specific for this organelle

(Dunham and Bryant, 1983). Similarly estimations of DNA and RNA give only indications of the purity of a nuclear fraction. A RNA–DNA ratio lying between 0.15 and 0.35 is typical for the nuclei of many animals and plants (Franke, 1974a; Philipp et al., 1976).

3.2.2. The Nuclear Envelope

The nuclear envelope is a membrane type with distinct similarities to the ER in both in terms of marker enzymes and composition. However, it differs significantly from the ER in that DNA and RNA remain attached to the membrane even after high salt treatment. The DNA is not completely removable with pancreatic DNase nor is it possible to cause a separation by flotation in 3 or 4 *M* CsCl solutions (Philipp et al., 1976).

The lipid composition of the NE from higher plant preparations has been determined on several occasions (Kemp and Mercer, 1968; Philipp et al., 1976; Stavy et al., 1973) and appears to be characterized by a high (relative to that in animal cells) proportion of nonpolar lipids, especially sterols. The phospholipids which are present are similar in composition to those of the ER, so that their presence and type cannot be used as a specific marker for the NE. Similarly, NADH–cytochrome *c* and NADPH–cytochrome *c* reductase activities, which are usually regarded as selective for the ER (see Chapter 2), are also present in NE preparations (Philipp et al., 1976).

3.3. SEPARATION

Of all the endomembrane organelles in plant cells, the nucleus is at one and the same time both the easiest and the most difficult to isolate. As a result of its size, it can be easily separated from other organelles by short centrifugations at low speeds, but this feature also makes it more susceptible to damage upon homogenization. Nuclear isolation has thus mainly been restricted to meristematic tissues where there are only primary walls and the nucleus–cytoplasm ratio is high, or to protoplasts. With whole-plant tissue a variety of mechanical homogenizers have been used including mixers, blenders, sonifiers, and mortar and pestle, although the highest yields appear to have been obtained by simply chopping with a razor blade (Dunham and Bryant, 1983; Spencer and Wildman,

1964). Even more effective is when the tissue is "softened" prior to homogenization by enzymic maceration (D'Alessio and Trim, 1968). With protoplasts a gentle homogenization with Potter or Dounce homogenizers normally gives adequate, careful breakage. In addition, squirting through a syringe (Tallman and Reeck, 1980) or centrifugation through a 20-μm nylon mesh (Hampp, 1980) have also been successfully employed.

As Quail (1979) has noted there are two types of nuclear isolations: one which employs Triton X-100 (<1%) in the homogenization medium and the other in which it is omitted. This detergent is added to remove contaminating membranes of other types and it is claimed (D'Alessio and Trim, 1968; Hughes et al., 1974) that the outer portion of the NE is also degraded leaving the inner intact. The electron microscopic evidence for this is, however, not convincing.

Homogenates after filtration through miracloth or muslin are usually centrifuged at 150–250 *g* for 10–20 min to provide a crude nuclear fraction (Dunham and Bryant, 1983). This can be further purified by repeated resuspension and centrifugation through buffered sucrose (Mascarenhas et al., 1974) or by high-speed centrifugation with a discontinuous gradient of sucrose (Kühl, 1964), metrizamide (Cheah and Osborne, 1977), or sorbitol (Ohyama et al., 1977). Continuous sucrose and Ludox (Percoll) gradients have also been employed (Mascarenhas et al., 1974; Trewavas, 1979). In isopycnic sucrose gradients the equilibrium density of nuclei is almost 1.32 $g{\cdot}cm^{-3}$ and in Percoll between 1.11 and 1.18 $g{\cdot}cm^{-3}$. The quality of such purified nucleus preparations is often exceedingly good, as seen in Figure 3.1.

Isolation of the NE from the nucleus can be achieved in one of two ways:

1. By resuspending the crude nuclear pellet in homogenizing medium followed by rapid dilution (Franke, 1966). This leads to "nuclear ghosts" which may be purified by layering over 62% sucrose and centrifuging at 3000*g* for 30 min.
2. By interval (10 × 2 sec) sonications followed by pelleting (150*g*, 5 min) and layering onto 66% sucrose and centrifuging at high speed (75,000*g*, 90 min) (Philipp et al., 1976; Stavy et al., 1973).

In both cases the NE is recovered at the interface between the heavy sucrose solution and the overlay.

REFERENCES

Bell, P. R. (1972). *J. Cell Sci.* **11,** 739–755.

Berezney, R. (1979). In *The Cell Nucleus* (H. Busch, ed.), Vol. 7, pp. 413–456. Academic Press, New York.

Bonner, W. M. (1978). In *The Cell Nucleus* (H. Busch, ed.), Vol. 6, pp. 97–148. Academic Press, New York.

Bouck, G. B. (1971). *J. Cell Biol.* **50,** 362–384.

Cheah, K. S. E. and Osborne, D. J. (1977). *Biochem. J.* **163,** 141–144.

d'Alessio, G. and Trim, A. R. (1968). *J. Exp. Bot.* **19,** 831–839.

Dunham, V. L. and Bryant, J. A. (1983). In *Isolation of Membranes and Organelles from Plant Cells* (J. L. Hall and A. L. Moore, eds.), pp. 237–275. Academic Press, London.

Dwyer, N. and Blobel, G. (1976). *J. Cell Biol.* **70,** 581–591.

Ely, S., Arcy, S., and Jost, E. (1978). *Exp. Cell Res.* **116,** 325–331.

Esau, K. and Gill, R. H. (1969). *Can. J. Bot.* **47,** 581–591.

Falk, H. and Kleinig, H. (1968). *Arch. Mikrobiol.* **61,** 347–362.

Franke, W. W. (1966). *J. Cell Biol.* **31,** 619–623.

Franke, W. W. (1974a). *Int. Rev. Cytol.* **4,** 71–236.

Franke, W. W. (1974b). *Phil. Trans. Roy. Soc. London B* **268,** 67–93.

Franke, W. W. (1974c). In *Biochemische Zytologie der Pflanzenzelle* (G. Jacobi, ed.), pp. 15–40. G. Thieme Verlag, Stuttgart.

Franke, W. W. and Scheer, U. (1974). in *The Cell Nucleus* (H. Busch, ed.), Vol. 1, pp. 219–347. Academic Press, New York.

Franke, W. W., Scheer, U., and Herth, W. (1974). *Progr. Bot.* **36,** 1–20.

Franke, W. W., Scheer, U., Krohne, G., and Jarasch, E.-D. (1981). *J. Cell. Biol.* **91,** 39s–50s.

Gerace, L., Blum, A., and Blobel, G. (1978). *J. Cell Biol.* **79,** 546–566.

Gibbs, S. P. (1962). *J. Cell. Biol.* **14,** 433–444.

Hampp, R. (1980). *Planta* **150,** 219–298.

Harris, J. R. (1978). *Biochem. Biophys. Acta* **515,** 55–104.

Hepler, P. K. (1980). *J. Cell Biol.* **86,** 490–499.

Hughes, B. G., Hess, W. M., and Smith, M. A. (1974). *Protoplasma* **93,** 267–274.

Jordan, E. G. (1974). *Protoplasma* **79,** 31–40.

Jordan, E. G. and Chapman, J. M. (1973). *J. Exp. Bot.* **24,** 197–214.

Kemp, R. J. and Mercer, E. T. (1968). *Biochem. J.* **110,** 119–125.

Kessel, R. G. (1968). *J. Ultr. Res.* **10,** 1–82.

Krohne, G., Franke, W. W., and Scheer, U. (1978). *Exp. Cell Res.* **116,** 85–102.

Krohne, G., Franke, W. W., Eli, S., d'Arcy, A., and Jost, E. (1978). *Cytobiologie* **18,** 22–38.

Krutrachue, M. and Evert, R. F. (1978). *Ann. Bot.* **42,** 15–21.

Kubai, D. F. (1975). *Int. Rev. Cytol.* **43,** 167–227.

Kühl, L. (1964). *Z. Naturforsch.* **19b,** 525–532.

Leedale, G. F. (1969). *Brit. Phycol. J.* **4,** 159–180.

Mascarenhas, J. P., Berman-Kurtz, M.,and Kulikowski, R. R. (1974). *Meth. Enzymol.* **31,** 558–565.

Maul, G. G. (1977). *Int. Rev. Chytol.* **6,** 75–186.

Moses, M. J. (1956). *J. Biophys. Biochem. Cytol.* **2,** 215–218.

Ohyama, K., Pelcher, L. E., and Horn, D. (1977). *Plant Physiol.* **60,** 179–181.

Philipp, E.-I., Franke, W. W., Keenan, T. W., Stadler, J., and Jarasch, E.-D. (1976). *J. Cell. Biol.* **68,** 11–29.

Quail, P. H. (1979). *Ann. Rev. Plant Physiol.* **30,** 425–484.

Quatrano, R. (1972). *Exp. Cell Res.* **70,** 1–12.

Scheer, U. and Franke, W. W. (1972). *Planta* **107,** 145–159.

Spencer, D. and Wildman, S. G. (1964). *Biochemistry* **3,** 954–959.

Stavy, R., Ben-Shaul, Y., and Galun, E. (1973). *Biochim. Biophys. Acta* **323,** 167–177.

Tallman, G. and Reeck, G. R. (1980). *Plant Sci. Lett.* **18,** 271–275.

Trewavas, A. (1979). In *Recent Advances in the Biochemistry of Cereals* (D. L. Laidman and R. G. Wyn Jones, eds.), pp. 175–208. Academic Press, London.

Wischnitzer, S. (1970). *Int. Rev. Cytol.* **27,** 65–100.

Woodcock, C. L. F. and Miller, G. J. (1973). *Protoplasma* **77,** 331–341.

4

MICROBODIES

Microbodies are an ubiquitous organelle in eukaryotic cells. First discovered in the 1950s [see Beevers (1980) and Tolbert and Essner (1981) for historical aspects] they belong to the best characterized of organelles from both animal and plant cells. As might be expected the term *microbody* was introduced at a time when the functional and biochemical aspects of this organelle were not, or only poorly, understood. Nowadays it is more appropriate to speak of peroxisomes or glyoxysomes, although they are morphologically identical structures.

The term *peroxisome* was introduced by De Duve (1965) to emphasize the fact that in these organelles hydrogen peroxide, H_2O_2, is metabolized. The name peroxisome does not, however, imply the existence of peroxidases but signifies instead the peroxidative activity of the enzyme catalase, which is responsible for the breakdown of H_2O_2. By definition, at least one H_2O_2-producing enzyme must also be present, but this is not always so [e.g., the "microperoxisomes" in some animal cells (Novikoff and Novikoff, 1972)]. The responsible oxidases in plant cells differ from those of animal sources (α-hydroxyacid), although flavin nucleotides are always involved as coenzymes. In the case of plants we recognize glycolate oxidase and β-acyl-CoA dehydrogenase. The former enzyme is responsible for the breakdown of glycolate formed as a result of photorespiration (Tolbert and Yamazaki, 1969), the microbody in question being termed *leaf peroxisome*. The latter is characteristic of the β-oxidation pathway for converting fatty acids to acetyl-CoA (Cooper and Beevers, 1969) and is present in the glyoxysomal type of microbody. In addition to these two types of microbody there are others present in a number of plant cells and tissues in which neither photorespiration nor the catabolism of fats can be claimed to be significant processes. According to Huang and Beevers (1971), these are called *nonspecialized microbodies* and usually have uricase or glycolate oxidase as the obligatory oxidase.

In many textbooks photorespiration is described as a phenomenon associated with plants showing the C_3-type of photosynthesis. As a result it is often falsely assumed that C_4 plants do not possess microbodies. Although it is true that photorespiration in C_4 plants cannot be measured by conventional gas exchange techniques (Chollet and Ogren, 1975; Zelitch, 1971), more refined measurements including $^{18}O_2$ exchange experiments (Black, 1973) do indicate that photorespiration does take place in C_4 plants *in situ*. Since the enzymes involved in the synthesis and metabolism of glycolate are present in homogenates of leaves from C_4 plants

(Chollet, 1974; Huang and Beevers, 1972; Rehfeld et al., 1970), it is clear that they possess the capacity to produce glycolate, even though a CO_2 release is not measurable. Thus peroxisomes have not only been detected ultrastructurally in the leaves of C_4 plants (Black and Mollenhauer, 1971; Frederick and Newcomb, 1971; Laetsch, 1974), but they have also been isolated (Gutierrez et al., 1974; Rathmann and Edwards, 1975).

There is an extensive literature dealing with microbodies from plant cells, so much so that I cannot hope to do justice to all of it. Of the newer reviews those of Beevers (1979, 1980), Tolbert (1980), Tolbert and Essner (1981), and Vigil (1984) are recommended to the reader for further details. In addition, Gerhardt wrote an excellent monograph (in German), in 1978 and Huang et al. wrote a recent one in 1983.

4.1. MORPHOLOGY

4.1.1. Form, Size, and Content

Microbodies are surrounded by a single membrane implicating their affiliation to endomembranous organelles rather than to mitochondria and chloroplasts. The membrane is superficially similar to the ER in that a clear tripartite structure is not as easy to discern, as for example with the PM, but is easily differentiated from it through the lack of ribosomes. Microbodies of the leaf peroxisome type are usually spherical with a diameter of up to 1.5 μm (Figure 4.1a). In contrast, glyoxysomes can often assume a pleomorphic form which, depending on the angle of section, give rise to dumbbell-shaped structures (Figure 4.2c) or cytoplasmic invaginations [see Figure 11.14 and Gruber et al. (1970)].

In general, the content of microbodies stains more intensely than the surrounding cytoplasm after conventional glutaraldehyde–osmium tetroxide fixation. A variety of inclusions are found in the matrix. Some are amorphous (Frederick et al., 1975), some are fibrillar [particularly in grasses (Frederick and Newcomb, 1971; Gruber et al., 1972)] and others are crystalline in nature (Hilliard et al., 1971; Vigil, 1973). In some cases all three types may be seen in one and the same microbody (Figure 4.1b). Values varying between 8.5 and 22 nm for the periodicity in microbody crystals have been given in the literature [see Gerhardt (1978) and Vigil (1973) for summaries] and probably reflect differing sectional angles

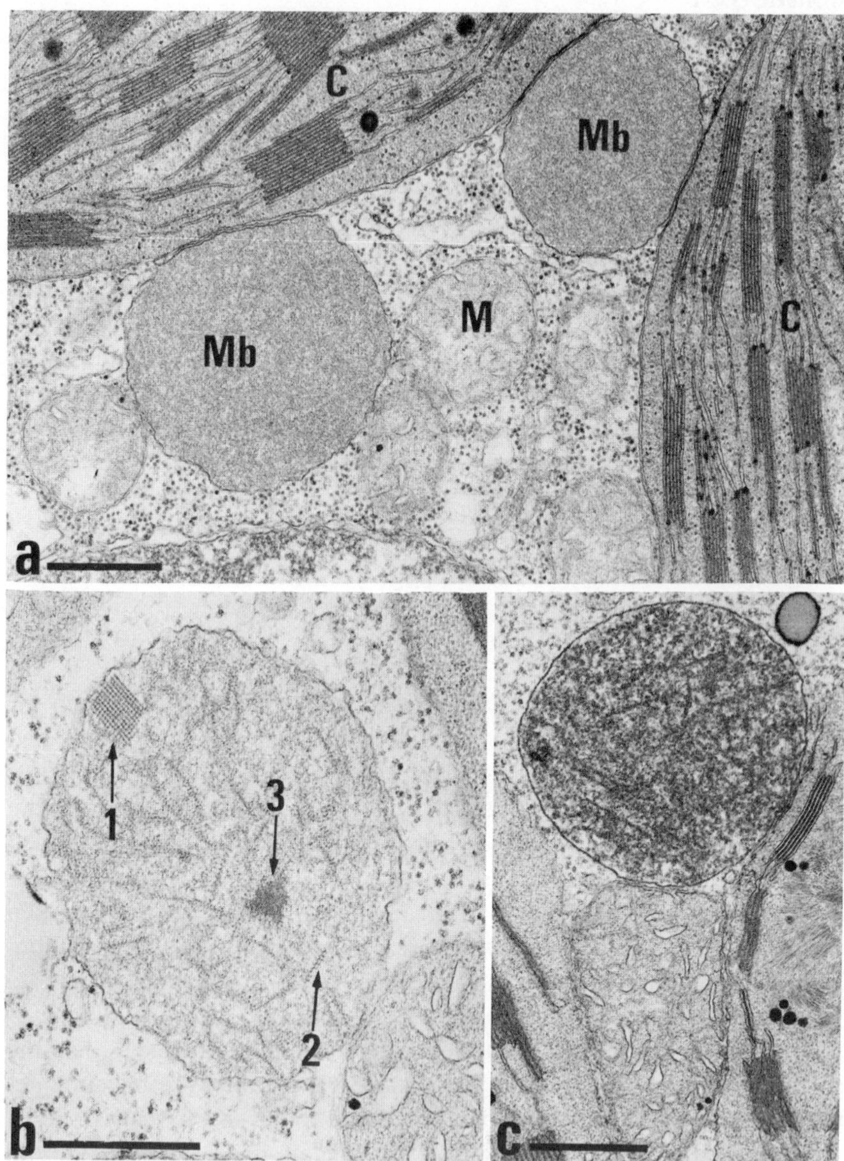

Figure 4.1. (a) Thin section through a tobacco leaf mesophyll cell. Microbodies (Mb) of the leaf peroxisomal type are seen wedged between chloroplasts (C) and mitochondria (M), their metabolic partners. Published micrograph of Frederick and Newcomb (1969). Bar = 0.5 μm. (b) Leaf peroxisome from a mesophyll cell of *Avena sativa* showing three different types of microbody inclusion (crystal, arrow 1; fibrillar, arrow 2; amorphous, arrow 3). Published micrograph of Gruber et al. (1972). Bar = 0.5 μm. (c) Leaf peroxisome from a mesophyll cell of *Avena sptiva* stained according to the DAB procedure. Published micrograph of Frederick and Newcomb (1970). Bar = 0.5 μm.

through the same structure. The crystal in peroxisomes from animal tissues appears to consist principally of uricase (Tsukuda et al., 1968, 1971) and therefore does not react positively with the DAB method (see Section 4.2). Plant microbody crystals in contrast stain strongly with DAB [see Vigil (1973) for references] indicating the presence of catalase.

4.1.2. Relationships to Other Organelles

Microbodies in plant cells are always found in close proximity to their metabolic partners. In the case of leaf peroxisomes, these are chloroplasts and mitochondria (Figure 4.1a); the former delivering glycolate as a result of photorespiration and receiving glycerate, the latter receiving glycine and delivering serine. On the other hand, glyoxysomes are intimately associated with lipid bodies (Figure 4.2), which provide fatty acids, and mitochondria, which receive malate and succinate as a result of β-oxidation and glyoxylate cycle reactions, respectively, and shuttle back aspartate [see Gerhardt (1978), Mettler and Beevers (1980), and Tolbert (1980) for details of the biochemistry involved].

There are numerous examples in the literature for an association between microbodies and ER (e.g., Frederick et al., 1968, 1975; Gruber et al., 1973). Perhaps the most extreme example is that seen in cells of the spadix from *Sauromatum guttatum* [see Figure 4.3 and Berger and Schnepf (1970)]. Generally the contact region of the ER is free of ribosomes, but parts more distant are often studded with ribosomes. Less frequent, however, are micrographs depicting direct continuities between ER and microbody membranes [for higher plants, Frederick et al (1968), Goeckermann and Vigil (1975), Matsushima (1972), Vigil (1970, 1973), and Wanner et al. (1982); for algae, Nilshammer and Walles (1974) and Silverberg and Sawa (1973)]. The significance of these latter observations in terms of the biogenesis of microbodies is dealt with in Section 11.7.

4.2. CYTOCHEMISTRY

4.2.1. The DAB Method

Upon oxidation 3,3-diaminobenzidine (DAB) gives rise to a polymerization product which is very osmiophilic (Seligman et al., 1968). Because

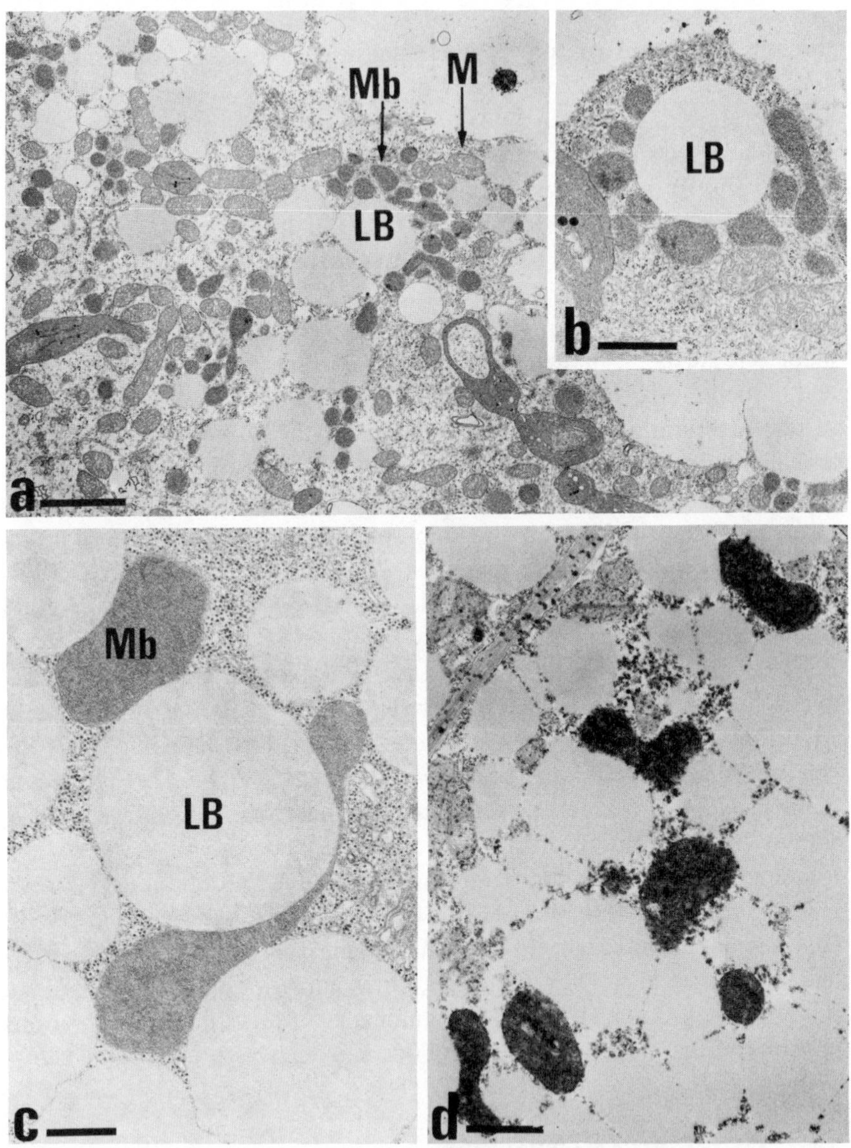

Figure 4.2. (a) Part of an endosperm cell from a germinating seedling of castor bean (*Ricinus communis*). Microbodies (Mb) of the glyoxysomal type are seen associated with lipid bodies (LB) and mitochondria (M). Bar = 0.25 μm. (b,c) High-magnification view of the "close association" of lipid bodies and glyoxysomes. Bars = 1 μm; 0.5 μm. (d) Glyoxysomes (Mb) of germinating cucumber (*Cucumis sativus*) cotyledons stained with the $Cu_2[Fe(CN)_6]$ technique to reveal the presence of malate synthase. Published micrographs of (a,b) Vigil, 1970; (c) Gruber et al., 1970; (d) Trelease et al., 1974. Bar = 1 μm.

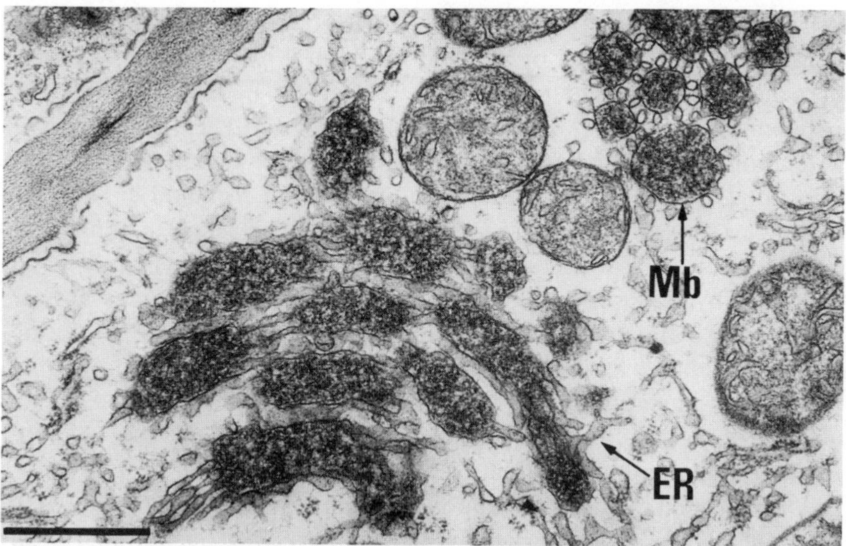

Figure 4.3. Section through a cortical parenchyma cell of the spadix of *Sauromatum guttatum* showing the close association between microbodies (Mb) and endoplasmic reticulum (ER). Published micrograph of Berger and Schnepf (1970). Bar = 1 μm.

this electron-dense precipitate is not removed during the usual washing and dehydrating procedures of the electron microscopist, DAB would appear to be a useful substance for localizing peroxidatic activity when H_2O_2 is present as an electron acceptor. However, one should be aware of the fact that DAB can donate electrons nonenzymically to a number of biologically important molecules—for example, haem proteins such as cytochrome *c* (Roels et al., 1975; Shnitka and Seligman, 1971) and also the photosynthetic electron transport chain (Chua, 1972). Catalase is also a haem protein and reacts with H_2O_2 both oxidatively and peroxidatively (Sies, 1974), although it is unclear which mechanism prevails *in situ* [see Gerhardt (1978) for a discussion]. For its cytochemical localization with DAB, care must be taken to differentiate it from other peroxidases.

A number of procedures have been developed to enable one to do this with varying degrees of specificity. Negative control experiments are, for example, those in which the reaction is carried out in the presence of substances which inhibit catalase (aminotriazole; Margoliash and Novogrodsky, 1958), or in which the catalytic rather than the peroxidative activity of catalase is favored [high, 1–2%, concentrations of H_2O_2 (Fahimi, 1969; Vigil, 1970)], or by choosing neutral pH conditions whereby peroxidases but not catalase are active (Vigil, 1969). Positive control experiments are those involving the inclusion of 1 m*M* KCN which inhibits cytochrome *c*-oxidase based mitochondrial DAB oxidation but is without effect on catalase (Graham and Karnovsky, 1966; Novikoff et al., 1972) or by carrying out the reaction at pH 6 in the presence of Mn^{2+} ions (Poux, 1972). A most useful property of catalase is that its activity is not destroyed by glutaraldehyde fixation, indeed the latter enhances activity. Herzog and Fahimi (1974) have shown that purified catalase has an optimal peroxidative activity after fixation with 6% glutaraldehyde, a treatment which in addition leads to an irreversible inhibition of the catalytic activity of the enzyme. In contrast, peroxidases generally lose their activity after glutaraldehyde fixation (Roehls et al., 1975).

When carried out under conditions optimal for the peroxidative activity of catalase, the DAB reaction product is almost exclusively found in microbodies (Figure 4.1c), confirming its localization as measured on isolated organelles (Figure 4.4). This is true for the peroxisomes and glyoxysomes of higher plants, but with algae considerable difficulty with the DAB method has been encountered. Although for many algae catalase activity has been measured *in vitro* (see Section 4.6), there are numerous examples where a positive DAB reaction in microbodies has not been obtained (Bibby and Dodge, 1973; Gerhardt and Berger, 1971; Silverberg, 1975). In some of these cases a positive reaction can be achieved but only after considerable changes in the usual reaction conditions have been made (Silverberg, 1975). Only in *Euglena gracilis* has a negative DAB reaction been correlated with the lack of measurable catalase activity as determined *in vitro* (Brown et al., 1975; Graves et al., 1971, 1972).

4.2.2. Other Methods

Another reaction successfully adapted for cytochemical studies on microbodies is the reduction of ferricyanide to ferrocyanide which, in the presence of copper sulfate, results in the production of the insoluble, electron-dense copper ferrocyanide, $Cu_2[Fe(CN)_6]$. This has been employed by Trelease and Becker (1972) and Trelease et al. (1974) for the localization of the glyoxysomal enzyme malate synthase. The electron donor for the reaction is reduced coenzyme A released upon the formation of malate from glyoxylate and acetyl-CoA in the glyoxylate cycle. Glyoxysomes stain very strongly using this method (Figure 4.2c) but less specifically than with DAB due to the presence of deacylases elsewhere in the cell which also produce reduced CoA.

In addition to methods which employ the properties of an enzymatic reaction, there is now the possibility of identifying the enzyme itself with the help of immunological methods (both through fluorescence- or ferritin-binding techniques). Particularly since such methods allow a distinction to be made between isoenzymes, they are clearly the methods of choice for visualizing microbody enzymes *in situ*. It has thus been possible, in the case of malate dehydrogenase, to differentiate between glyoxysomal, mitochondrial, and cytosolic isoenzymes (Sautter and Hock, 1982).

4.3. ISOLATION

4.3.1. Markers

Microbodies can be detected in sucrose density gradients by testing for enzymes involved in the various metabolic events which are supposed to occur in them. Catalase is prominent in both glyoxysomes and leaf peroxisomes; in fact, according to Gerhardt (1978, p. 62), the rate of H_2O_2 production and catalase content of plant microbodies is about 10–100 times higher than for liver peroxisomes.

Huang (1975) has analyzed the enzymic content of glyoxysomes from a number of different germinating seeds rich in fat reserves and has established similar proportional activities for catalase, the enzymes of the glyoxylate pathway, glycolate oxidase, and alkaline lipase (Table 4.1). In addition, uricase and D-amino acid oxidase have been measured in glyoxysomes (Theimer and Beevers, 1971; Tajima and Yamamoto, 1975). Leaf peroxisomes, while sharing some of the enzymes found in glyoxysomes (e.g., catalase, malate dehydrogenase), possess the two aminotransferases: serine-glyoxylate and glutamate-glyoxylate and hydroxypyruvate reductase together with considerably higher levels of glycolate oxidase (Gerhardt, 1978; Tolbert, 1980).

The separation of membrane from content through osmotic rupture or sonification can easily be carried out (Huang and Beevers, 1973). By washing with 0.2 *M* KCl, peripheral and unspecifically adsorped proteins

TABLE 4.1. ENZYME ACTIVITIES IN GLYOXYSOMES FROM DIFFERENT FATTY SEEDLINGS

Enzymes	Castor Bean Endosperm	Watermelon Cotyledon	Peanut Cotyledon	Cucumber Cotyledon	Pine Megagametophyte
	(nmole/min/mg protein)				
Catalase	13,000,000	9,900,000	11,000,000	11,000,000	9,500,000
Malate dehydrogenase	38,000	55,000	34,000	80,000	60,000
Fatty acyl-CoA dehydrogenase	6,500	7,100	6,200	5,600	6,400
Malate synthetase	2,100	1,100	1,300	1,600	2,100
Isocitrate lyase	930	520	300	330	480
Citrate synthetase	840	700	330	800	350
Glycolate oxidase	95	70	58	89	Not checked
Alkaline lipase	19	4	7	2	8

Source: From Huang, 1975.

can be removed from these "ghost" preparations. Several glyoxysomal marker enzymes are known to be peripherally associated with the limiting membrane, for example, citrate and malate synthases and some of the enzymes responsible for β-oxidations (Biegelmayer et al., 1973; Huang and Beevers, 1973; Kagawa and Gonzalez, 1981). The others (e.g., catalase, uricase, isocitrate lyase, and possibly malate dehydrogenase) are matrix proteins as well [see Table 7.13 in Gerhardt (1978)]. For leaf peroxisomes a similar situation exists with glycolate oxidase, hydroxypruvate reductase, the amino transferases, and catalase being soluble matrix enzymes (Donaldson et al., 1972; Huang and Beevers, 1973). A recent report (Harson et al., 1983) has shown that those proteins which bind specifically to the luminal surface of the glyoxysomal membrane are attached through the oligosaccharide side chains of membrane glycoproteins. Although glyoxysomal enzymes do not bind to ER membranes, similarities do exist between ER and glyoxysomal membranes in terms of their phospholipid composition and in the presence of NADH-dependent antimycin-A-insensitive CCR activity in both (Donaldson and Beevers, 1977; Donaldson et al., 1972).

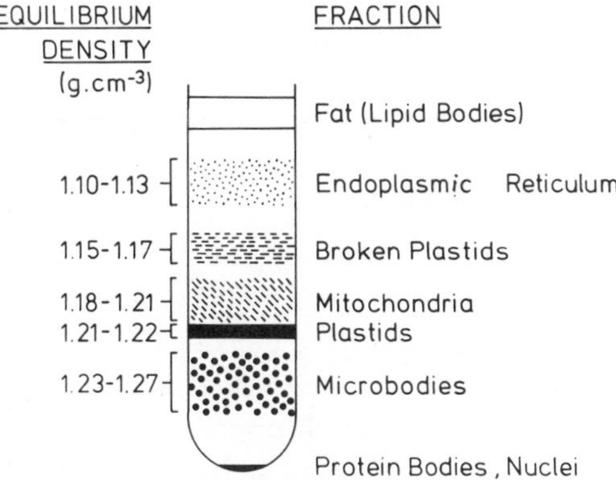

Figure 4.4. Schematic portrayal of the distribution of organelles from cotyledon or endosperm tissue of germinating fat-rich seedlings in an isopycnic sucrose density gradient.

4.3.2. Separation

The methodology of microbody isolation has been described on several occasions (Beevers et al., 1974; Tolbert, 1974; Vigil, 1983) and has mainly involved the separation of organelles by isopycnic sucrose density gradient centrifugation. As a rule, due to the higher cytoplasm–vacuole ratio, the yield of glyoxysomes from germinating seeds is higher than that for peroxisomes from leaves. In order to avoid damaging the microbodies, homogenization methods employing low shear are preferred. As a measure of intactness the "latency" of putative microbody fractions is tested (see Chapter 1). Unfortunately, as Vigil (1983) stresses, there is a good possibility that pleomorphic microbodies of the glyoxysomal type are disrupted and rapidly reanneal giving rise to an artificial population. Indirect confirmation of this comes from the demonstration of direct connections between glyoxysome and lipid body (Wanner et al., 1982; see also Section 11.7 and accompanying figures). These connections are not preserved, as witnessed by the fact that lipid bodies and glyoxysomes are always well separated on sucrose gradients.

After isopycnic conditions have been attained, microbodies usually band around 1.25 $g \cdot cm^{-3}$ in linear sucrose gradients (Figure 4.4). There are conflicting values in the literature for the relative sedimentation velocities of mitochondria and microbodies. According to Rocha and Ting (1970), mitochondria from spinach leaves sediment slower than microbodies, but the converse is given for microbodies from liver (De Duve, 1969) and yeasts (Avers, 1971). In the case of the alga *Polytomella caeca,* Gerhardt (1971) has shown that while equilibrium conditions for mitochondria are attained after 5 hr at 60000*g*, a further 10 hr are required for the microbodies. Equilibrium densities lower than those given above have been recorded for microbodies from several different suspension-cultured higher plant cells. Working with soybean cells, Moore and Beevers (1974) found that the maximum equilibrium densities for mitochondria and microbodies are obtained with stationary phase cultures. Catalase profiles for cells in the lag phase of growth appreciably overlap mitochondrial CCO profiles. A similar distribution was also obtained for sugar cane cells (Robinson and Glas, 1983). Kudielka et al. (1981) in contrast reported peaks at densities corresponding to ER fractions (around 1.13 $g{\cdot}cm^{-3}$) for anise cultures grown in sucrose-containing medium. Cultures starved of sucrose and then grown in acetate medium gradually develop a catalase profile with a peak at densities around 1.22 $g{\cdot}cm^{-3}$. This effect is analogous to that studied more frequently in algae and fungi (see Section 4.5).

One of the main reasons for the discrepancies in equilibrium density just mentioned is that there is in fact a sucrose uptake by microbodies, the extent of which can vary from species to species. In attempting to overcome this problem, centrifugations with Percoll (see also Chapter 1) have recently been introduced (Mettler and Beevers, 1980; Riezman and Becker, 1981; Schwitzguebel et al., 1981; Sautter et al., 1982). With this substance, mitochondria and microbodies are obtained in excellent structural condition, although their densities are much lower (1.07–1.09 $g{\cdot}cm^{-3}$ and 1.08–1.10 $g{\cdot}cm^{-3}$, respectively).

4.4. GLYOXYSOME–PEROXISOME TRANSITION

Glyoxysomes are abundant in the cotyledons of germinating seedlings rich in fat. In the early stages of germination the cotyledons are essentially deprived of light and gradually become exposed to it. This event results in a decline of measurable activity for glyoxylate cycle and β-oxidation pathway enzymes. On the other hand, enzymes associated with photorespiration (e.g., glycolate oxidase, hydroxypyruvate reductase) appear. This glyoxysome–peroxisome transition has attracted great interest over the last 15 years with opinion being divided over "one-" or "two-population models".

According to the one-population hypothesis (Trelease et al., 1971), the enzyme complement of the glyoxysome is exchanged for that of the peroxisome. In the two-population hypothesis championed by Beevers and colleagues (Kagawa and Beevers, 1975; Beevers, 1979) the transition phase is characterized by the simultaneous destruction of glyoxysomes and the *de novo* synthesis of leaf peroxisomes. Implicit in the one-population hypothesis is the posttranslational insertion of the enzymes of photorespiration, although such a phenomenon was unknown at the time it was proposed. With the demonstration that both peroxisomal and glyoxysomal enzymes are synthesized on cytosolic rather than ER-bound polysomes (see Kindl, 1982, and Section 11.7, this volume) much of the conceptual difficulty in accepting this hypothesis is done away with. In this vein, recent work by Kindl's group (Gerdes et al., 1982) on the synthesis of glycolate oxidase may be taken as strong, though not unequivocal, support for the one-population hypothesis. They have shown that during the greening of cucumber cotyledons a considerable increase in mRNA for this enzyme is seen and small oligomeric forms (<5S) are detectable in the cytosol before their insertion and assembly into larger complexes (8S–11S) in the microbody. Furthermore, at a stage when the matrix is becoming characterized by leaf peroxisomal enzymes (glycolate oxidase, β-hydroxypyruvate reductase, serine–glyoxylate transaminase), the membrane of the transition microbody still possesses glyoxysomal enzymes (e.g., malate synthase).

The decline in glyoxylate cycle enzyme activity as a consequence of fat depletion has been made synonymous with a disappearance of the glyoxysome (Beevers, 1979), although there is no evidence for this from ultrastructural studies (Burke and Trelease, 1975; Gruber et al., 1970; Trelease et al., 1971). Unfortunately, an abrupt loss of lipid bodies and a subsequent proplastid–chloroplast development is not seen, which makes the interpretation of these papers somewhat subjective. Nevertheless, a population of smaller microbodies in the transition phase during greening of the cotyledons, which might be expected on the basis of the destruction of the glyoxysomes and the budding-off of leaf peroxisomes from the ER, is not apparent. Furthermore, one must ask what kind of mechanism is to be envisaged for the destruction of the glyoxysomes. Although there are no electron micrographs available which depict this situation, Hock (personal communication) has demonstrated by immunofluorescence the presence of glyoxysomal malate dehydrogenase in vacuoles of dark-grown watermelon seedlings. If one considers the vacuole as being the lysosomal compartment of the cell (see Chapter 12), this might indicate at least the turnover of glyoxysomal contents.

It is sometimes assumed that leaf peroxisomes do not contain the enzymes associated with fatty acid β-oxidation and glyoxylate cycle (Vigil, 1983; Tolbert, 1980), although there are reports to the contrary (Kindl and Majunke, 1973; Gerhardt, 1981). The changes in enzymic complement in the glyoxysome–peroxisome transition are therefore not absolute but relative and, according to Gerdes et al. (1982), reflect the availability of soluble pool precursors, since "microbodies probably cannot select if several different, but appropriate precursors are at their disposal" (Kindl, 1982).

4.5. MICROBODIES IN ALGAE AND FUNGI

Depending on the nature of the carbon source, microbodies of a glyoxysomal or leaf peroxisomal character may be induced in *Euglena* (Collins et al., 1975). When grown heterotrophically on acetate medium, glyoxysomes are present. In contrast, microbodies in photoautotrophically grown cells have a leaf peroxisomal nature. Under photoheterotrophic conditions, both the enzymes of the glyoxylate cycle and glycolate metabolism are present. Unfortunately, this nice analog to the situation during greening in higher plants is not universally true for algae. Thus, in the Chlorophyceae several of the key enzymes of the glyoxylate cycle appear to be localized in the cytosol (see Gerhardt, 1978, chapter 9). The oxidation of glycolate in algae can proceed with the help of glycolate oxidase or with glycolate dehydrogenase. In those algae where the former is the case the activity and that of hydroxypyruvate reductase is detectable in a microbody fraction (Codd et al., 1973; Stabenau, 1975). In other algae, particularly the chlamydomonads, glycolate oxidase is localized in the mitochondria (Stabenau, 1974; Stabenau and Beevers, 1974) rather than in microbodies, whereby the oxidation (without H_2O_2 production) is probably coupled to the electron transport chain.

In contrast to microbodies from algae and higher plants, those from fungi possess a lower equilibrium density (1.17–1.20 $g \cdot cm^{-3}$; see Chapter

10, Gerhardt, 1978). Catalase as a microbody marker is not measurable as a rule in cells cultivated with glucose or sucrose but develops when ethanol, acetate, or amino acids are substituted as the C source. Under these conditions considerably higher activities of the glyoxylate cycle enzymes isocitrate lyase and malate synthase are observed (Haarasilta and Oura, 1975). While two enzymes are principally found in the microbodies of *Neurosposa crassa* (Flavell and Woodward, 1971; Theimer et al., 1978), there are conflicting reports (cf. Avers, 1971, with Parish, 1975) as to their cytosolic location in *Saccharromyces cerevisiae*. Since the two enzymes citrate synthase and malate dehydrogenase are localized in the mitochondria of *N. crassa* (Kobr and Vanderhaeghe, 1973) and *Candida tropicalis* (Kawamoto et al., 1977), it is clear that the entire glyoxylate cycle is not housed in the microbodies.

REFERENCES

Avers, C. J. (1971). *Sub-Cell Biochem.* **1,** 25–37.

Beevers, H. (1979). *Ann. Rev. Plant Physiol.* **30,** 159–193.

Beevers, H. (1980). In *The Biochemistry of Plants* (P. K. Stumpf, ed.), Vol. 4, pp. 117–130. Academic Press, New York.

Beevers, H., Theimer, R. R., and Feierabenc, J. (1974). In *Biochemische Cytologie der Pflanzenzelle. Ein Praktikum* (G. Jacobi, ed.), pp. 127–146. Thieme Verlag, Stuttgart.

Berger, C. and Schnepf, E. (1970). *Protoplasma* **69,** 237–251.

Bibby, B. T. and Dodge, J. D. (1973). *Planta* **112,** 7–16.

Biegelmayer, C., Graf, J., and Ruis, H. (1973). *Eur. J. Biochem.* **37,** 553–562.

Black, C. C. (1973). *Ann. Rev. Plant Physiol.* **24,** 253–286.

Black, C. C. and Mollenhauer, H. H. (1971). *Plant Physiol.* **47,** 15–23.

Brown, R. H., Collins, N., and Merrett, M. J. (1975). *Plant Physiol.* **55,** 1123–1124.

Chollet, R. (1974). *Arch. Biochem. Biophys.* **163,** 521–529.

Chollet, R., and Ogran, W. L. (1975). *Botan. Rev.* **41,** 137–180.

Chua, N.-H. (1972). *Biochim. Biophys. Acta.* **267,** 179–189.

Codd, G. A., Schmid, G. H., and Kowallik, W. (1973). *Arch. Mikrobiol.* **92,** 21–38.

Collins, N., Brown, R. H., and Merrett, M. J. (1975). *Plant Physiol.* **55,** 1018–1022.

Cooper, T. G. and Beevers, H. (1969). *J. Biol. Chem.* **244,** 3514–3520.

de Duve, C. (1965). *J. Cell. Biol.* **27,** 25A.

de Duve, C. (1969). *Proc. Roy. Soc. B* **173,** 71–83.

Donaldson, R. P. and Beevers, H. (1977). *Plant Physiol.* **59,** 259–263.

Donaldson, R. P., Tolbert, N. E., and Schnarrenberger, C. (1972). *Arch. Biochem. Biophys.* **152,** 199–215.

Fahimi, H. D. (1969). *J. Cell Biol.* **42,** 257–288.

Flavell, R. B. and Woodward, D. O. (1971). *J. Bacterial.* **105,** 200–210.

Frederick, S. E. and Newcomb, E. H. (1969). *J. Cell Biol.* **43,** 343–353.

Frederick, S. E. and Newcomb, E. H. (1971). *Planta* **96,** 152–174.

Frederick, S. E., Gruber, P. J., and Newcomb, E. H. (1975). *Protoplasma* **84,** 1–29.

Frederick, S. E., Gruber, P. J., Vigil, E. L., and Wergin, W. P. (1968). *Planta* **81,** 229–252.

Gerdes, H.-H., Behrends, W., and Kindl, H. (1982). *Planta* **156,** 572–578.

Gerhardt, B. (1971). *Arch. Mikrobiol.* **80,** 205–218.

Gerhardt, B. (1978). *Cell Biology Monographs,* Vol. 5. Springer Verlag, Vienna.

Gerhardt, B. (1981). *FEBS Lett.* **126,** 71–73.

Gerhardt, B. and Berger, C. (1971). *Planta* **100,** 155–166.

Goeckermann, J. A. and Vigil, E. L. (1975). *J. Histochem.* **23,** 957–973.

Graham, R. C. and Karnovsky, M. J. (1966). *J. Histochem. Cytochem.* **14,** 291–302.

Graves, L. B., Hanzely, L., and Trelease, R. N. (1971). *Protoplasma* **72,** 141–152.

Graves, L. B., Trelease, R. N., Grill, A., and Becker, W. M. (1972). *J. Protozool.* **19,** 527–532.

Gruber, P. J., Becker, W. M., and Newcomb, E. H. (1972). *Planta* **105,** 114–138.

Gruber, P. J., Becker, W. M., and Newcomb, E. H. (1973). *J. Cell Biol.* **56,** 500–518.

Gruber, P. J., Trelease, R. N., Becker, W. M., and Newcomb, E. H. (1970). *Planta* **93,** 269–288.

Gutierrez, M., Huber, S. C., Ku, S. B., Kanai, R., and Edwards, G. E. (1974). In *Proceedings of the Third International Congress on Photosynthesis Research* (M. M. Avron, ed.), pp. 1219–1320. Elsevier, Amsterdam.

Haarasilta, S. and Oura, E. (1975). *Eur. J. Biochem.* **52,** 1–18.

Harson, M. M., Cooper, M. J., and Lord, J. M. (1983). *Planta* **157,** 143–149.

Herzog, V. and Fahimi, H. D. (1974). *J. Cell Biol.* **60,** 303–310.

Hilliard, J. H., Gracen, V. E., and West, S. H. (1971). *Planta* **97,** 93–105.

Huang, A. H. C. (1975). *Plant Physiol.* **55,** 870–874.

Huang, A. H. C. and Beevers, H. (1971). *Plant Physiol.* **48,** 637–641.

Huang, A. H. C. and Beevers, H. (1972). *Plant Physiol.* **50,** 242–248.

Huang, A. H. C. and Beevers, H. (1973). *J. Cell Biol.* **58,** 379–389.

Huang, A. H. C., Trelease, R. N., and Moore, T. S. (1983). *Plant Peroxisomes,* Academic Press, New York.

Kagawa, T. and Beevers, H. (1975). *Plant Physiol.* **55,** 258–264.

Kagawa, T. and Gonzalez, E. (1981). *Plant Physiol.* **68,** 845–850.

Kawamoto, S., Tanaka, A., Yamamura, M., Terainski, Y., Fukui, S., and Osumi, M. (1977). *Arch. Microbiol.* **112,** 1–8.

Kindl, H. (1982). *Int. Rev. Cytol.* **80,** 193–229.

Kindl, H. and Majunke, G. (1973). *Hoppe Seyler's Z. Physiol. Chem.* **354,** 999–1005.

Kobr, M. J. and Vanderhaeghe, F. (1973). *Experientia* **29,** 1221–1223.

Kudielka, R. A., Kock, H., and Theimer, R. R. (1981). *FEBS Lett.* **136,** 8–12.

Laetsch, W. M. (1974). *Ann. Rev. Plant Physiol.* **25,** 27–52.

Margoliash, E. and Novogrodsky, A. (1958). *Biochem. J.* **68,** 468–475.

Matsushima, H. (1972). *J. Electron Microscopy* **21,** 293–299.

Mettler, I. J. and Beevers, H. (1980). *Plant Physiol.* **66,** 555–560.

Moore, T. S. and Beevers, H. (1974). *Plant Physiol.* **53,** 261–265.

Nilshammer, M. and B. Walles, (1974). *Protoplasma* **79,** 317–332.

Novikoff, P. M. and Novikoff, A. B. (1972). *J. Cell Biol.* **53,** 532–560.

Novikoff, A. B., Novikoff, P. M., Davis, C., and Quintana, N. (1972). *J. Histochem. Cytochem.* **20,** 1006–1023.

Parish, R. W. (1975). *Arch. Bicrobiol.* **105,** 187–192.

Poux, N. (1972). *J. Microscopie* **14,** 183–201.

Rathnam, C. K. and Edwards, G. E. (1975). *Arch. Biochem. Biophys.* **171,** 214–225.

Rehfeld, D. W., Randall, D. D., and Tolbert, N. E. (1970). *Can. J. Bot.* **48,** 1219–1226.

Riezman, H. and Becker, W. M. (1981). *Plant Physiol.* **65,** *Suppl.* 32.

Robinson, D. G. and Glas, R. (1983). *J. Exp. Bot.* **34,** 668–675.

Rocha, V. and Ting, I. P. (1970). *Arch. Biochem. Biophys.* **140,** 398–407.

Roels, F., Wisse, E., de Prest, B., and Meulen, J. v. d. (1975). *Histochem.* **41,** 281–312.

Sautter, C. and Hock, B. (1982). *Plant Physiol.* **70,** 1162–1168.

Sautter, C., Kollmannsberger, E., and Hock, B. (1982). *GIT Fachzeit. Lab.* **26,** 550–555.

Schwitzguebel, J. P., Möller, I. M., and Palmer, T. M. (1981). *J. Gen. Microbiol.* **120,** 289–296.

Seligman, A. M., Karnovsky, M. J., Wasserkrug, H. L., and Hanker, J. S. (1968). *J. Cell. Biol.* **38,** 1–14.

Shnitka, T. K. and Seligman, A. M. (1971). *Ann. Rev. Biochem.* **40,** 375–396.

Sies, H. (1974). *Angew. Chem.* **86,** 789–801.

Silverberg, B. A. (1975). *Protoplasma* **83,** 269–295.

Silverberg, B. A. and Sawa, T. (1973). *Can. J. Bot.* **5,** 2025–2032.

Stabenau, H. (1974). *Plant Physiol.* **54,** 921–924.

Stabenau, H. (1975). *Ber. Dtsch. Bot. Ges.* **88,** 469–471.

Stabenau, H. and Beevers, H. (1974). *Plant Physiol.* **53,** 866–869.

Tajima, S. and Yamamoto, Y. (1975). *Plant Cell Physiol.* **16,** 271–282.

Theimer, R. R. and Beevers, H. (1971). *Plant Physiol.* **47,** 246–251.

Theimer, R. R., Wanner, G., and Anding, G. (1978). *Cytobiol.* **18,** 132–144.

Tolbert, N. E. (1974). In *Methods in Enzymology* (S. Fleischer and L. Packer, eds.), Vol. XXXI, Part A, pp. 734–746. Academic Press, New York.

Tolbert, N. E. (1980). *Biochem. of Plants* **1,** 359–388.

Tolbert, N. E. and Essner, E. (1981). *J. Cell. Biol.* **91,** 271s–283s.

Tolbert, N. E. and Yamazaki, R. K. (1969). *Ann. NY Acad. Sci.* **168,** 425–441.

Trelease, R. N. and Becker, W. M. (1972). *J. Cell Biol.* **55,** 262a.

Trelease, R. N., Becker, W. M., and Burke, J. H. (1974). *J. Cell Biol.* **60,** 483–495.

Trelease, R. N., Becker, W. M., Gruber, P. J., and Newcomb, E. N. (1971). *Plant Physiol.* **48,** 461–475.

Tsukuda, H., Koyama, S., Gotoh, M., and Tadano, H. (1971). *J. Ultr. Res.* **36,** 159–175.

Tsukuda, H., Mochizuki, Y., and Konishi, T. (1968). *J. Cell Biol.* **37**, 231–243.

Vigil, E. L. (1969). *J. Histochem. Cytochem.* **17**, 425–428.

Vigil, E. L. (1970). *J. Cell Biol.* **46**, 435–454.

Vigil, E. L. (1973). *Sub. Cell Biochem.* **2**, 237–285.

Vigil, E. L. (1983). In *Isolation of Membranes and Organelles from Plant Cells* (J. L. Hall and A. L. Moore, eds.), pp. 211–236. Academic Press, New York.

Wanner, G., Vigil, E. L., and Theimer, R. R. (1982). *Planta* **156**, 314–325.

Zelitch, I. (1971). *Photosynthesis, Photorespiration and Plant Productivity*. Academic Press, New York.

5

THE GOLGI APPARATUS

The Golgi apparatus (GA) is the collective name for the complement of dictyosomes and secretory vesicles in a eukaryotic cell. In contrast to the literature on plant cells one encounters the word *dictyosome* much less frequently in description of the animal Golgi apparatus (see, e.g., Farquhar and Palade, 1981). This is not simply a difference in the terminologies used by botanists and zoologists (an example of this is the much more frequent use of the word *Plasmalemma* instead of *plasma membrane* by the former group) but unintentionally reflects differences in form and function of the GA in these two major phyla.

It is universally acknowledged that the GA is involved in secretion. If one therefore compares the GA in electron micrographs of, for example, root tip cells with glandular cells such as liver or pancreas, one can immediately appreciate two major morphological differences. The first is the much more discrete nature of the dictyosome in the plant cell and the second is that there are usually more of them per cell than in animal cells.

There are numerous descriptions of the plant GA in the literature [for recent reviews see Mollenhauer and Morré (1980), Morré (1976), and Robinson and Kristen (1982)]. Nevertheless, although the quality of electron micrographs, which depict the salient features of this organelle, have hardly been improved upon over the last 15 years, there has been an unfortunate tendency during this period to overinterpret. As a result the dictyosome has often achieved a uniform morphology, characteristic for all eukaryotic cells and faithfully followed in most textbooks. Generalizations of this type are misleading and certainly do not reflect the great versatility in form and function of this organelle.

5.1. THE BASIC UNIT: THE CISTERNA

The lowest level of organization of the GA, be it plant or animal in origin, is the cisterna. This consists of two parts: a disclike central plate region together with peripheral elaborations extending from it. The central plate region when sectioned is essentially a flattened vesicle usually with a diameter of 0.5–1 μm, although in some algae greater sizes may be encountered. Because of its continuous nature central plate sizes are often mistakenly given for those of the complete cisterna or for that matter the dictyosome. A look at negatively stained cisternae or flat sections through a cisternal stack (Figure 5.1a–d) demonstrates how dangerous such un-

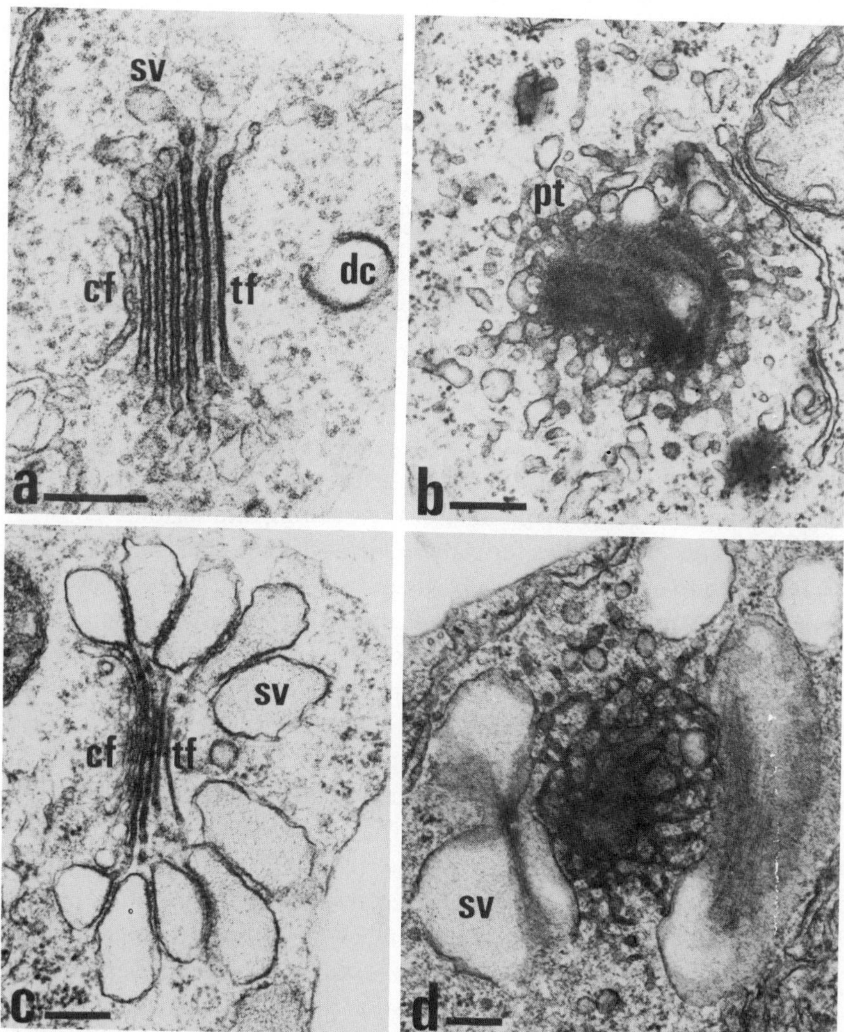

Figure 5.1. Dictyosomes of higher plants (a,b—suspension-cultured sycamore cells; c,d—maize root cap cells) sectioned perpendicular (a,c) and plane (b,d) with respect to the cisternae of the stack (from Robinson 1980; enlarged): cf—*cis*-face; dc—degenerating cisterna; ie—intercisternal elements; pt—peripheral tubules; sv—secretory vesicle; tf—*trans*-face. Bar = 0.2 μm.

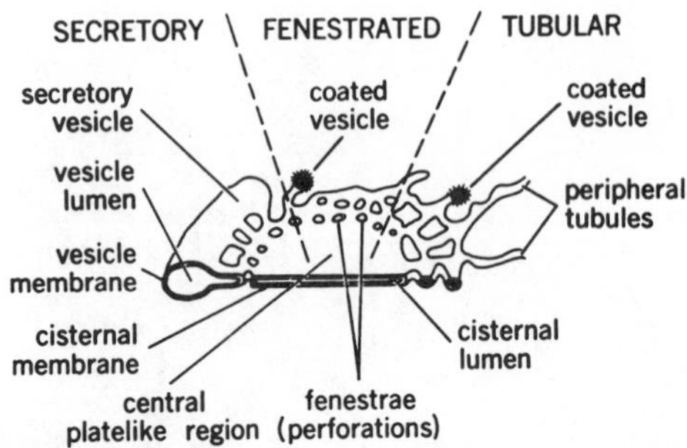

Figure 5.2. Composite diagram of the various types of higher plant GA cisternae. The central plate region is common to all, but the peripheral elaborations can be quite varied. (From Mollenhauer and Morré, 1980).

derstimates really are. It is difficult to generalize, but elaborations at the periphery of the central plate may bring the total extent or diameter of the cisterna to at least double that of the central plate region.

The various possibilities for elaboration at the periphery of the central plate region are diagrammatically given in Figure 5.2. For all cisternal types a fenestration or perforation of the central plate is the immediate peripheral elaboration. In some types this gives way to an extensive network of anastomosing tubules (30–50 nm in diameter). Anastomoses between peripheral tubules from adjacent cisternae also seem possible (Kristen et al., 1984). In addition, direct connections between GA cisternae and elements of the ER via such tubular protuberances have been recorded (e.g., Poux, 1973; see also "GERL" in Section 5.4.). With respect to higher plants this is certainly not the rule (Juniper et al., 1982; Robinson, 1980) and appears to be restricted to gland cells secreting protein rather than polysaccharide (Juniper et al., 1982; Kristen, 1980) or to cells in germinating seeds involved in the mobilization of protein reserves (Harris and Oparka, 1983). There is no question that direct connections of this type are difficult to visualize both in thin and thick sections (Figure 5.3) and may only be of a transitory nature. By reducing secretion through cold treatment (4 °C; several or more hours) it has been claimed that such connections in maize root cap cells become much clearer (Mollenhauer et al., 1975). These observations, based in part on $KMnO_4$ fixations, have not been confirmed in reinvestigations using the same tissue (Robinson, 1981a).

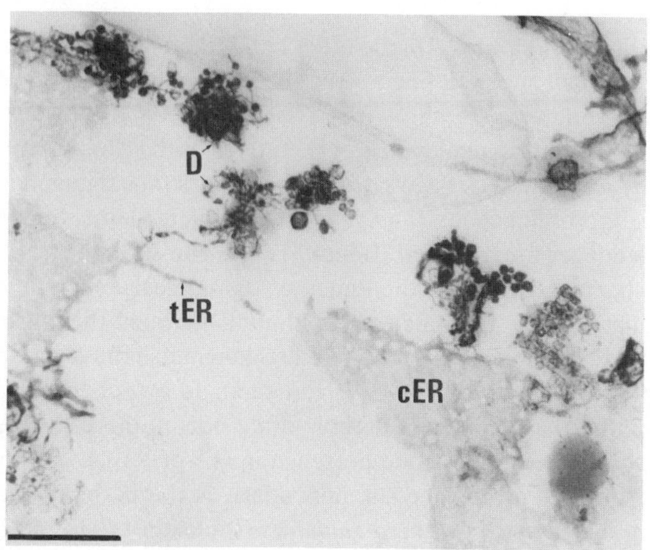

Figure 5.3. 1-μm thick section through a developing wheat endosperm cell (20 days postanthesis) visualized by high-voltage (1-MV) electron microscopy after zinc iodide–osmium impregnation. Possible tubular connections between dictyosomal and ER cisternae are indicated by arrows (from Parker and Hawes, 1982): cER—cisternal ER; D—dictyosome; tER—tubular ER. Bar = 1 μm.

Single tubular extensions may end in either coated (see Section 5.6) or smooth-membraned secretory vesicles. Particularly in the mantle cells of monocotyledon root tips the secretory vesicles attain an appreciable size (Figure 5.1c,d). This is sometimes described as the hypertrophied condition (Rougier, 1981). Here the secretory vesicles are usually kidney shaped and are attached through several tubules to the cisterna (Robinson, 1980). Cisternae of this type illustrate most clearly the fact that the vesicles at their periphery are not all of the same type, since numerous attached coated vesicles are seen between paired hypertrophied secretory vesicles. Furthermore, vesicles are not always of the same type during cellular development. Thus at least five different types of GA vesicles are envisaged during the life cycle of the desmid *Micrasterias* (Kiermayer, 1981).

For the most part, attached secretory vesicles in higher plant cells have a diffuse content which has on numerous occasions (Roland and Sandoz, 1969; Ryser, 1979; Rougier, 1971) been shown to react positively with stains for polysaccharides. On the other hand, there are a number of algae in which a much more defined structural component is present in the secretory vesicle or cisterna.

Many algae, but particularly representatives of the Chrysophyceae and Xanthophyceae, possess flagella which bear hairs (mastigonemes). These consist of a proteinaceous tubular shaft to which filamentous carbohydrate appendages are attached (Bouck, 1972). The site of synthesis of the shaft has been shown to be the lumen of the nuclear envelope *per se* or its continuation (the "perinuclear continuum") around the plastids (Leedale et al., 1970; Bouck, 1971). The mastigoneme shafts are also seen in the peripheral regions of the GA, and there is cytochemical evidence indicating that the addition of carbohydrate occurs there (Mignot et al., 1972). Another example for the presence in GA cisternae of an extracellular, presumably proteinaceous, appendage is the barbs in the chlorococcalean alga *Acanthosphaera zacharasi* (Schnepf et al., 1982). These barbs are 200–300 nm long and are attached to the outer surface of long cellulose spikes that emanate from particular positions on the cell surface. During cell division (i.e., correlated with new spike formation) barbs are seen attached to the membranes and project into the lumen of inflated ER as well as peripheral regions of GA cisternae.

More interesting, perhaps because of their architectural complexity, are the scales secreted through the GA by various marine phytoplankton. These scales may have a microfibrillar (Haptophyceae) or nonmicrofibrillar skeleton and in at least one case (*Pleurochrysis scherffelii*) contain cellulose (Brown and Romanovicz, 1976). In addition, silica or calcite may be deposited onto these scales. The cytology of scale structure and production has been intensively studied, particularly by Manton and Brown and their co-workers [see Gunning and Steer (1975), Robinson (1981b), and Romanovicz (1981) for reviews]. In these organisms a clear distinction between central plate and peripheral portions to the cisterna is not present and the scales are more or less centrally located. The variability in synthetic capability as far as scale type is concerned is great. There are thus cells in which one scale per cisterna is produced with either all cisternae making the same kind of scale or different cisternae synthesizing different scales. Alternately, there are forms in which several scales per cisterna are assembled and the scales are either identical or different from one another. The scales usually have a dorsiventral nature and are oriented accordingly at the cell surface after their release. This

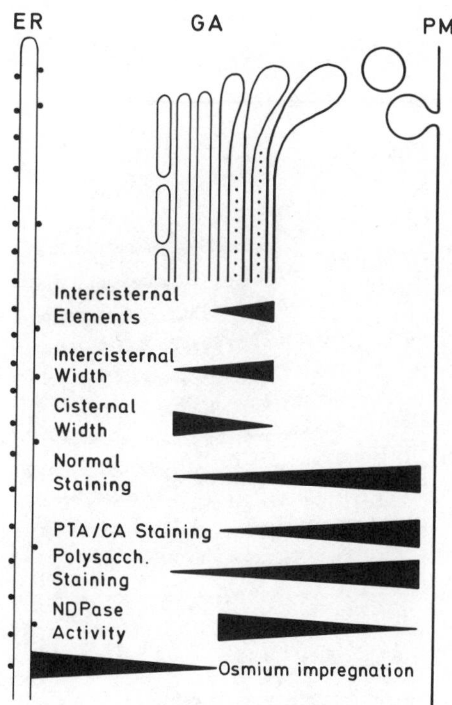

Figure 5.4. Diagrammatic representation of the various parameters contributing to polarity in the higher plant dictyosome. Gradients are marked accordingly as arrowheads. (From Robinson and Kristen, 1982.)

feature is reflected by their similar orientation in the cisternae and by the fusion of whole cisterna with the plasma membrane.

5.2. THE CISTERNAL STACK

Usually between 5 and 10 cisternae are stacked together to constitute the dictyosomes of higher plant cells, although in algae [e.g., the euglenoid and scaled flagellates (Dodge, 1975)] their number may be greater. The cisternae in an individual stack are not identical; instead there is a gradation in both structural and cytochemical terms from one side of the stack to the other. This is represented diagrammatically in Figure 5.4. It must be stressed, however, that this is a summation of all known polarity indicators. There are certainly cases (see, e.g., Juniper et al., 1982) where a gradient is unclear or abrupt or, perhaps due to poor fixation, not discernible. As shown by Mollenhauer and Morré (1978), however, a gradient

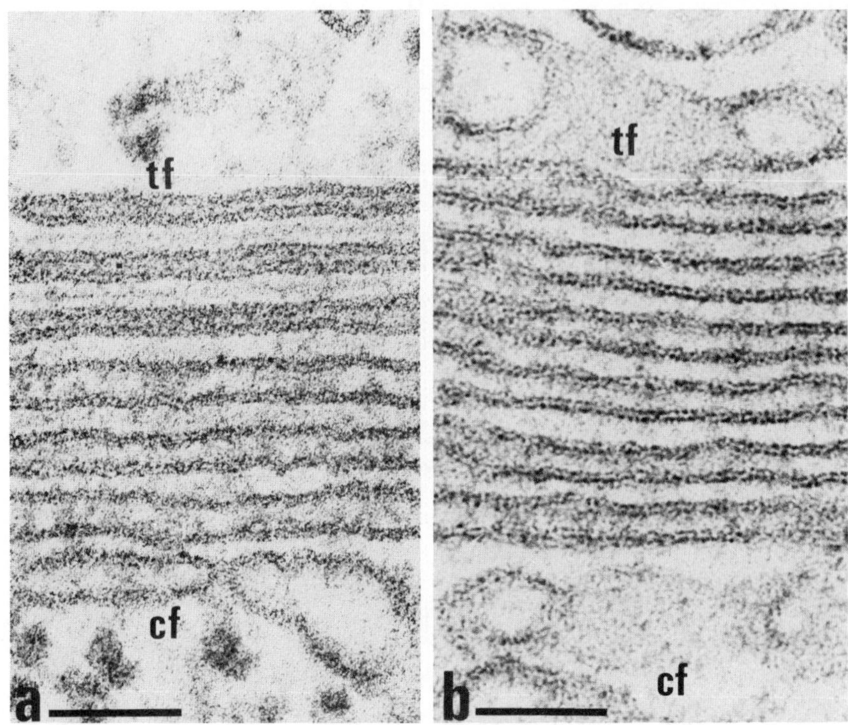

Figure 5.5. A comparison of stacked cisternae in the higher plant (a—from a red poppy stigmatic cell) and animal (b—rat epididymal cell) GA. The structural gradient is much clearer in the former than in the latter (from Mollenhauer and Morré, 1978): cf—*cis*-face; tf—*trans*-face. Bar = 0.05 μm.

across the cisternal stack in the GA from animal cells is much more difficult to recognize (Figure 5.5a,b).

In the plant literature it has been hitherto customary to designate the poles of the cisternal stack as forming and maturing or secretory faces, respectively. However, not just simply to comply with current terminology in the animal literature (e.g., Farquhar and Palade, 1981; Rothman, 1981) but because of their more neutral meaning, I have decided to adopt here the corresponding terms *cis* (forming) and *trans* (maturing).

From the cis to trans face the spacing between the cisternae (intercisternal width) increases while the spacing within the cisternae (intracisternal width) decreases. This contradistinction in morphological gradient has been recorded numerous times in the literature (e.g., Amelunxen and Gronau, 1966; Kristen, 1978; Mollenhauer and Morré, 1975, 1978, 1980; Ryser, 1979; Robinson, 1980). In a study involving a variety of higher plant cell types Mollenhauer and Morré (1978) have given average values for the spacing between terminal *cis*- and *trans*-face cisternae as 7.8 and 13.5 nm, respectively. Corresponding values for changes in intracisternal width are not given by these authors, indeed they have seldom been recorded. As a rule, however, the intracisternal width of terminal *cis* cisternae are usually 2–3 times those for terminal *trans* cisternae.

Correlated with the increase in intercisternal width is the increased presence of so-called intercisternal elements (Turner and Whaley, 1965; Mollenhauer and Morré, 1972). They are fibrous in nature (about 4 nm in diameter) and lie oft parallel to one another between the central plate-like portion of the cisternae. Although a role in cross-linking the cisternae has been proposed for them (Franke et al., 1972), their real function has yet to be defined since other substances appear to be involved in maintaining the cohesion of the cisternal stack. These are structurally less defined and are readily extracted by chaotropic agents (Mollenhauer et al., 1973). In those cells where large slime vesicles are seen attached at the periphery of the cisterna (e.g., in the root cap cells of grasses) the intercisternal fibers are restricted to this region. According to Mollenhauer and Morré (1975), they serve to organize the form and position of these secretory vesicles. Clearly, this negates the suggestion of Kristen (1978) whereby the intercisternal fibers are considered to effect a compression of the central portion of the stack, thereby allowing the production of vesicles at the periphery.

Membrane thickness and the intensity of staining in normal fixed-stained material (i.e., glutaraldehyde–osmium fixed, uranyl acetate–lead citrate poststained) are also indicators of polarity. Once again we must thank Morré and his co-workers (Grove et al., 1968; Morré and Ovtracht, 1977), for documenting these features. Thus, both membrane thickness and staining in the cisternal stack represent a gradient lying between those values for the ER (thin, lightly stained) and PM (thicker, more heavily stained).

Various special staining procedures also confirm the polar nature of the cisternal stack. Osmium impregnation (Morré and Mollenhauer, 1974) and zinc oxide–osmium staining (Dauwalder and Whaley, 1973; Juniper et al., 1982; Parker and Hawes, 1982; Shannon et al., 1982) both give

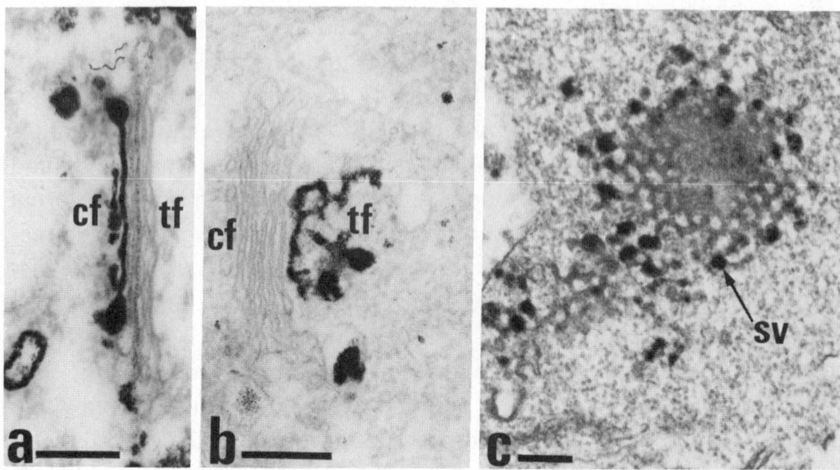

Figure 5.6. Cytochemistry of the higher plant GA. Euphorbic root tip dictyosomes stained (a) with zinc iodide–osmium tetroxide, (b) for acid phosphatase activity. Lepidium root tip dictyosome stained (c) for inosine diphosphatase activity [from Marty (1973, 1978) and Ellinger (unpublished)] cf—*cis*-face; sv—secretory vesicle; tf—*trans*-face. Bar = 0.25 μm.

strong reactions for the ER and *cis*-face cisternae. In contrast, staining for polysaccharides with the Thiéry test (Rougier, 1971; Dexheimer, 1981) is mainly associated with trans cisternae but more particularly mature and released secretory vesicles. Staining for nucleoside disphosphatase activity (Dauwalder et al., 1969; Zaar and Schnepf, 1969; Dexheimer, 1978) also shows a localization of the reaction in *trans* cisternae and especially the secretory vesicles (Figure 5.6a,c).

In addition to information obtained by thin-sectioning, evidence for polarity in the cisternal stack has also been provided by freeze-fracturing. Thus, a progressive increase in intramembranous particles (IMPs) from *cis*- to *trans*-face cisternae has been noted for dictyosomes from both algae (Staehelin and Kiermayer, 1970) and higher plants (Vian, 1974).

5.3. THE *CIS*-FACE

In a number of very persuasive articles Morré, Mollenhauer, and coworkers have laid great emphasis on the homology in form and function of the GA in plant and animal cells (Morré and Mollenhauer, 1974; Morré and Ovtracht, 1977; Morré, Mollenhauer, and Bracker, 1971; Morré, Franke, Deumling, Nyquist, and Ovtracht, 1971; Morré et al., 1979). The grand scheme of the "endomembrane concept," as it has been put forward in these publications, is appealing, but in my opinion significant differences between the systems described tend to have been played down somewhat. We have already seen that in the case of the higher plant dictyosome, a structural gradient within the cisternal stack is much more marked than in the animal GA. Similarly, there are deviations from the "norm" with respect to the *cis*-face in higher plant dictyosomes.

According to Mollenhauer and Morré (1980), "Presumably, vesicles from endoplasmic reticulum juxtaposed to the forming pole of the dictyosome transfer membrane to the Golgi apparatus. Vesicles from the endoplasmic reticulum accumulate approximately 100 Å from the forming pole of the dictyosome where they appear to fuse to form new cisternae." Although the authors are quick to add: "However, no unequivocal evidence is available to show product or membrane transfer between endoplasmic reticulum and Golgi apparatus in plants," one might be forgiven for thinking that it is only a question of time before the necessary proof is at hand. From the structural point of view, pertinent information on plants has been available for some time. While there is no doubt that many algae do indeed show a clear relationship between a specific segment of ER (or NE) and the GA (Figure 5.7) and that a traffic of small transition vesicles between the two occurs (e.g., Falk, 1976; Fraser and Gunning, 1973; Goodenough and Porter, 1968; Massalski and Leedale, 1969; Pickett-Heaps, 1971), the situation is not universally so in the plant kingdom.

As early as 1965 Sievers spoke of the cis face as the "überwiegende Vesikel-freie Seite des Dictyosoms" (predominantly vesicle-free face of the dictyosome) in a paper on the *Chara* rhizoid. In oft published micrographs of Turner (given as Abb. 2 in Schnepf, 1969, and as Figure 19 in Whaley, 1975) the absence of ER and of an accumulation of ER-derived transition vesicles at the *cis*-face of the *Nitella* dictyosome is strikingly apparent. It may be a coincidence, but that which is shown by the above two members of the Characeae, which is to be regarded as perhaps the most advanced algal family Pickett-Heaps, 1975), is also shared by many higher plants.

That a close and specific relationship between ER and GA in higher plants is not often seen has been noted by Gunning and Steer (1975, p.

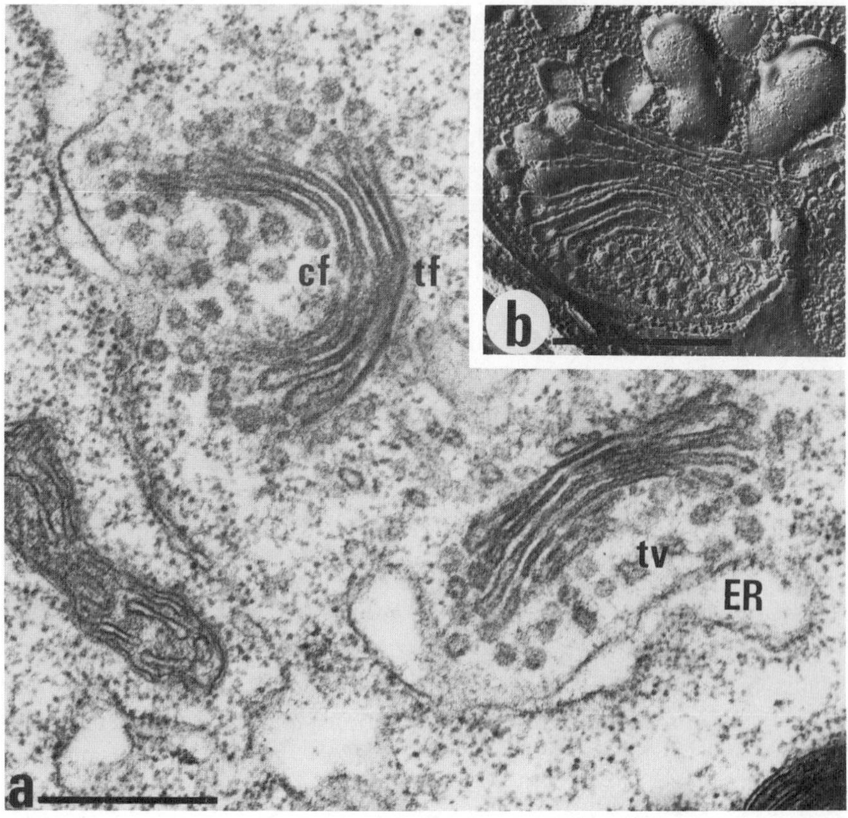

Figure 5.7. The GA in *Chlamydomonas smithii*. In many algae, particularly flagellates, a clear relationship between dictyosome and ER can be seen. This spatial association is often accompanied by the easily discernible traffic of transition vesicles between the two (unpublished micrograph of the author): cf—*cis*-face; tf—*trans*-face; tv—transition vesicle. Bar = 0.5 μm.

85) and may well be related to a much more vigorous cytoplasmic streaming in these cells as opposed to specialized animal secretory cells or flagellates where such relationships are easily seen. Cytoplasmic streaming might also be one reason why transition vesicles do not normally accumulate at the *cis*-face of higher plant dictyosomes. This feature has been occasionally remarked upon (e.g., Gunning and Steer, 1975, plate 28, p. 238) but was highlighted in a publication in 1980 by Robinson. Dictyosomes from suspension-cultured sycamore cells and maize root caps cells

were subjected to a serial section analysis. No evidence for a collection of transition vesicles at the *cis*-face waiting to fuse with one another could be found in either of the two cell types examined. This observation on maize root cap dictyosomes has in the meantime been confirmed by both normal transmission electron microscopy (Shannon et al., 1982) and high-voltage electron microscopy of thick sections (Hawes, 1982, personal communication). In addition Shannon et al. (1982) have also claimed the absence of transition vesicles at the *cis*-face of dictyosomes from digestive gland cells of the insectivorous plant *Dionaea*.

In his 1980 publication Robinson pointed out that, depending on the nature of the forming face and the angle of section with respect to the cisternal stack, the illusory appearance of a set of transition vesicles at the *cis*-face can be obtained. Thus the tubular peripheral extrusions of a cisterna when cross-sectioned may give rise to the appearance of a set of vesicles. The extent to which a *cis*-face cisterna is fenestrated is variable from one organism to another. In the case of the maize root cap dictyosome there are diverging opinions. In the majority of published micrographs [see, e.g., Figures 6 and 8 in Robinson (1980) and Figure 1 from Mollenhauer and Morré (1974)] as well as in unpublished high-voltage electron micrographs (Hawes, personal communication) the first *cis* cisterna has a significant central plate portion. Shannon et al. (1982) maintain in contrast that the *cis*-face cisterna is extensively fenestrated. Whatever the outcome of this controversy, it is clear that the deceptive appearance of a set of transition vesicles at the *cis*-face will occur more frequently with highly fenestrated *cis*-face cisternae.

Claims that transition vesicles are absent from the *cis*-face of many higher plant dictyosomes must be qualified. Whereas an accumulation of transition vesicles at the central plate portion of the *cis*-face cisterna now seems unlikely, it is still possible that small numbers of ER-derived vesicles may fuse at the periphery of these cisternae. As suggested by Wienecke et al. (1982), the fusion of such vesicles cannot easily be differentiated from the early stages of secretory vesicle production at the same cisterna. Nor can conventional electron microscopy of thin sections say much about the number of transition vesicles which might be released from the ER or the frequency with which they fuse at any particular dictyosome. It is difficult, at the moment, to see how these problems might be solved, but perhaps the application of immunocytochemical methods in relation to the synthesis and secretion of cell wall proteins may be helpful.

5.4. THE *TRANS*-FACE

Although it cannot be ruled out that secretory vesicles might also be released from cisternae within the stack (Robinson, 1980), the general consensus of opinion is that their release is restricted to the terminal *trans*-face cisterna (Juniper et al., 1982; Mollenhauer and Morré, 1980; Shannon et al., 1982; Schnepf, 1969). One of the favorite study objects for demonstrating the salient features of *trans*-face cisternae are the mantle cells of germinating maize root tips (Juniper and Roberts, 1966; Juniper et al., 1982; Mollenhauer and Morré, 1966; Mollenhauer et al., 1961). In these cells the secretory vesicles are restricted in number to either one or two per cisterna. The vesicles are large and kidney shaped and their contents are usually stained somewhat more than the surrounding cytoplasm. Mollenhauer (1971) drew attention to the fact that in these cells both the terminal trans cisterna and the attached secretory vesicle become separated from the dictyosome before the connection between the vesicle and the cisterna is severed. The vesicle then moves towards the PM and the cisterna begins to round up. Micrographs depicting these events are shown in Figure 5.8. What happens to the cisterna after it has rounded up is not at all clear apart from the knowledge that it is no longer recognizable in thin sections. Presumably, it is broken down, but there is no evidence for the participation of lysosomelike vesicles in this autolytic process.

Not all higher plant dictyosomes show this feature so markedly as in the maize root cap cell, but there are other examples e.g., in suspension-cultured sycamore cells (Figure 5.1a). Perhaps part of the difficulty in its recognition lies in the speed with which detachment and rounding up of the terminal trans cisterna takes place. Furthermore, the rounded-up cisterna, being much smaller than the "parent" dictyosome, may often not lie in the plane of the section. Yet another possibility is that it signifies an active process (i.e., membrane flow through the dictyosome—see Chapters 15–18) and not all sectioned dictyosomes are "active" in this respect at the moment of fixing and sectioning. A separation of cisterna from secretory vesicle at the trans face is not always the case. Some higher plant dictyosomes [see, e.g., Plant 28a in Gunning and Steer (1975)] and

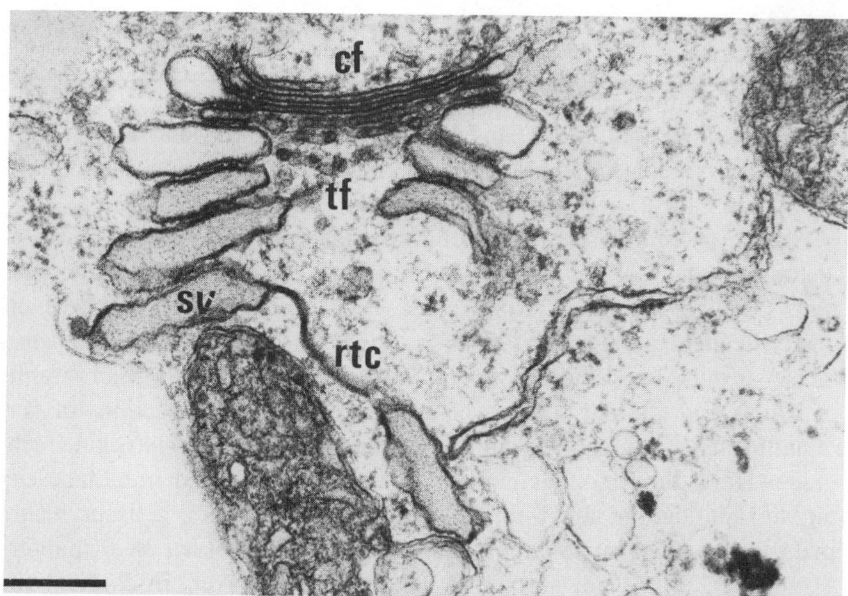

Figure 5.8. Sloughing-off of the terminal *trans*-face cisterna from the stack in a maize root cap dictyosome. Separation of the secretory vesicle from the cisterna occurs subsequently (unpublished micrograph of the author): cf—*cis*-face; rtc—released terminal *trans* cisterna; sv—secretory vesicle; tf—*trans*-face. Bar = 0.25 μm.

many algae (e.g., the scale-producing flagellates) do not differentiate their dictyosomal membrane in this manner. Instead the entire cisterna becomes the secretory vesicle and is gradually released from the trans pole of the dictyosome.

Because of the functional similarity between vacuole genesis in plants (see Matile, 1975) and lysosome production in animals, it is logical to try to impose a structural analogy. Thus, in his studies on vacuole initiation in *Euphorbia* root tip cells, Marty (1978) has sought to use the term *GERL* (Golgi associated–endoplasmic reticulum–lyosomal complex). Coined originally by Novikoff (1964, 1976), this is best described as a specialized region of ER which lies at the *trans*-face of a dictyosome and which together with the *trans*-face cisterna is involved in the production of primary lysosomes. Whereas it has previously been thought possible to differentiate cytochemically GA and ER elements in GERL from one another (Hand, 1980), this is not always the case. Thus there are now strong claims for excluding ER from *trans*-face activities of the GA in some animal cells (Ono, 1979; Pavelka and Ellinger, 1983). Nevertheless, GERL has been well studied in a number of different animal cell types, particularly neurons (Novikoff et al., 1971) and hepatocytes (Novikoff and Yam, 1978), and the component membranes react characteristically positive towards acid phosphatase.

Marty (1978) has shown that the ramification of the tubular network at the periphery of *trans*-face cisternae in *Euphorbia* dictyosomes also react positively in the acid phosphatase test (Figure 5.6b). If, however, one critically examines Marty's micrographs, the presence of ER at *trans*-face cisterna is not very convincing. More convincing are the micrographs of Harris and Oparka (1983) who have examined thick sections of germinating bean cotyledon cells with the zinc oxide–osmium tetroxide technique. Here direct connections between tubular ER and trans face cisternae ramifications are well demonstrated. Since this is a tissue where hydrolytic or lysosomal events are particularly expressed (see Chapters 11 and 12), this is not surprising. One notes, however, in contrast the failure to demonstrate such connections in cells of developing wheat endosperm (Parker and Hawes, 1982), a tissue where hydrolytic activities are minimal.

5.5. ISOLATION OF CISTERNAE AND DICTYOSOMES

Because of its organization the isolation of the plant GA can be considered at the level of the cisterna or the dictyosome. In the latter case particular precautions are usually taken to preserve the cisternal stack, which is then much easier to identify morphologically in fractions. Unstacking, which takes place during homogenization, can largely be overcome by employing low-shear breakage methods (see Chapter 1), by the inclusion of special chemicals such as dextran (Morré and Mollenhauer, 1964; Morré, 1971) or glutaraldehyde (Morré et al., 1965), or by combining both (Ray et al., 1976, Robinson, 1977). Clearly, care must be taken when using glutaraldehyde as a "stabilizing agent" because of its well-known properties as a fixative. Too high a concentration in the homogenizing medium can lead to abnormal sedimentation properties due to the unspecific coagulation of membranes from different organelles and to a loss of measurable enzyme activity. Working with internodal tissue from etiolated pea seedlings, Ray et al. (1976) found that an end concentration of 0.3% (v/v) glutaraldehyde was sufficient to preserve dictyosomal integrity, while still allowing the measurement of some marker enzyme activity. Inclusion of glutaraldehyde in the gradient media, in addition to homogenization media, is not recommended.

In animal cells the most frequently used marker enzymes for elements of the GA are thiamine pyrophosphatase (TPPase) and UDP-galactose–*N*-acetylglucosamine galactosyl-transferase (Gal-Tase) (Fleischer et al., 1969; Morré, 1971; Bretz et al., 1980). By contrast, these enzymes are hard to measure in GA fractions from plant tissues. Instead, inosine diphosphatase (IDPase) and other glycosyl transferases, particularly glucose, are used (Ray et al., 1969). IDPases have been shown by cytochemical means to be present primarily in the plant GA (Dauwalder et al., 1969; Goff, 1973) but not exclusively localized there. Thus, the vacuole appears to possess considerable activity (Poux, 1967) and probably contributes to the "soluble activity" measured after homogenization. This is, however, normally separated from the membrane-associated activity by pelleting or gel filtration (see Chapter 1) prior to gradient centrifugation.

Originally the IDPase assay was carried out on plant cell fractions which had been maintained in the cold (4 °C) for about 4 days. Sometimes longer cold storage periods are required (Leonard et al., 1973; Leonard and Van der Woude, 1976), and in some cases "latent" activity is still not detectable even after 8 days in the refrigerator (Koehler et al., 1976; Nagahashi and Beevers, 1978). Nevertheless, a GA-localized IDPase activity measured in this way has been recorded for a large number of different tissues and cell types—for example, carrot root discs (Gardiner and Chrispeels, 1975), maize roots (Bowles and Northcote, 1972), onion stems (Morré et al., 1977), and suspension-cultured soybean cells (Moore and Beevers, 1974). Some authors have, however, presented data showing IDPase activity localized in ER and PM fractions from maize coleoptiles and bean hypocotyls (Hendriks, 1978; Bowles and Kauss, 1976; M'Voula-Tsieri et al., 1981). A multisite distribution of IDPase activity has been recorded in the animal literature (Wattiaux-De Comninck and Wattiaux, 1969; Tulkens et al., 1974; Eppler and Morré, 1982) and NDPase isoenzymes are known from plant tissues (Troyer et al., 1977), but poor or incomplete resolution of GA elements in gradients could also be responsible for these unusual observations in plant cells. Thus, when the same material (maize coleoptiles) was used in a recent investigation, IDPase activity was found associated with GA membranes only (Mandala et al., 1982).

An alternate, quicker, and probably more reliable method for the measurement of IDPase activity, involves the inclusion of detergents, particularly Triton X-100 in the assay medium. This has been known for a long time in animal systems (Ernster and Jones, 1962; Novikoff and Heus, 1963) and is also applicable to plant cells (Klohs and Goff, 1980; Nagahashi and Kane, 1982). Although detergents interfere with the determination of the released phosphate in some methods, their presence enables the analysis of freshly collected fractions. The action of the detergent has been assumed to be one of increasing accessibility of the substrate to a luminal site of the enzyme in the vesicles or cisternae (Kuhn and White, 1977; Kuriyama, 1972) but it is by no means clear whether the latent and detergent IDPase activities represent one and the same enzyme (Little et al., 1976; Klohs and Goff, 1980; Nagahashi and Nagahashi, 1982). In contrast there seems to be more agreement as to the role of the IDPase. Glycosyl transferases are inhibited by small concentrations of the reaction product UDP (Ray et al., 1969; Staver et al., 1978; Gooday and De-Roussett-Hall, 1975), and it is logical that nucleoside diphosphatases must be present to prevent this.

Glucosyl transferase activities using uridine diphosphoglucose as substrate have been known to be associated with "particulate fractions" of plant cells for almost two decades (Barber et al., 1964). Using μM substrate concentrations, Ray et al. (1969) showed, after a combination of rate zonal and isopycnic centrifugation techniques, that this activity resides in a fraction containing the GA. A subsequent demonstration by Van der Woude et al. (1974) of a second glucosyl transferase activity measured at mM substrate concentrations and associated with the plasma membrane (see also Chapter 7) has led to the two activities being termed *glucan synthase I* (μM) and *II* (mM). Although GS I and GS II activities equilibrate at the same position in isopycnic sucrose gradients of homogenates from the frequently used leguminosae systems (e.g., pea stems, bean hypocotyls), there is no doubt that these activities are carried out on different organelles as judged by their behavior in relation to Mg^{2+} concentration (Robinson et al., 1982).

Until recently these activities have been considered specific for GA and PM membranes, respectively. However, the demonstration that the two activities are measurable in both GA and PM fractions from suspension-cultured grass-type cells (Henry et al., 1983; Robinson and Glas, 1983) questions this generalization for all higher plant cells.

For so important a marker enzyme it is surprising how irregular the assay conditions for GS I are in the literature. This is certainly the reason for the occasional disparity in the results obtained. Thus, some do not include a carrier at the end of the incubation (e.g., Henry and Stone, 1982), others add powdered cellulose (e.g., Hendriks, 1978; Pierce and Hendrix, 1979; Shore and MacLachlan, 1975; Van der Waude et al., 1974), and still others add heat-denatured microsomal membranes (Ray et al., 1976; Robinson et al., 1982) as carriers. The majority of workers separate the product from the unused substrate by centrifuging and washing, but there are some who have applied instead filtration through glass fiber discs (e.g., Pierce and Hendrix, 1979; Henry and Stone, 1982). These are perhaps minor technical points but more important is the wash procedure, since this determines the nature of the product whose radioactivity is measured. Some workers have washed their products with aqueous (60–70%) ethanol alone (Ray, 1973; Robinson et al., 1982) or with absolute ethanol and acetone (Henry and Stone, 1982). Others have employed wash procedures which have included hot water, 2:1 (v/v), chloroform–methanol, and absolute methanol (Hendriks, 1978; Pierce and Hendrix, 1979; Van der Woude et al., 1974). In contrast there are two early papers from MacLachlan's group in which organic solvents were not employed (Shore and MacLachlan, 1975; Shore et al., 1975). Instead washing was carried out with hot water and hot 1 *N* NaOH. As a result, peaks for GS I and GS II activity were obtained in the ER fraction in addition to the mixed GA–PM fraction in isopycnic gradients from pea stem material. Subsequent work (Dürr et al., 1979; Hopp et al., 1979) has demonstrated the presence of glucosyl transferases in ER fractions from this tissue which are capable of the glycosylation of endogenous polyphenyl phosphates, and it would appear prudent always to employ a wash procedure in GS I assay capable of removing such lipidlike compounds.

Whether or not homogenates have been stabilized with glutaraldehyde, the positions of peaks for IDPase and GS I activity in isopycnic linear sucrose gradients remain unchanged (Ray et al., 1976), indicating the same equilibrium density for free or stacked cisternae. These activities usually occupy the region of 27–34% (1.11–1.15 g·cm^{-3}) in sucrose gradients,

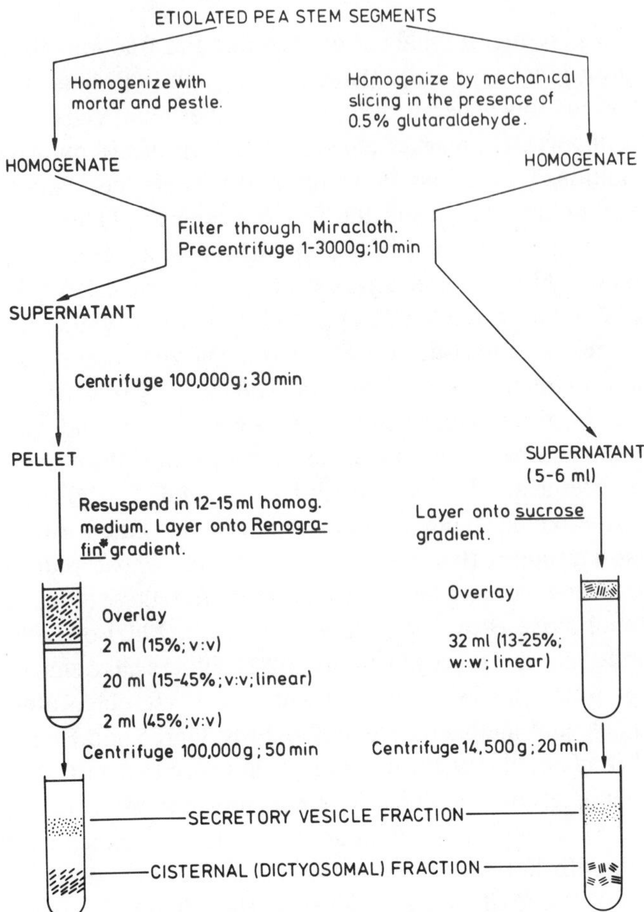

Figure 5.9. The use of rate zonal centrifugation techniques to separate secretory vesicles from GA cisternae (left: according to Taiz et al., 1983) or from dictyosomes (right: according to Ray et al., 1976).

with the GS I activity tending to have a narrower and somewhat deeper distribution than the IDPase. In step gradients (e.g., Green and Northcote, 1979; Morré, 1970; Morré et al., 1974) or flotation gradients (e.g., Dashek, 1970; Morré, 1971) using sucrose, these activities are found at corresponding positions.

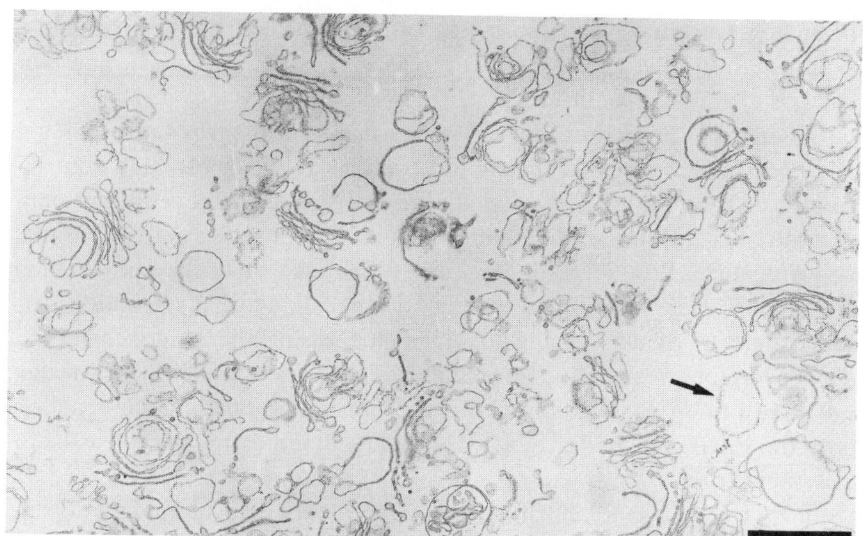

Figure 5.10. Dictyosomes isolated from glutaraldehyde-stabilized pea stem homogenates. The dictyosome fraction obtained in the scheme shown in Figure 5.9 was subjected to a further purification step by centrifugation isopycnically. A small amount of contamination due to ER vesicles (arrow) is apparent. (From Ray et al., 1976). Bar = 0.5 μm.

When glutaraldehyde is used, dictyosomes can be separated from other membranes by a short rate zonal centrifugation (Ray et al., 1976). The dictyosomes, being heavier than individual, free cisternae, sediment more rapidly (Figure 5.9). By collecting the fractions corresponding to he GS I activity from such a gradient and then subjecting them to a subsequent isopycnic centrifugation, a very pure dictyosomal fraction can be obtained (Figure 5.10 and Robinson, 1977).

5.6. ISOLATION OF SECRETORY VESICLES

A subfractionation of the GA into so-called heavy and light fractions corresponding to *cis*- and *trans*-face cisternae, respectively, has been achieved for liver cells (Bergeron et al., 1978). Although this is not yet possible for plant cells, the separation of secretory vesicles from the cisternae of the GA in higher plant cells has been carried out. In one case use was made of a system with an extremely high proportion of naturally occurring secretory vesicles. In the other two cases differences in the sedimentation characteristics of secretory vesicles as against single or stacked cisternae was utilized.

Pollen tubes are a much investigated system for tip growth [see Picton and Steer (1982) for references] and contain an extremely high concentration of secretory vesicles which are transported to the tip and concentrate there. Using a combination of millipore filtration and differential centrifugation, Van der Woude et al. (1971) have succeeded in isolating secretory vesicles from germinating Lily pollen. Particular caution was taken here to avoid damage to organelles during the homogenization. Their secretory vesicle fraction sedimented at 58,000*g* and passed through a 0.45-μm filter. It was identified cytochemically by the intense reaction with the PASH reagent of Picket-Heaps (1968) indicating the presence of polysaccharide.

In their rate zonal centrifugation experiments on glutaraldehyde-stabilized pea stem homogenates Ray et al. (1976) provided evidence for the presence of secretory vesicles in a nondictyosomal fraction. This fraction was a mixed microsomal fraction containing fragments of ER, TP, PM, as well as the secretory vesicles. By recentrifuging isopycnically on a linear sucrose gradient, at least the ER components could be separated out of this fraction. Since the position of the secretory vesicles in this second centrifugation was recognized alone through the presence of radioactive polysaccharides which had been synthesized *in vivo* prior to homogenization, contamination through PM fragments cannot be ruled out. Indeed, since the peak position for this radioactivity was around 1.13 $g \cdot cm^{-3}$, which is also coincident with the equilibrium density for the PM in this system (see above and Chapter 7), PM contamination of this fraction is most likely. The position of TP fragments was not determined in this study, but based on their equilibrium density values (around 1.10 $g \cdot cm^{-3}$, see Chapter 6) they probably did not contaminate this secretory vesicle fraction. Of course these experiments are only of relative use: they showed that secretory vesicles can be separated from their cisternal origin, but the use of glutaraldehyde, necessary for the preservation of dictyosomal integrity, precludes a thorough biochemical characterization.

The separation of secretory vesicles from GA cisternae without using glutaraldehyde has recently been achieved by Taiz et al. (1983). Their technique has also involved rate zonal centrifugation but employed reno(uro)grafin instead of sucrose as the gradient medium (see Figure

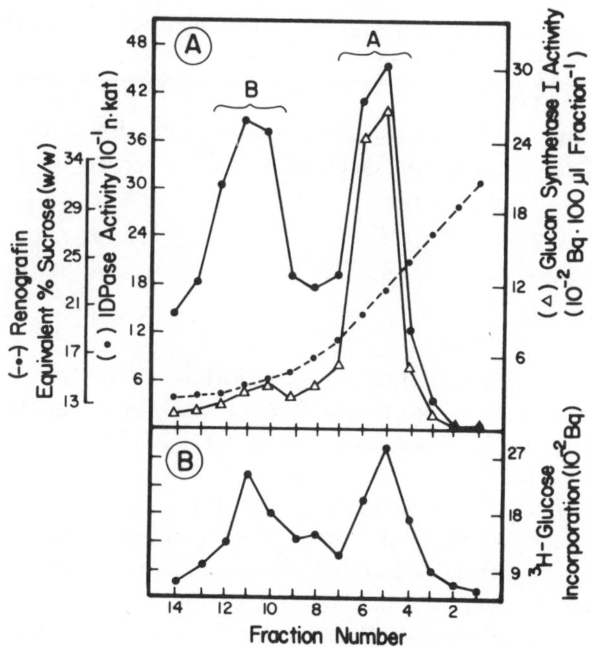

Figure 5.11. Profiles for GA marker enzyme activities on rate zonal renografin gradients prepared as shown in Figure 5.9. In addition, the incorporation of ^{3}H-glucose into polysaccharides *in vivo* after a 5-min pulse is given. (From Taiz et al., 1983.)

5.9). Secretory vesicles can be separated in this way either directly from a crude microsomal fraction or, by running an initial isopycnic centrifugation on a linear sucrose gradient, from partially purified GA fractions. If the latter is carried out, the secretory vesicle fraction subsequently obtained appears essentially free from contamination from other membranes, as judged by the lack of appropriate marker enzyme activity. Even PM vesicles present in an isopycnic sucrose GA fraction appear to sediment quicker than the secretory vesicles with this procedure. The cisternae do not remain stacked together but nevertheless clearly possess sedimentation characteristics different from the secretory vesicles with respect to renografin.

With this method secretory vesicles were recognized through the presence of IDPase activity, which as already discussed (see above 5.2.) tends to be associated with *trans*-face cisternae and attached secretory vesicles. Renografin gradients prepared in this way show two peaks of IDPase activity (Figure 5.11). The faster traveling one also bears GS I activity and has been shown electron microscopically to contain cisternae. Although the secretory vesicle fraction surprisingly did not show GS I or GS II activities, a clear pulse-chase relationship with respect to *in vivo* synthesized radioactive polysaccharides was observed for the cisternal and secretory vesicle fractions.

REFERENCES

Amelunxen, F. and Gronau, G. (1966). *Z. Pflanzenphysiol.* **55,** 327–336.

Barber, G. A., Elbein, A. D., and Hassid, W. Z. (1964). *J. Biol. Chem.* **239,** 4056–4061.

Bergeron, J. J. M., Borts, D., and Cruz, J. (1978). *J. Cell Biol.* **76,** 87–97.

Bouck, G. B. (1971). *J. Cell Biol.* **50,** 362–384.

Bouck, G. B. (1972). *Adv. Cell. Molec. Biol.* **2,** 237–271.

Bowles, D. J. and Kauss, H. (1976). *Biochem. Biophys. Acta* **443,** 360–374.

Bowles, D. J. and Northcote, D. H. (1972). *Biochem. J.* **130,** 1133–1145.

Bretz, R., Bretz, H., and Palade, G. E. (1980). *J. Cell. Biol.* **84,** 87–101.

Brown, R. M. and Romanovicz, D. K. (1976). *Applied Polym. Symp.* **28,** 537–585.

Dashek, W. V. (1970). *Plant Physiol.* **46,** 831–838.

Dauwalder, M. and Whaley, W. G. (1973). *J. Ultr. Res.* **45,** 279–296.

Dauwalder, M., Whaley, W. G., and Kephart, J. E. (1969). *J. Cell Sci.* **4,** 455–497.

Dexheimer, J. (1978). *Z. Pflanzenphysiol.* **86,** 189–201.

Dexheimer, J. (1981). *Z. Pflanzenphysiol.* **101,** 333–346.

Dodge, J. D. (1975). *The Fine Structure of Algal Cells.* Academic Press, New York.

Dürr, M., Bailey, D. S., and MacLachlan, G. A. (1979). *Eur. J. Biochem.* **97,** 445–453.

Eppler, C. M. and Morré, D. J. (1982). *Eur. J. Cell Biol.* **29,** 13–23.

Ernster, L. and Jones, L. C. (1962). *J. Cell Biol.* **15,** 563–577.

Falk, H. (1976). *Arch. Mikrobiol.* **58,** 222–227.

Farquhar, M. G. and Palade, G. E. (1981). *J. Cell. Biol.* **91,** 775–1035.

Fleischer, B., Fleischer, S., and Ozawa, H. (1969). *J. Cell Biol.* **43,** 59–79.

Fraser, T. W. and Gunning, B. E. S. (1973). *Planta* **113,** 1–19.

Gardiner, M. and Chrispeels, M. J. (1975). *Plant Physiol.* **55,** 536–541.

Goff, C. W. (1973). *Protoplasma* **78,** 397–416.

Gooday, G. W. and Rousett-Hall, A. de. (1975). *J. Gen. Microbiol.* **89,** 137–145.

Goodenough, U. and Porter, K. R. (1968). *J. Cell Biol.* **38,** 403–425.

Green, J. R. and Northcote, D. H. (1979). *J. Cell Sci.* **40,** 235–244.

Grove, S. N., Bracker, G. E., and Morré, D. J. (1968). *Science* **161,** 171–173.

Gunning, B. E. S. and Steer, M. W. (1975). *Ulstructure and the Biology of Plant Cells.* Edward Arnold, London.

Hand, A. (1980). *J. Histochem. Cytochem.* **28,** 82–86.

Harris, N. and Oparka, K. J. (1983). *Protoplasma* **114,** 93–102.

Hendriks, T. (1978). *Plant Sci. Lett.* **11,** 261–274.

Henry, R. J. and Stone, B. A. (1982). *Plant Physiol.* **69,** 632–636.

Henry, R. H., Schibeci, A., and Stone, B. A. (1983). *Biochem. J.* **209,** 627–633.

Hopp, H. E., Romero, P., and Pont Lezica, R. (1979). *Plant Cell Physiol.* **20,** 1063–1069.

Juniper, B. E. and Roberts, R. M. (1966). *J. Roy. Micr. Soc.* **85,** 63–72.

Juniper, B. E., Hawes, C. R., and Horne, J. C. (1982). *Bot. Gaz.* **143,** 135–145.

Kiermayer, O. (1981). In *Micrasterias in Cytomorphogenesis in Plants* (O. Kiermayer, ed.), pp. 147–189. Springer Verlag, Berlin.

Klohs, W. D. and Goff, C. W. (1980). *Eur. J. Cell Biol.* **20,** 228–233.

Koehler, D. E., Leonard, R. T., van der Woude, W. J., Linkins, A. E., and Lewis, L. N. (1976). *Plant Physiol.* **58,** 324–330.

Kristen, U. (1978). *Planta* **138,** 29–33.

Kristen, U. (1980). *Europ. J. Cell Biol.* **23,** 16–21.

Kristen, U., Lockhausen, J., and Robinson, D. G. (1984). *J. Exp. Bot.* (in press).

Kuhn, N. J. and White, A. (1977). *Biochem. J.* **168,** 423–433.

Kuriyama, Y. (1972). *J. Biol. Chem.* **247,** 2979–2988.

Leedale, G. F., Leadbeater, B. S. C., and Massalski, A. (1970). *J. Cell. Sci.* **6,** 701–723.

Leonard, R. T. and van der Woude, W. J. (1976). *Plant Physiol.* **57,** 105–114.

Leonard, R. T., Hansen, D., and Hodges, T. K. (1973). *Plant Physiol.* **51,** 749–754.

Little, J. S., Thiers, D. R., and Widnell, C. C. (1976). *J. Biol. Chem.* **251,** 7821–7825.

Mandala, S., Mettler, I. J., and Taiz, L. (1982). *Plant Physiol.* **70,** 1743–1747.

Marty, F. (1978). *Proc. Nat. Acad. Sci.* **75,** 852–856.

Massalski, A. and Leedale, G. F. (1969). *Brit. Phycol. J.* **4,** 159–180.

Matile, P. (1975). In *Cell Biology Monographs,* Vol. 1. Springer Verlag, Berlin.

Mignot, J. P., Brugerolle, G., and Metenier, G. (1972). *J. Microscopie* **14,** 327–342.

Mollenhauer, H. H. (1971). *J. Cell Biol.* **49,** 212–214.

Mollenhauer, H. H. and Morré, D. J. (1966). *J. Cell Biol.* **29,** 373–376.

Mollenhauer, H. H. and Morré, D. J. (1972). *What's New in Plant Physiology* **4,** 1–4.

Mollenhauer, H. H. and Morré, D. J. (1975). *J. Cell Sci.* **19,** 231–237.

Mollenhauer, H. H. and Morré, D. J. (1978). *J. Cell Sci.* **32,** 357–362.

Mollenhauer, H. H. and Morré, D. J. (1980). In *The Biochemistry of Plants* (P. K. Stumpf and E. E. Conn, eds.), Vol. 1, pp. 437–488. Academic Press, New York.

Mollenhauer, H. H., Morré, D. J., and Totten, C. (1973). *Protoplasma* **78,** 443–459.

Mollenhauer, H. H., Morré, D. J., and Van Der Woude, W. J. (1975). *Mikroskopie* **31,** 257–272.

Mollenhauer, H. H., Whaley, W. G., and Leech, J. H. (1961). *J. Ultrastruct. Res.* **5,** 193–200.

Moore, T. S. and Beevers, H. (1974). *Plant Physiol.* **53,** 261–265.

Morré, D. J. (1970). *Plant Physiol.* **45,** 791–799.

Morré, D. J. (1971). *Meth. Enzymol.* **22,** 130–148.

Morré, D. J. (1976). In *International Cell Biology* (B. R. Brinkley and K. R. Porter, eds.), pp. 293–303. Rockefeller University Press, New York.

Morré, D. J. and Mollenhauer, H. H. (1964). *J. Cell Biol.* **23,** 295–305.

Morré, D. J. and Mollenhauer, H. H. (1974). In *Dynamic Aspects of Plant Ultrastructure* (A. Robards, ed.), pp. 84–137. McGraw-Hill, London.

Morré, D. J. and Ovtracht, L. (1977). *Int. Rev. Cytol. Suppl.* **5,** 61–188.

Morré, D. J., Kartenbeck, J., and Franke, W. W. (1979). *Biochem. Biophys. Acta* **559,** 72–152.

Morré, D. J., Lembi, C. A., and Van Der Woude, W. J. (1977). *Cytobiologie* **16,** 72–81.

Morré, D. J., Mollenhauer, H. H., and Bracker, C. E. (1971). In *Results and Problems in Cell Differentiation. II. Origin and Continuity of Cell Organelles* (J. Reinert and H. Ursprung, eds.), pp. 82–126. Springer Verlag, Berlin.

Morré, D. J., Mollenhauer, H. H., and Chambers, J. E. (1965). *Exp. Cell Res.* **38,** 672–675.

Morré, D. J., Franke, W. W., Deumling, B., Nyquist, S. E., and Ovtracht, L. (1971). *Biomembranes* **2,** 95–104.

M'Voula-Tsieri, M., Hartmann-Bouillon, M. A., and Benveniste, P. (1981). *Plant Sci. Lett* **20,** 379–386.

Nagahashi, J. and Beevers, L. (1978). *Plant Physiol.* **16,** 451–459.

Nagahashi, J. and Kane, A. P. (1982). *Protoplasma* **112,** 167–173.

Nagahashi, J. and Nagahashi, S. L. (1982). *Protoplasma* **112,** 174–180.

Novikoff, A. B. (1964). *Biol. Bull.* **127,** 358.

Novikoff, A. B. (1976). *Proc. Nat. Acad. Sci.* **73,** 2781–2787.

Novikoff, A. B. and Heus, M. (1963). *J. Biol. Chem.* **238,** 710–716.

Novikoff, P. M. and Yam, A. (1978). *J. Cell Biol.* **76,** 1–11.

Novikoff, P. M., Novikoff, A. B., Quintana, N., and Hauw, J. J. (1971). *J. Cell Biol.* **50,** 859–886.

Ono, K. (1979). *Histochemistry* **62,** 113–124.

Parker, M. L. and Hawes, C. R. (1982). *Planta* **154,** 277–283.

Pavelka, M. and Ellinger, A. (1983). *Eur. J. Cell Biol.* **29,** 253–261.

Pickett-Heaps, J. D. (1968). *J. Cell Sci.* **3,** 55–63.

Pickett-Heaps, J. D. (1971). *Protoplasma* **72,** 275–314.

Pickett-Heaps, J. D. (1975). *Green Algae*. Sinnauer Assoc., Massachusetts.

Pierce, W. S. and Hendrix, D. L. (1979). *Planta* **146,** 161–169.

Poux, N. (1967). *J. Microsc.* **6,** 1043–1058.

Poux, N. (1973). *Compt. Rend. Acad. Sci.* (*Paris*) **276,** 2163–2166.

Ray, P. M. (1973). *Plant Physiol.* **51,** 601–608.

Ray, P. M., Eisinger, W. R., and Robinson, D. G. (1976). *Ber. Dtsch. Bot. Ges.* **89,** 121–146.

Ray, P. M., Shininger, T. L., and Ray, M. M. (1969). *P.N.A.S.* **64,** 605–612.

Robinson, D. G. (1977). *Adv. Bot. Res.* **5,** 89–151.

Robinson, D. G. (1980). *Eur. J. Cell Biol.* **23,** 22–36.

Robinson, D. G. (1981a). In *Cell Walls '81* (D. G. Robinson and H. Quader, eds.), pp. 43–52. Wissenschaftliche Verlagsgesellschaft, Stuttgart, FRG.

Robinson, D. G. (1981b). *Encycl. Plant Physiology* **13B,** 317–332.

Robinson, D. G. and Glas, R. (1984). *J. Exp. Bot.* **34,** 668–675.

Robinson, D. G. and Kristen, U. (1982). *Int. Rev. Cytol.* **77,** 89–127.

Robinson, D. G., Eberle, M., Hafemann, C., Wienecke, K., and Graebe, J. E. (1982). *Z. Pflanzenphysiol.* **105,** 323–330.

Roland, J.-C. and Sandoz, D. (1969). *J. Microscopie* **8,** 263–268.

Romanovicz, D. (1981). In *Cytomorphogenesis in Plants* (O. Kiermayer, ed.), Vol. 8, pp. 27–62. Springer Verlag, Berlin.

Rothman, J. E. (1981). *Science* **213,** 1212–1219.

Rougier, M. (1971). *J. Microscopie* **10,** 67–82.

Rougier, M. (1981). *Encycl. Plant Physiology* **13B,** 542–574.

Ryser, U. (1979). *Protoplasma* **98,** 223–239.

Schnepf, E. (1969). *Protoplasmatologia* **8**(8), 8–22.

Schnepf, E., Deichgräber, G., and Herth, W. (1982). *Protoplasma* **110,** 203–214.

Shannon, T. M., Henry, Y., Picton, J. M., and Steer, M. W. (1982). *Protoplasma* **112,** 189–195.

Shore, G. and MacLachlan, G. A. (1975). *J. Cell Biol.* **64,** 557–571.

Shore, G., Raymond, Y., MacLachlan, G. S. (1975). *Plant Physiol.* **56,** 34–38.

Sievers, A. (1965). *Z. Pflanzenphyiol.* **53,** 193–213.

Staehelin, L. A. and Kiermayer, O. (1970). *J. Cell. Sci.* **7,** 787–792.

Staver, J. J., Glick, K., and Baisted, D. J. (1978). *Biochem. J.* **169,** 297–303.

Taiz, L., Murry, M., and Robinson, D. G. (1984). *Planta* **758,** 534–539.

Troyer, G. D., Goff, C. W., and Klohs, W. D. (1977). *J. Histochem. Cytochem.* **25,** 1247–1253.

Tulkens, P., Beaufay, H., and Trouet, A. (1974). *J. Cell. Biol.* **63,** 383–401.

Turner, F. R. and Whaley, W. G. (1965). *Science* **147,** 1303–1304.

Van Der Woude, W. J., Lembi, C. A., Morré, D. J., Kindinger, J. I., and Ordin, L. (1974). *Plant Physiol.* **54,** 333–340.

Van Der Woude, W. J., Morré, D. J. and Bracker, A. E. (1971). *J. Cell Sci.* **8,** 337–357.

Vian, B. (1974). *Comptes Rend. Acad. Sci., D.* **278,** 1483–1486.

Wattiaux-De Coninck, S. and Wattiaux, R. (1969). *Biochem. Biophys. Acta* **183,** 118–128.

Whaley, W. G. (1975). *Cell Biol. Monographs* Vol. 2, Springer Verlag, pp. 1–190.

Wienecke, K., Glas, R., and Robinson, D. G. (1982). *Planta* **155,** 58–63.

Zaar, K. and Schnepf, E. (1969). *Planta* **88,** 224–232.

6

TONOPLAST AND VACUOLES

Together with the existence of plastids and a cell wall the vacuole is one of the three distinguishing characteristics of plant cells. Indeed for differentiated higher plant cells it is the largest single organelle, occupying up to 90% of the total cell volume. The limiting membrane of the vacuole is called the *tonoplast* (TP), a term introduced by De Vries in the last century to express a "tension" in vacuolar precursors. As we now know, because of its osmotic properties, the vacuole plays an integral role in the maintenance of "tissue tension," or more appropriately turgor, and as a result is a primary factor in controlling cell shape. The vacuole also allows the plant to increase its total surface area without involving a net synthesis of cytoplasm, thus enabling a better uptake of mineral nutrients, which is necessary for a successful autotrophic existence (Wiebe, 1968).

There are numerous contributions from the pre-EM period [for summaries the reader is referred to Guilliermond (1941) and Zirkle (1937)] dealing with vacuole structure and function; some are only of marginal interest but many are quite valuable and deserve amplification with the techniques now at our disposal. As is the case with several other organelles, the advent of the EM and cell fractionation techniques has led to a renaissance of research into what might appear on the surface to be an uninteresting organelle. Starting with Matile's classic monograph of 1975 there have appeared a number of useful reviews on various aspects of the higher plant vacuole over the last 8 years. Of these the work of Boller (1982), Matile (1978, 1982), Marty et al. (1980), Leigh (1983), and Wagner (1982, 1983) are to be recommended.

6.1. STRUCTURE AND MORPHOLOGY

The TP is, like the PM, in section a tripartite structure free of ribosomes. Freeze-fracture studies (Fineran, 1970a,b; Leigh and Branton, 1976; Marty, 1982) have shown that the two fracture faces of the TP are not identical. As a rule the protoplasmic face of the membrane (outer bimolecular leaflet viewed from within the vacuole, PF—see Figure 6.1c) has a higher density of IMPs (intramembranous particles) than that of the luminal face of the membrane (inner bimolecular leaflet viewed from outside of the vacuole, EF—see Figure 6.1d). The IMPs, however, tend to be only slightly different in size from one another.

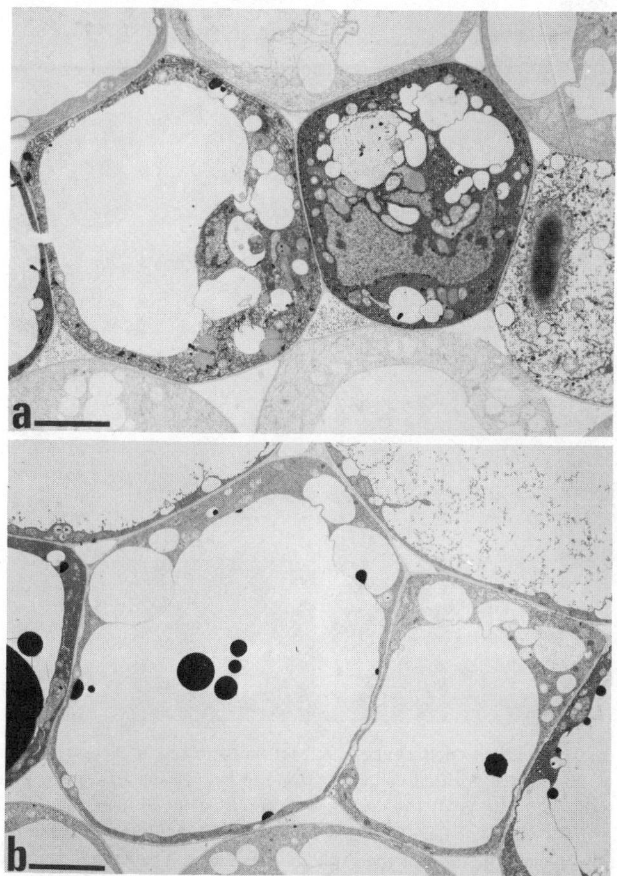

Figure 6.1. (a) Thin section through phenolic-containing cells in the roots of cotton seedlings 3 mm from the tips. Conventional fixation with glutaraldehyde and osmium tetroxide. (b) The same cells as in Figure 6.1a but fixed with 1% caffeine in the primary fixation and wash medic. The irregular blackening of the cytoplasm in Figure 6.1a is replaced by aggregates in the vacuole, the cytoplasm instead being "normally" fixed. Bar = 5 μm. Published micrographs of Mueller and Greenwood (1978).

Vacuoles can vary tremendously in size and shape; even during the life of a single cell there can be appreciable differences in the number and morphology of this organelle. In meristematic cells vacuoles are usually small, although it is not always the case that in differentiated cells they are large and few in number. As every young botany student will witness,

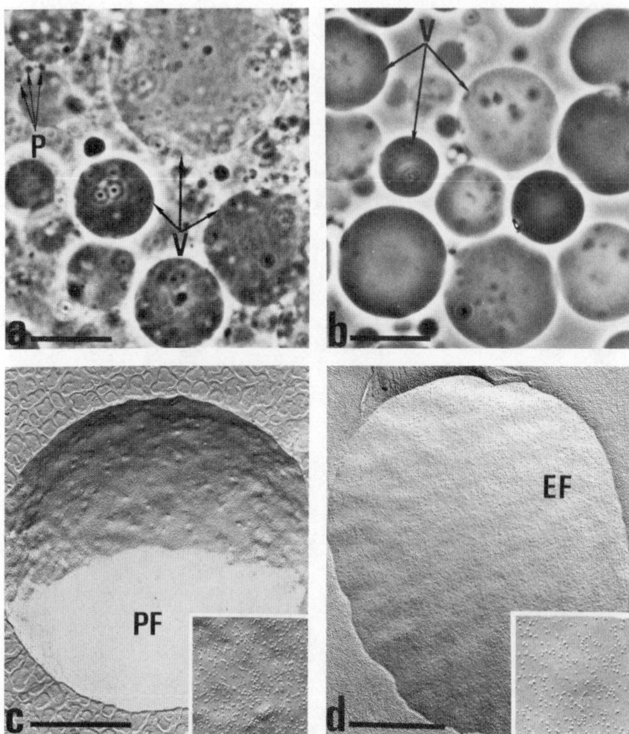

Figure 6.2. (a) Crude (2000*g* pellet) vacuole fraction from red beet tissue. p—plasma; v—vacuola. Bar = 20 μm. (b) Purified vacuoles from red beet tissue as seen by phase-contrast microscopy. Bar = 20 μm. (c,d) Freeze-fracture replicas of red beet vacuoles. PF—practoplasmic face; EF—vacuolar face (see text for further explanation of these descriptions). Insets: High-power magnification of the replicas in a and d. The particles are the so-called intramembrane particles (IMPs). Bar = 1 μm. Published micrographs of Leigh and Branton (1978).

vacuoles in differentiated cells are often traversed by strands of cytoplasm. These strands vary in size and position.

Many plant cells store phenolics in their vacuoles. Under normal fixation conditions such cells leak their phenolics into the cytoplasm during the glutaraldehyde fixation, resulting in a blackening of the cytoplasm during the subsequent osmification. This problem is overcome to a great extent by the inclusion of caffeine (0.1–1%w/v) in the glutaraldehyde and wash media (Mueller and Greenwood, 1978). The phenolics are coagulated into large droplets which tend to remain in the vacuole (Figure 6.2).

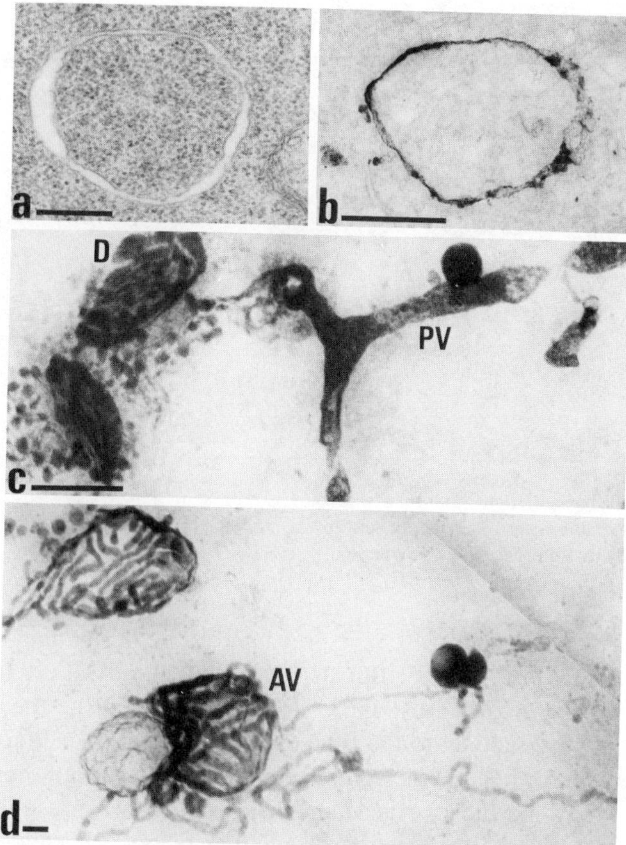

Figure 6.3. (a,b) Ringlike autophagic vacuoles in meristematic root tip cells of *Euphorbia characics*. (a) Normal fixation; (b) cytochemical demonstration of thiolacetic acid esterase activity. (c,d) Thick (μm) sections from wheat (*Triticum sativum* L.) root tips prepared by the zinc iodide–osmium tetroxide fixation technique and viewed at 2-MeV accelerating voltage. (c) Tubular provacuoles (PV) emanating from the peripheral region of a dictyosome (D). (d) Provacuolar tubes leading into a cagelike autophagic vacuole (AV). Bar = 0.5 μm. Published micrographs of Marty (1978, 1980).

6.2. THE ORIGIN OF VACUOLES

This section is concerned with the question "How do vacuoles originate in meristematic cells?" Problems related to their maintenance and growth are dealt with in Chapters 11 and 17.

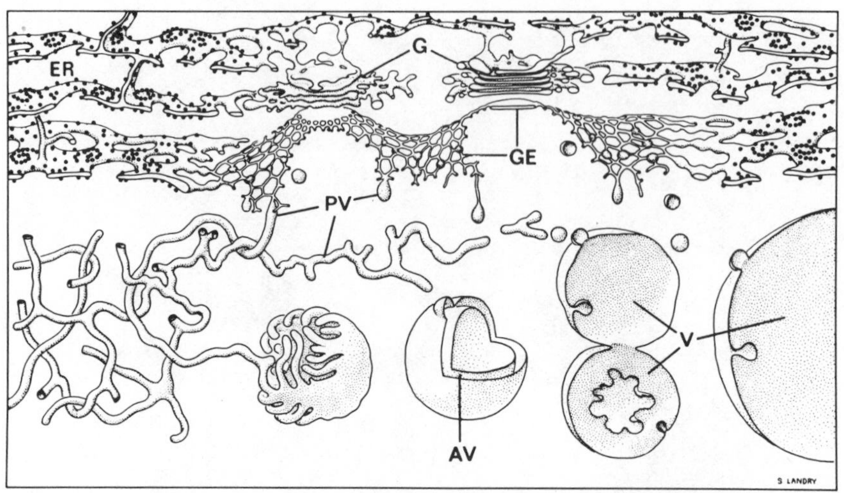

Figure 6.4. Stages in the formation of vacuoles as represented by Marty (1978). Essentially the developmental series begins at the left (lower) with the production of a "lysosomal cage" and proceeds to small vacuoles (lower right).

For many years vacuoles were presumed to be derivatives of the ER (Berjak, 1972; Bowes, 1965; Buvat, 1971; Matile and Moor, 1968; Mesquita, 1969) or GA (Marinos, 1963; Ueda, 1966). With the recognition of the vacuole as an analog to the lysosome of the animal cell (see Chapter 12) attempts have been made at determining whether or not the structural basis for lysosome biogenesis, namely GERL, is also present in meristematic plant cells. The question of whether or not ER elements are in close proximity to or fuse with peripheral tubules of *trans*-face GA cisternae has already been discussed (Chapter 5). Here I will concern myself with the consequences of a "plant GERL."

According to Marty (1973a,b, 1978, 1980), who has looked at zinc-oxide–osmium-tetroxide-stained semithin sections with a high-voltage EM, tubelike "provacuolar" structures emanate from the GERL complex (Figure 6.3c). Starting as a pear-shaped bleb, these tubular structures (diameter ~0.1 μm) branch and ramify throughout the cell. These provacuolar tubes then wrap themselves around pieces of cytoplasm or organelles (Figure 6.3d) and fuse laterally engulfing completely the object in question. This stage is visualized in Figure 6.3a and finds confirmation in the cytochemical demonstration of acid esterase activity in the lumen of such "lysosomal cages" (Figure 6.3b).

One notes that a reticulate provacuolar system corresponding to these EM observations has in fact already been described by light microscopists who have applied the vacuolar dye neutral red to root tips (Guilliermond et al., 1933; Dangeard, 1956). Marty has proposed that the inner membrane of the lysosomal cage or shell is gradually digested together with the engulfed contents. Possible micrographs depicting this event are seen in Symillides and Marty (1977). These authors also demonstrate the fusion of vacuolar cages which are apparently at different developmental stages. Finally, the various smaller vacuoles fuse with one another to produce the large central vacuole.

The various events just described are summarized in Figure 6.4. While quite appealing, Marty's observations suffer from not having been corroborated by other authors and also having been made principally on only one object: the root tips of *Euphorbia characias* L. One hopes that in the next years scientists with access to high-voltage EMs will be attracted to this area of plant cytology.

6.3. ISOLATION

6.3.1. Tonoplast Markers

A couple of years ago it would not have been possible to write this section; a situation reflected in Quail's 1979 review. In the intervening period numerous papers attest to the existence of a TP-situated electrogenic ATPase which is responsible for pumping protons. The majority of the publications involve measurements carried out on mechanically isolated vacuoles or microsomal TP vesicles, but find confirmation in studies employing intact vacuoles obtained from protoplasts. Apart from the well-known mitochondrial ATPase, recent research has also led to the recognition of another electrogenic proton-pumping ATPase at the PM (see also Chapter 7), making it necessary to compare carefully the properties of the TP ATPase with the other two. This is done with help of Table 6.1.

TP, PM, and mitochondrial ATPase activities are seemingly ATP specific in contrast to soluble phosphatases. They are also not inhibited by ammonium molybdate, which is the case for the latter (Leigh and Walker, 1980). Although the mitochondrial ATPase also has an alkaline pH op-

TABLE 6.1. A COMPARISON OF THE PROPERTIES OF TONOPLAST (TP), PLASMA MEMBRANE (PM), AND MITOCHONDRIAL (MITO) ASSOCIATED ATPASE ACTIVITIES

Property	TP	PM	Mito
1. pH optimum	7–8[a] (6.5)[b]	6.5[c]	8–9[d]
2. divalent cation effects	Mg^{2+}, Mn^{2+} requirement Ca^{2+} inhibitory[a,b]	Mg^{2+} requirement[e]	Mg^{2+} requirement
3. monovalent cation effect	insensitive to K^+ [a,b]	K^+ stimulation[e]	K^+ stimulation
4. anion effects	Cl^- stimulation[a,f] NO_3^- inhibitory	—	—
5. Effect of ouabain	no effect[g]	no effect[g]	no effect[g]
6. Effect of oligomycin and azide	no effect[e]	no effect[e]	inhibitory[e]
7. Effect of *N*,*N*′-dicyclohexyl carbodiimide (DCCD)	inhibitory[a,b]	inhibitory[e]	no effect[h]
8. Effect of diethylstilbesterol (DES)	inhibitory[a,b]	inhibitory[c]	no effect[i]
9. Effect of orthovanadate	no effect[a,b]	inhibitory[j]	no effect[j]

[a] Mettler et al., 1982.
[b] Du Pont et al., 1982.
[c] Briskin and Poole, 1983.
[d] Hodges and Leonard, 1972.
[e] Leonard and Hodges, 1973.
[f] Walker and Leigh, 1981.
[g] Hodges, 1976.
[h] Leonard and Hotchkiss, 1976.
[i] Balke and Hodges, 1979.
[j] Du Pont et al., 1981.

timum, the inhibitory effects of oligomycin and azide easily set it apart from the other two. Lack of stimulation by K^+ ions and the failure of vanadate to inhibit its activity are the distinguishing characters of the TP ATPase.

Isolated vacuoles, while maintaining large concentration gradients of some solutes are apparently incapable of doing the same for protons (Komor et al., 1982b; Lin et al., 1977), a fact which has been explained by the lack of an energy source for a TP proton pump under these conditions (Komor et al., 1982a; Marin et al., 1981). When measured *in vitro*, both TP and PM ATPases pump protons inwardly; that is, the luminal pH of the membrane vesicle in question drops. Vectorially speaking, this is in agreement with that which would have been expected from the *in vivo* situation for the TP but not for vesicles derived from the PM (see Chapters 7 and 13). The luminal acidification of TP and PM vesicles can be measured in a variety of ways: through a change in the absorption spectrum of the pH indicator neutral red (Hager et al., 1980), through the quenching of the fluorescence of quinacrine (Du Pont et al., 1982; Mettler et al., 1982) or acridine orange (Vianello et al., 1982), or through the uptake of radioactively labeled anions [e.g., imidazole (Stout and Cleland, 1982), methylamine (Mettler et al., 1982), or thiocyanate (Rasi-Caldongo et al., 1981)]. In the latter case it is thought that these anions diffuse into the lumen of the vesicle in response to the higher (inside positive) membrane potential generated by the proton pumping ATPase.

Another distinguishing feature of the TP ATPase is its stimulation through Cl^- ions (Walker and Leigh, 1981). Although Cl^- ions are known to decrease the membrane potential in vesicle preparations, it appears that Cl^- ions can also directly stimulate ATP hydrolysis (Du Pont et al., 1982; Stout and Cleland, 1982). Conversely, ATP hydrolysis stimulates Cl^- uptake into the vesicles, as shown in experiments with ^{36}Cl (Mettler et al., 1982). Thus Cl^- ions play a double role. Confirmation of the idea that the inwardly directed moments of protons and Cl^- ions are coupled in TP vesicle preparations is provided by experiments employing ionophores. Both gramicidin (which nonspecifically facilitates the transport of protons and K^+ or Na^+ ions) and carbonyl cyanide *p*-trifluoromethoxyphenyl hydrazone (FCCP), a protonophore, dissipate gradients in membrane potential and pH and also prevent the accumulation of Cl^- and methylamine in TP vesicles (Du Pont et al., 1982; Mettler et al., 1982).

6.3.2. Content Markers

It is now generally recognized that a large number of hydrolytic enzymes are localized in the vacuole of higher plant cells (Matile, 1978, 1982). In a recent review Boller (1982) has conveniently summarized the appropriate literature so that it is not necessary to go into great detail here. Enzymes found in the vacuole up to now are as follows (given alphabetically according to their common names): acid phosphatase, acid proteinase, α-galactosidase, α-mannosidase, β-fructosidase, β-glucosidase, carboxypeptidase, DNase, invertase, *N*-acetyl glucosaminidase, peroxidase, phosphodiesterase, phytase, protease, RNase.

According to Boller (1982) and Martinoia et al. (1981), α-mannosidase is probably the best universal vacuolar content marker. It is easy to measure and is probably localized exclusively in the vacuole. Depending on the tissue involved and whether the vacuoles were isolated directly or indirectly (see Sections 6.3.3. and 6.3.4.), the proportions of cellular enzyme activity varies between 50 and 100% (with the majority of values nearer 100 than 50%). Critical for the validity of this statement is that the degree of nonvacuolar contamination should be much smaller than those values for the vacuole preparation in question. Thus, Boller (1982) has quite correctly pointed out the equivocal nature of some claims for vacuolar enzyme localization where measurements of extravacuolar markers were either neglected (e.g., Lancaster and Collin, 1981) or were of the same order as the enzyme in question [cf. Leigh and Branton (1976) with Leigh and Walker (1980)].

During the homogenization of plant tissue it is unavoidable that many vacuoles will burst. Much of their contents is then liberated into the cytosol and only a small amount will be preserved by the rapid reannealing of the TP into small vesicles. Measurement of vacuole content markers on density gradients made from total cell homogenates can therefore lead to false conclusions. As Boller and Kende (1979) have shown, at least one vacuolar enzyme, phosphodiesterase, exists in oligomeric form and behaves as a 35S particle. By prolonged centrifugation (e.g., overnight at 80,000*g*), this enzyme leaves the overlay or sample volume at the top of the gradient and begins to move into the gradient where it can mistakenly be allocated to a "foreign" membranous organelle.

Although there is no evidence that this may also apply to α-mannosidase, recent results of Mandala et al. (1982) and Taiz, Murry, and Robinson (1983) show activity distributions on sucrose density gradients which are not compatible with that of a soluble enzyme. Speaking against the oligomeric nature of this enzyme are the relatively short centrifugation times involved in these experiments. Instead the fact that the distribution of this enzyme mirrors the general protein distribution in the gradients in question suggests that a nonspecific binding of this enzyme to membranes may take place.

6.3.3. Direct Isolation through Homogenization

There have been few attempts at isolating intact vacuoles directly from plant tissue. Although Gregory and Cocking (1965) reported the release of intact vacuoles from tomato tissue, this claim has not been subjected to rigorous biochemical tests. In 1976 Leigh and Branton successfully isolated vacuoles from the storage tissue of beetroots. The crude vacuole preparation is released from the tissue by a special "chopper" and is collected in a buffered sorbitol medium. Purification of the vacuoles is carried out by various centrifugation steps and finally through a metrizamide step gradient (Figure 6.5). The disadvantage of this method, namely the relatively poor yield, is offset by the large amounts of starting material available so that significant numbers of vacuoles can be obtained.

A modification of the Leigh and Branton (1976) procedure has been introduced by Willenbrink's and Matile's groups (Doll et al., 1979; Grob and Matile, 1979, 1980; Willenbrink and Doll, 1979) and involves the plasmolysis of the tissue prior to homogenization. Again separation of the vacuoles was also achieved via step gradients using either dextran (which because of its noninhibitory effects on sucrose uptake was adopted instead of metrizamide) or renografin.

Apart from the generally low yields of vacuoles obtained as a result of the mechanical disruption of cells, there are two other potential drawbacks. One is that this procedure appears to be restricted to firm tissues; for instance, there have been no reports of its successful application to suspension-cultured cells. Another problem is the possibility that the vauoles finally obtained may have been subjected to a breakage and resealing before the purification steps are begun. Speaking against this fear are the results of Leigh, Rees, Fuller, and Banfield (1979) who included nonpermeating ^{3}H-inulin in the collection medium. Determination of the ratio of betanin (the red pigment in beetroot vacuoles) to ^{3}H was the

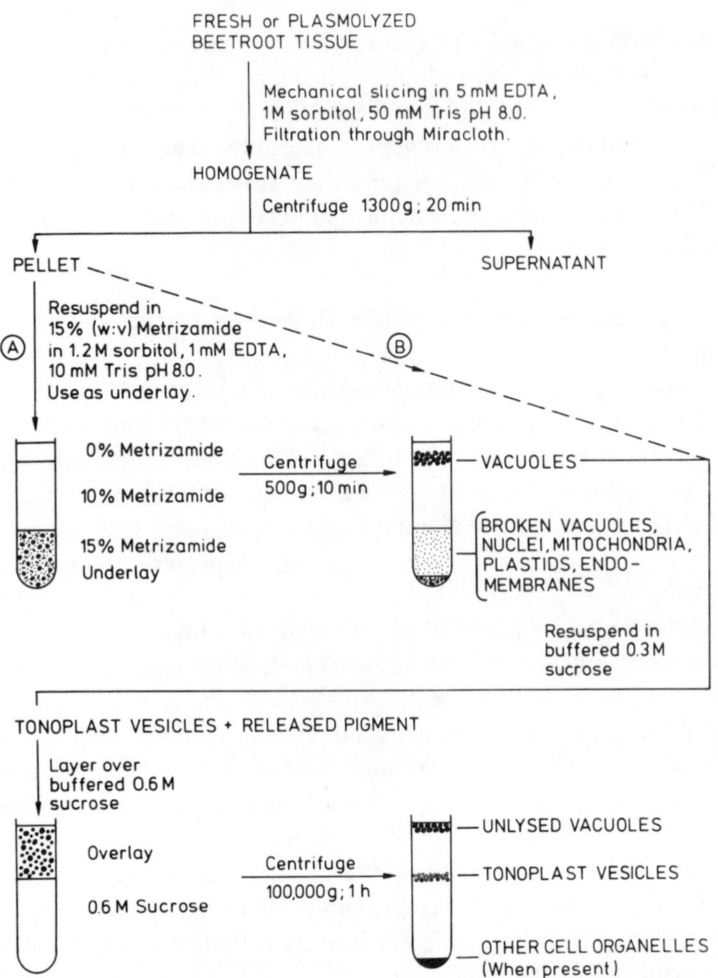

Figure 6.5. Procedures developed for the preparation and purification of vacuoles and tonoplast vesicles from red beet tissue (brought together from published details in Marty et al., 1980; Leigh et al., 1979b).

highest in the purified vacuole preparation, indicating that at least when the vacuoles enter the collection medium they are sealed. Just what happens immediately upon cell breakage is, however, a matter of conjecture. Certainly the larger the vacuole, the greater the chance that it will burst upon mechanical disruption of the cell. It is, therefore, not surprising that the best yields are obtained from meristematic regions [e.g., root tips (Pohl, 1981)].

If intact vacuoles were obtained it would have been relatively easy to separate the membrane (TP) from the content (see, e.g. Figure 6.5). Because contamination from other membranes is largely removed during the preparation of the vacuoles, recognition of TP vesicles after this method is relatively easy. This, of course, is not normally the case for homogenates from nonmeristematic plant tissue. However, with the aid of the Cl^--stimulated Mg^{2+}-ATPase (see above) it is now possible to locate the position of TP vesicles on isopycnic sucrose gradients. According to Mandala et al. (1982), they band at a density of 1.11 $g \cdot cm^{-3}$ in gradients prepared from maize coleoptiles under low Mg^{2+} conditions. This puts them in the same region as that of the ER, but in contrast to the ER, which shifted to a higher density in response to high Mg^{2+} concentrations, the peak for the ATPase activity remained constant. A distinction between ER and TP vesicles was also obtained by centrifuging isopycnically on a linear 1–12% dextran gradient.

6.3.4. Indirect Isolation via Protoplasts

The preparation of protoplasts is becoming routine in a number of areas of plant research and the field of vacuolar physiology is no exception, where it is now the method of choice. Vacuoles may be released from protoplast preparations, equilibrated in 0.5–0.7 *M* mannitol or sorbitol solutions, in one of the following ways:

1. *By osmotic lysis.* Here there are two methods in use. Either the protoplasts are rapidly resuspended in 0.1–0.2 *M* K-phosphate buffer at pH 8 (Boller and Kende, 1979; Wagner and Siegelmann, 1975; Wagner, 1983), or the osmolarity of the bathing polyol is reduced (Mettler and Leonard, 1979; Sasse et al., 1979; Saunders and Conn, 1978). In the latter case the osmolarity is usually reduced by about 40–50% although the best yields (~50%) have been obtained with a respective change from 0.7 to 0.5 *M* mannitol at ice temperature (Kringstad et al., 1980).
2. *By high-shear forces under isotonic conditions.* This is achieved by either high-speed centrifugation through a sucrose or Ficoll cushion (see Figure 6.6; Guy et al., 1979; Komor et al., 1982a,b;

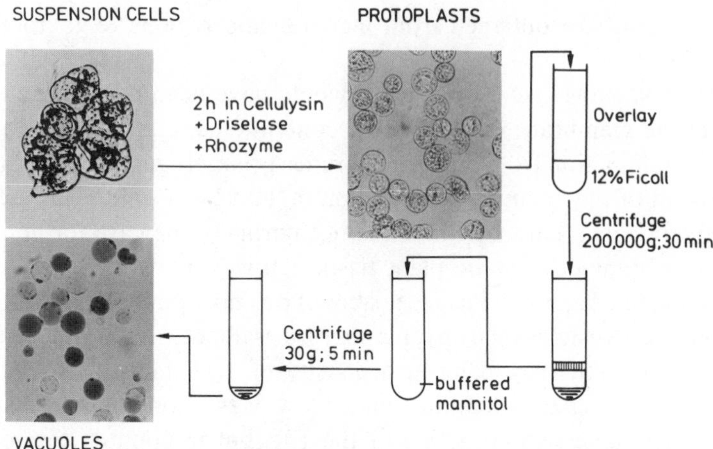

Figure 6.6. Vacuole isolation from protoplasts. Protoplasts prepared enzymatically from sugar cane cells are burst open by high-speed centrifugation onto a Ficoll cushion. The vacuoles are stained with neutral red. Published drawing of Komar et al. (1982a).

Lorz et al., 1976) or by extrusion through a fine needle (Martinoia et al., 1981).

3. *By polybase-induced lysis.* Here DEAE–dextran in buffered mannitol is added to the protoplast suspension at a concentration equivalent to 10–50 pg· protoplast^{-1}. Excess unbound polybase is removed by addition of dextran sulfate resulting in the lysis of the protoplasts after several minutes (Boudet et al., 1982; Buser and Matile, 1977; Schmidt and Poole, 1980).

Separation of the vacuoles from the rest of the protoplast homogenate is variable depending in part on the method of lysis. A step gradient usually involving Ficoll or Metrizamide is often used, but the concentrations and centrifugation speeds and duration are often quite different. Thus there are papers describing vacuole isolation with low speeds (≤1000*g*) and short duration (≤60 min) (e.g., Boller and Kende, 1979; Mettler and Leonard, 1979) and there are those where high speeds (100,000*g*) and long duration (2 hr) have been employed (e.g., Lin and Wittenbach, 1981; Saunders and Conn, 1978). In addition, flotation rather than sedimentation has also been used as a method of centrifugal separation (e.g., Martinoia et al., 1981).

REFERENCES

Balke, N. E. and Hodges, T. K. (1979). *Plant Physiol.* **63,** 48–52.

Berjak, P. (1972). *Ann. Bot. NS* **36,** 73–81.

Boller, T. (1982). *Physiol. Veg.* **20,** 247–257.

Boller, T. and Kende, H. (1979). *Plant Physiol.* **63,** 1123–1132.

Boudet, A. M., Canut, H., and Alibert, G. (1982). *Plant Physiol.* **68,** 1354–1358.

Bowes, B. G. (1965). *Cellule* **65,** 357–364.

Briskin, D. P. and Poole, R. J. (1983). *Plant Physiol.* **71,** 350–355.

Buser, C. and Matile, P. (1977). *Z. Pflanzenphysiol.* **82,** 462–466.

Buvat, R. (1971). In *Origin and Continuity of Cell Organelles* (J. Reinert and H. Ursprung, eds.) pp. 127–157. Springer Verlag, Berlin.

Dangeard, P. (1956). *Protoplasma* **3,** 1–41.

Doll, S., Rodier, F., and Willenbrink, J. (1979). *Planta* **144,** 407–411.

Du Pont, F. M., Burke, L. L., and Spanswick, R. M. (1981). *Plant Physiol.* **67,** 59–63.

Du Pont, F. M., Giorgi, D. L., and Spanswick, R. M. (1982). *Plant Physiol.* **70,** 1694–1699.

Fineran, B. A. (1970a). *Protoplasma* **70,** 457–478.

Fineran, B. A. (1970b). *J. Ultr. Res.* **33,** 574–586.

Gregory, D. W. and Cocking, E. C. (1965). *J. Cell Biol.* **24,** 143–146.

Grob, K. and Matile, P. (1979). *Plant Sci. Lett.* **14,** 327–335.

Grob, K. and Matile, P. (1980). *Z. Pflanzenphysiol.* **98,** 235–243.

Guilliermond, A. (1941). In Chronica Botanica Co., Waltham, MA.

Guilliermond, A., Mangenot, G., and Plantefol, L. (1933). *Traité de Cytologie Végétale*. Le Francois, Jouve, Paris.

Guy, M., Reinhold, L., and Michaeli, D. (1979). *Plant Physiol.* **64,** 61–64.

Hager, A., Frenzel, R., and Laible, D. (1980). *Z. Naturforsch.* **35,** 783–793.

Hodges, T. K. (1976). Encycl. Plant Physiol. Vol. IIA, pp. 260–283. Springer Verlag, Berlin.

Hodges, T. K. and Leonard, R. T. (1972). *Meth. Enzym.* **32,** 392–406.

Komor, E., Thom, M., and Maretzki, A. (1982a). *Physiol. Veg.* **20,** 277–287.

Komor, E., Thom, M., and Maretzki, A. (1982b). *Plant Physiol.* **69,** 1326–1330.

Kringstad, R., Kenyon, W. H., and Black, C. C. (1980). *Plant Physiol.* **65,** 379–382.

Lancaster, J. E. and Collin, H. A. (1981). *Plant Sci. Lett.* **22,** 169–176.

Leigh, R. A. (1983). *Physiol. Plant.* **57,** 390–396.

Leigh, R. A. and Branton, D. (1976). *Plant Physiol.* **58,** 656–662.

Leigh, R. A. and Walker, R. R. (1980). *Planta* **150,** 222–229.

Leigh, R. A., Rees, T. A. P., Fuller, W. A., and Banfield, J. (1979). *Biochem. J.* **178,** 539–547.

Leigh, R. A., Branton, D., and Marty, F. (1979). In *Plant Organelles* (E. Reid, ed.), Vol. 9, Methodological Surveys in Biochemistry, pp. 69–80. Ellis Harwood Ltd., Chichester.

Leonard, R. T. and Hodges, T. K. (1973). *Plant Physiol.* **52,** 6–12.

Leonard, R. T. and Hotchkiss, C. W. (1976). *Plant Physiol.* **58,** 331–335.

Lin, W. and Wittenbach, V. A. (1981). *Plant Physiol.* **67,** 969–972.

Lin, W., Wagner, G. J., Siegelman, H. W., and Hind, G. (1977). *Biochim. Biophys. Acta* **465,** 110–117.

Lorz, H., Harms, C. T., and Potrykus, I. (1976). *Biochem. Physiol. Pflanz.* **169,** 617–620.

Mandala, S., Mettler, I. J., and Taiz, L. (1982). *Plant Physiol.* **70,** 1743–1747.

Marin, B., Marin-Lanza, M., and Komor, E., (1981). *Biochem. J.* **198,** 365–372.

Marinos, N. G. (1963). *J. Ultr. Res.* **9,** 177–185.

Martinoia, E., Heck, U., and Wiemken, A., (1981). *Nature* **289,** 292–294.

Marty, F. (1973a). *C. R. Acad. Sci. Paris D* **276,** 1549–1552.

Marty, F. (1973b). *C. R. Acad. Sci. Paris D* **277,** 2681–2684.

Marty, F. (1978). *Proc. Nat. Acad. Sci.* **75,** 852–856.

Marty, F. (1980). *J. Histochem. Cytochem.* **28,** 1129–1132.

Marty, F. (1982). In *Plasmalemma and Tonoplast: Their Functions in the Plant Cell* (D. Marmé, E. Marré, and R. Hertel, eds.), pp. 179–188. Elsevier Biomedical Press, Amsterdam.

Marty, F., Branton, D., and Leigh, R. A., (1980). *Biochem. Plants* **1,** 625–658.

Matile, P. (1975). In *Cell Biology Monographs,* Vol. 1. Springer Verlag, pp. 1–183. Berlin.

Matile, P. (1978). *Ann. Rev. Plant Physiol.* **29,** 193–213.

Matile, P. (1982). *Physiol. Vég.* **20,** 303–310.

Matile, P. and Moor, H. (1968). *Planta* **80,** 159–175.

Mesquita, J. F. (1969). *J. Ultr. Res.* **26,** 242–250.

Mettler, I. J. and Leonard, R. T. (1979). *Plant Physiol.* **64,** 1114–1120.

Mettler, I. J., Mandala, S., and Taiz, L. (1982). *Plant Physiol.* **70,** 1738–1742.

Mueller, W. C. and Greenwood, A. D. (1978). *J. Exp. Bot.* **29,** 757–764.

Pohl, U. (1981). *Ber. Dtsch. Bot. Ges.* **94,** 127–134.

Quail, P. H. (1979). *Ann. Rev. Plant Physiol.* **30,** 425–484.

Rasi-Caldongo, F., De Michelis, M. I., and Pugliarello, M. C. (1981). *Biochim. Biophys. Acta* **642,** 37–45.

Sasse, F., Backs-Hüsemann, D., and Barz, W. (1979). *Z. Naturforsch.* **35c,** 848–853.

Saunders, J. A. and Conn, E. E. (1978). *Plant Physiol.* **61,** 154–157.

Schmidt, R. and Poole, R. J. (1980). *Plant Physiol.* **66,** 25–28.

Stout, R. G. and Cleland, R. E. (1982). *Plant Physiol.* **69,** 798–803.

Symillides, Y. and Marty, F. (1977). *C. R. Acad. Sc. Paris D* **56,** 721–724.

Taiz, L., Murry, M., and Robinson, D. G. (1983). *Planta* **158,** 534–539.

Ueda, K. (1966). *Cytologia* **31,** 461–472.

Vianello, A. Dell'Antone, P., and Macrì, F. (1982). *Biochim. Biophys. Acta* **689,** 89–96.

Wagner, G. J. (1982). *Rec. Adv. Phytochem.* **14,** 145.

Wagner, G. J. (1983). In *Isolation of Membranes and Organelles from Plant Cells* (A. L. Moore and J. L. Hall, eds.), pp. 83–118. Academic Press, New York.

Wagner, G. J. and Siegelman, H. W. (1975). *Science* **190,** 1298–1299.

Walker, R. R. and Leigh, R. A. (1981). *Planta* **153,** 140–149.

Wiebe, H. H. (1968). *Bioscience* **28,** 327–331.

Willenbrink, J. and Doll, S. (1979). *Planta* **147,** 159–162.

Zirkle, C. (1937). *Bot. Rev.* **3,** 1–30.

7

PLASMA MEMBRANE

Although the PM is a structure common to all eukaryotic cells, it has a number of functions in plant cells which clearly delineate it from its counterpart in animal cells. Ion transport properties (in higher plants at least—Lüttge and Higinbotham, 1979), for example, are in many ways different from those in animal cells, and microfibril synthesis (see Chapter 9) is a basic property of all plant cells which is not seen in animals. Coupled with their heterotrophic life-style animal cells regularly take up a variety of molecules from the cell exterior through endocytosis (see Chapter 8). The autotrophic nature of plant cells does not preclude this feature as judged by the pinocytotic capability of protoplasts (Cocking, 1970) and by the uptake of N_2-fixing bacteria in root cells of legumes (Goodchild and Bergersen, 1966), but may, because of the cell wall, be less frequent (Baker and Hall, 1973).

Despite its great importance and in contrast to the situation in animal cells (e.g., Wallach, 1972), there are few general reviews on the structure and isolation of the plant PM (e.g., Hall, 1983; Leonard and Hodges, 1980). Instead one has to look for the appropriate sections in books/reviews on cell walls and transport phenomena and so forth.

7.1. STRUCTURE AND MORPHOLOGY

Typically the PM is the thickest of all membranes (at least 7 nm; Morré and Bracker, 1976) and displays in section the clearest dark–light–dark profile (Grove et al., 1968). Sometimes the outermost of the dark lines stains more intensely than the inner one (Ledbetter and Porter, 1970). Freeze-fracture electron microscopy reveals the presence of numerous 7–15-nm particles, some of which are aggregated into complexes involved in microfibril biosynthesis (Chapter 9).

7.1.1. Increase of Surface Area

The cell can increase its surface area for transport purposes by an evagination or invagination of the PM. Evaginations of the PM are well known to every botanist in the form of the root hair (Esau, 1965) and are essentially restricted to this one case. Invaginations both in animal and plant cells are, in contrast, the more frequent option. Animal cells commonly resort to the microvillus (or collectively brush-border) to achieve this.

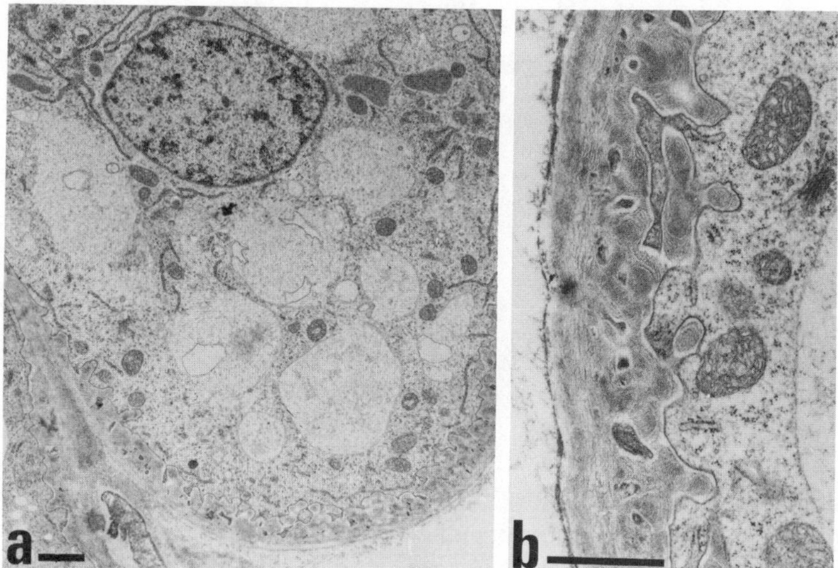

Figure 7.1. (a,b) Low and high magnifications of epidermal transfer cells in roots of *Helianthus annuus* induced by iron deficiency. Bar = 1 μm. Unpublished micrographs of D. Kramer.

Regular infoldings of this type are not present in plant cells, instead irregular projections of the cell wall ("labyrinth walls") are often seen (Figure 7.1). Cells possessing such surface elaborations are quite common in higher plants and are termed *transfer cells*, irrespective of their anatomical origin or type (Gunning and Pate, 1974). Infoldings which are not accompanied by cell wall development are in comparison quite rare but are found in salt gland cells of certain grasses (Levering and Thomson, 1971; Oross and Thomson, 1982). In these cases the invaginations are so deep that they traverse almost the whole cell.

7.1.2. Plasmodesmata

As a result of an incomplete separation of daughter protoplasts during cell division (see Chapter 14), higher plant cells remain in contact with one another. This is not like the contact forms in animal cells, where portions of the PM of adjacent cells touch (tight junctions) or are held together by special structures (gap junctions), but represents a direct con-

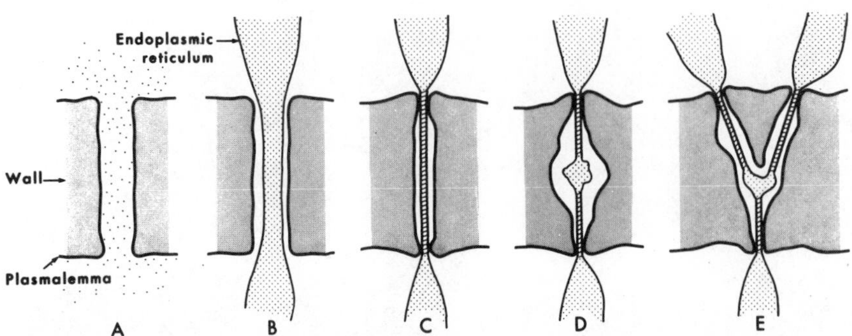

Figure 7.2. Types of plasmodesmata in higher plants. A—desmotubule type (PM constriction). B—open ER type (no PM constriction). C—median nodule type (PM constriction). D—anastomosing type (PM constriction). Published drawing of Robards (1975).

tinuity of the PM between the daughter cells. Since cell wall deposition (cell plate formation) is an integral part of cytokinesis in plant cells, this continuity assumes the form of a tube ("plasmodesma") which bridges the intervening cell wall. Thus the PM in higher plants is continuous throughout the whole organism and as a consequence creates two transport compartments: the symplast (the interconnecting protoplasts) and the apoplast (the interconnecting cell walls).

Two exceptions to this rule exist: one involves cell death (xylem cell differentiation) and the other reproduction (pollen formation, embryo development). In these cases either existing plasmodesmata are severed or, when sporophytic and gametophytic tissues are in contact with one another, are not developed.

Over the last 10 years there have appeared numerous articles on plasmodesmata, including several reviews (Gunning and Overall, 1982; Gunning and Robards, 1976a; Robards, 1975) and a book (Gunning and Robards, 1976b). It is, therefore, unnecessary to give here more than salient information on the structure of plasmodesmata. For a discussion of the formation of plasmodesmata the reader is referred to Chapter 14.

Although not always present, ER usually traverses the plasmodesma, (Figure 7.2) which shows that membrane continuity between adjacent cells is not just restricted to the PM. The plasmodesma in transverse section (Figure 7.3a,c) is often restricted, squeezing the ER into a structure termed *desmotubule*, which alludes to the tubular nature of the ER within the plasmodesma. In order to produce such a small-diameter struc-

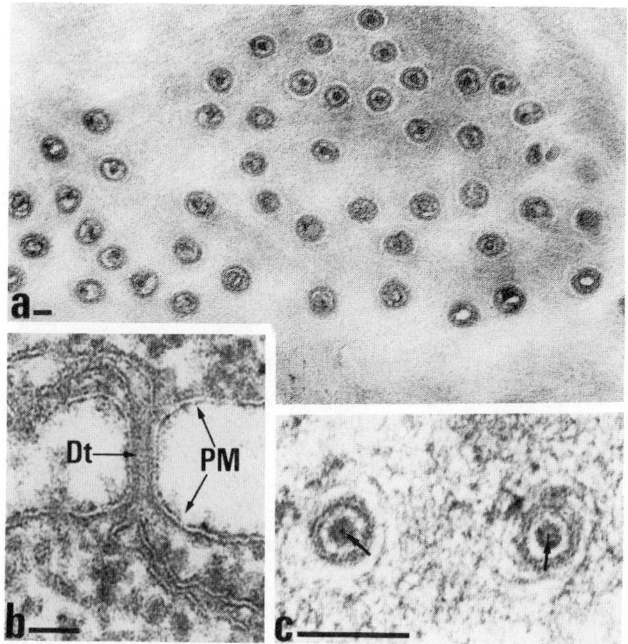

Figure 7.3. (a) Transverse-sectioned plasmodesmata in a primary pit field between mesophyll and bundle sheath cells of the bermuda grass leaf. (b) Longitudinal section through a plasmodesma from an Azolla root cell fixed in the presence of tannic acid. The continuity of ER at both sides of the plasmodesma is clearly seen (PM = plasma membrane; Dt = desmotubule). (c) Transversely sectioned plasmodesmata from an *Abutilon* nectary (arrows indicate the "central rod"). Bar = 0.1 μm. Published micrographs of Thomson (a), Gunning and Overall (b, 1982); Gunning and Hughes (c, 1976).

ture (16–20 nm), the desmotubule (according to Robards, 1968, 1971) represents a lipid-poor region of the ER. Robards's model envisages the desmotubule as a hollow entity which in the center of its lumen has a central rodlike structure, although the origin of the latter is unclear, Another feature of this model is that the PM constricts at the cytoplasmic openings.

In contrast to this conception of the architecture of the plasmodesma, an older model, that of Lopez-Saez et al. (1966), has a closed central cylinder without constrictions of the PM at the cytoplasmic openings. The advantage of this model lies in the easier explanation for the central rod, which was interpreted as representing a concentration of polar head

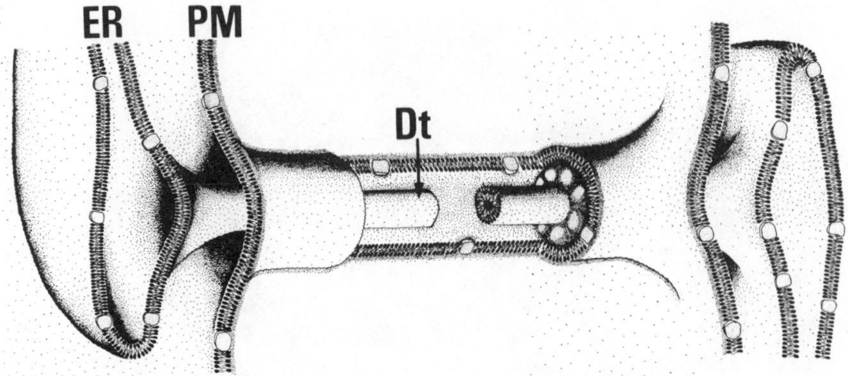

Figure 7.4. Diagram representing the molecular architecture of a plasmodesma. The space between desmotubule (Dt) and PM is interpreted as being occluded with particular material. Published drawing of Gunning and Overall (1982).

groups. The other set of electron-dense polar groups was thought to constitute the tubular structure later termed desmotubule by Robards.

Robards's difficulty in accepting an almost solid membrane cylinder in the Lopez-Saez et al. model has not found support in recent theoretical considerations of the molecular packing of lipids (Overall et al., 1982). Moreover, refined EM techniques (tannic acid and potassium ferricyanide fixations—Hepler, 1982; Overall et al., 1982) show convincingly that the lumen of the ER is constricted to such an extent that it becomes totally occluded leaving only the "central rod" visible. The region between the ER and the PM is also now thought to be partially occluded with particulate material (Figure 7.4). In some cells this occlusion is localized in the neck region and has been termed *sphincter* (Evert et al., 1977; Oleson, 1979).

Because of the solid desmotubule a luminal transport within the ER from one cell to another is no longer tenable. Furthermore, depending on the degree of occlusion of the extradesmotubular space, cytosolic symplastic transport cannot therefore occur as freely as previously imagined. Goodwin (1983) has recently determined the size limits for molecules passing through plasmodesmata by microinjecting fluorescent-labeled peptides of different molecular weight. The upper limits appear to be less than 10^3. The plasmodesma is thus a molecular filter of the same order of magnitude as that of the gap junction in animal cells (Flagg-Newton et al., 1979).

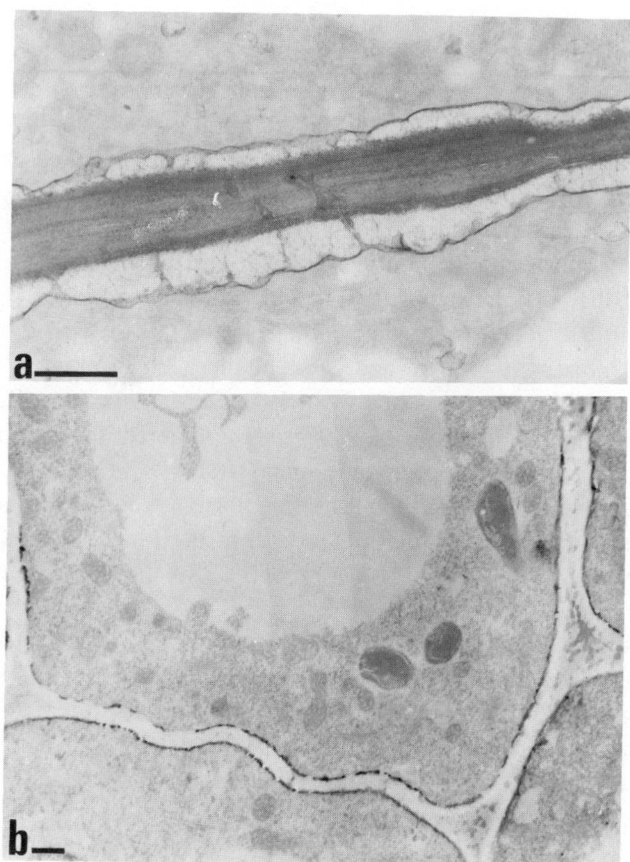

Figure 7.5. Cytochemistry of the PM. (a) Pea root cell stained with silicotungstate–chromate. (b) Wheat coleoptile epidermal cell stained for pH 7 $K^+(Mg^{2+})$ATPase activity. Bar = 0.5 μm. Unpublished (a) micrograph of Roland; published (b) micrograph of Hall et al. (1982).

7.2. CYTOCHEMISTRY

In 1972 Roland et al. introduced phosphotungstate-chromate (PTAC) as a selective stain for the PM of plant cells [see Figure 7.5 and Roland (1978) for details of the method]. It is certainly not a difficult method to use but "strict adherence to the experimental protocol" is to be recommended (Quail, 1979). Inadequate periodic acid treatment and too thin a section are pitfalls which should be avoided (Leonard and Van der Woude, 1976; Nagahashi et al., 1978). Nevertheless, even when carried out correctly, there are numerous reports in the literature where the selective nature of the stain has been called into question. On the one hand, there are cases where the PM does not stain (freshly isolated protoplasts in comparison to older ones—Mayo and Cocking, 1969; Taylor and Hall, 1979) and on the other there are micrographs showing a positive reaction for other membranes such as the TP (Hall and Flowers, 1976; Thom et al., 1975), ER (Thom et al., 1975), and prolamellar bodies (Quail and Hughes, 1977). Secretory vesicles which are about to fuse with the PM also react positively (Morré and Mollenhauer, 1983; Vian and Roland, 1972), but perhaps these can be regarded as PM *sensu lato*. Because of these deviating examples the use of this stain to detect the PM in gradient fractions (see below) has been cautioned (Hall, 1982; Quail, 1979). However, when one realizes that the other membrane types mentioned do not possess equilibrium densities (in sucrose) in the range where the PM is to be expected (1.13—1.18 $g \cdot cm^{-3}$) much of the caution is, I think, unjustified.

The widely accepted thesis that the PM contains ATPase activities that participate in ion transport [see Lüttge and Higinbotham (1979) and Spanswick et al. (1980) for references] has led to attempts to demonstrate this fact by cytochemical means. The method of choice is that of heavy-metal phosphate precipitation (see Figure 7.4b in Essner, 1973, for details), which has been applied successfully to phloem cells (e.g., Cronshaw, 1980) and to cells of nonconducting tissue (Hall et al., 1980, 1982). By choosing the appropriate incubation conditions, the deposition of reaction product at sites other than the PM can be kept to a minimum (Hall et al., 1982).

7.3. MARKERS FOR ISOLATION

7.3.1. Enzymes

Three types of enzyme activities have been used to recognize PM vesicles: glucosylation reactions, ATPases, and cellulases. Of these the first two are in wide use (Quail, 1979).

There is a great deal of evidence implicating the PM in the synthesis of glucans at the cell surface (Chrispeels, 1976; Delmer, 1977; Robinson, 1977). Since both cellulose and callose are deposited there, it is reasonable to expect that 1,4- and 1,3-β-glucan synthase activities are associated with the PM. As already mentioned (Section 5.5), the PM glucan synthase activity, GS II, is measured using m*M* concentrations of nucleotide sugar (Van der Woude et al., 1974). Until recently it has been regarded as a specific marker for the PM, but this appears to be true only for dicotyledenous plants. In monocots, however, GS II activity is also measurable in GA-rich fractions and this is not due to contaminating PM vesicles (Robinson and Glas, 1983). It has also been customary to carry out this assay with little or no Mg^{2+} present (Ray, 1977, 1979; Raymond et al., 1978), although the reasons for doing so are rarely clearly given. Both of these factors are known to influence the amount of radioactive glucose incorporated as well as the type of linkage involved. Although there are exceptions (see, e.g, Heiniger and Delmer, 1977), 1,3-β linkages are usually favored by high UDPG concentrations (Cook et al., 1980; Raymond et al., 1978; Smith and Stone, 1973; Tsai and Hassid, 1973). On the other hand, increased Mg^{2+} concentrations tend to a reduction in the proportion of 1,3-β linkages in the product (Fevre and Rougier, 1981; Tsai and Hassid, 1971) but to an increase in the total incorporation (Henry and Stone, 1982; Robinson et al., 1982). From the point of view of optimizing the assay for gradient fractions, without paying attention to the nature of the products formed, the inclusion of Mg^{2+} is therefore to be recommended. Another factor overlooked in the past is the stimulatory effect of sucrose on GS II activity (Kemp et al., 1978; Ray, 1979). Sucrose does not have to be present in the assay to achieve this effect: a pretreatment followed by centrifugation to remove the sucrose is sufficient (Ephritikhine et al., 1980). The mechanism appears to be an osmotic one since media such as Metrizamide and Ficoll do not elicit the same effect in this density range. It might therefore be worthwhile in terms of raising the levels of incorporation to remove such media by pelleting and to resuspend the membranes in sucrose at densities at least 1.10 $g \cdot cm^{-3}$.

A further glucosyl transferase reaction, namely that involving the glucosylation of sterols, can also be employed as a PM marker. In the early paper of Van der Woude et al. (1974) the PM was shown to be the major site of glycolipid synthesis *in vitro*. This has been confirmed on both monocotyledenous (Hartmann et al., 1977; Hartmann-Bouillon et al.,

1979) and dicotyledenous (Chadwick and Northcote, 1980) plants. The reaction however is not yet in general use as a standard PM marker.

Two types of ATPase reactions are in current use as PM markers and both require Mg^{2+} (Briskin and Poole, 1983; Gross and Marme, 1978). One is an electrogenic pump which is stimulated by K^+ ions and has a pH optimum at 6.5 (Leonard and Hodges, 1973). It is still not certain whether this pump is responsible for proton–K^+ exchange or whether K^+ is transported in response to an electrochemical potential gradient for protons generated by the ATPase (Leonard, 1982). This ATPase differs from a similar one localized at the TP (see Table 6.1, Chapter 6.) through its inhibition by vanadate (Du Pont et al., 1981; Mandala et al., 1982). At low (~5 μM) concentrations this substance is a potent inhibitor of PM ATPases in mammalian (Cantley et al., 1977) and fungal (Bowman and Slayman, 1979) cells. In plant PM preparations higher (50–100 μM) concentrations are usually required for an effective inhibition, which also inhibit soluble acid phosphatase activity (Gallagher and Leonard, 1982). The inhibition appears to be noncompetitive and increases with increasing K^+ concentrations. Nevertheless, the sensitivity of the ATPase toward vanadate can be used as a measure for the presence and position of PM vesicles in gradients (Taiz et al., 1981).

Another ATPase activity, which is apparently associated with the PM, transports Ca^{2+}. This activity is distinctly different from mitochondrial Ca^{2+} uptake in being unaffected by respiratory inhibitors such as KCN or antimycin A or by the uncoupler CCCP (Dieter and Marmé, 1980a). Whereas this Ca^{2+} transport ATPase is stimulated by calmodulin from both plant and animal sources, mitochondrial Ca^{2+} uptake is not (Dieter and Marré, 1980b). Unfortunately, the localization of this activity at the PM has been claimed (Dieter and Marmé, 1980a) but not proven. Interestingly, in a recent report Buckhout (1983) described the existence of a similar Ca^{2+} uptake at the ER. In contrast to the K^+–ATPase just described, Ca^{2+}–ATPase is not inhibited by vanadate (Dieter and Marmé, 1980a).

Ultrastructural observations on cellulytic activity suggest a localization at the PM–wall interface (Bal et al., 1976). Because of this, cellulase has been introduced as a PM marker (Koehler et al., 1976; Pierce and Hendrix, 1979). Perhaps because of the long (>10 hr) incubation periods necessary for the measurement of this enzyme, cellulase is not regularly used in gradient analysis.

7.3.2. Ligand Binding *in Vitro*

At present there are two ligand binding properties of the PM which are, dependent on the availability of radioactive ligand, sometimes used to locate the position of the PM in gradients. The more frequently used is the antiauxin naphthylphthalamic acid (NPA—Jacobs and Hertel, 1978; Lembi et al., 1971) but the fungal toxin fusicoccin, which is often used in studies on extension growth (Marré, 1979), has also been applied with equal success (Dohrmann et al., 1977). It is important in this technique to ensure that nonspecific binding is kept as low as possible. To this end, the tests are carried out in duplicate on resuspended gradient fractions, to which an excess of nonradioactive ligand is added to one of the samples.

Lectins are a special form of ligand which react with the saccharide portion of membrane glycoproteins of glycolipids and may cause the agglutination of animal cells or plant protoplasts (Larkin, 1978a,b). Binding sites for the lectin concanavalin A (Con A) at the surface of protoplasts have been demonstrated cytochemically on several occasions (Burgess and Linstead, 1976, 1977; Williamson, 1979; Williamson et al., 1976). As a result, the addition of radioactive- or ferritin-labeled Con A to isolated membrane fractions in order to recognize the PM has been successfully carried out (e.g., Hartmann-Bouillon et al., 1982; Travis and Berkovitz, 1980). However, the method suffers from two disadvantages. The first is that binding sites on the ER have also been detected (Berkovitz and Travis, 1981; Hartmann-Bouillon et al., 1982). Although not yet tested, it might be expected that sites will also be found on the GA and will, together with those at the ER, presumably represent synthesis and transport of the PM lectin receptor. Another problem is that in one and the same plant the number of binding sites per surface area of PM can vary according to the developmental stage of the cells involved (Berkovitz and Travis, 1982).

7.3.3. Surface Labeling *in Vivo*

With this technique the PM is tagged with a radioactive, fluorescent, or electron-dense label which can be followed during fractionation and isolation. Normally such methods are carried out with protoplasts to enable adequate binding and to rule out unspecific reactions with cell wall components or the edges of cut and damaged cells in tissue segments. There are several methods available:

1. *Iodination.* This can be carried out with $Na^{125}I$ either enzymically with either H_2O_2 or lactoperoxidase or chemically and involves the iodination of tyrosine groups in cell surface proteins. Although the method has often been used on animals (e.g., Phillips and Morrison, 1973) and has recently been successfully applied to protoplasts from ryegrass endosperm cells (Schibeci et al., 1982), it cannot always be recommended because of the frequent high levels of endogenous peroxidase in plant cells (Galbraith and Northcote, 1977; Hall and Sexton, 1972).
2. *Diazotized sulfanilic acid.* Originally developed for animal cells (Berg and Hirsh, 1975; Edwards et al., 1979), it has also been used with success on fungal cells (Scarborough, 1975). Soybean (Galbraith and Northcote, 1977) and maize leaf (Perlin and Spanswick, 1980) protoplasts have been effectively labeled using S- or I-labeled derivatives, but, in contrast, ryegrass endosperm protoplasts were found to be permeable to these (Schibeci et al., 1982), thus preventing their application in this system.
3. *Lectin binding.* There are only a few publications where labeled con A has been applied to protoplasts before their homogenization (e.g., Boss and Ruesink, 1979). Agglutination occurred, but the large sheetlike pieces of PM which can be obtained with fungi (Scarborough, 1975) were not achieved. Ryegrass protoplasts are not agglutinated with Con A but can be induced to do so with the 1,6-β-galactose-binding proteins axinellin and murine myeloma protein J539 (Schibeci et al., 1982). Upon cell rupture the PM in this case does not fragment but remains in large sheetlike pieces. These authors have suggested that galactose-binding lectins may be much more efficient than Con A, which is specific for mannose and glucose residues, in agglutinating plant protoplasts.
4. *Electron-dense metals.* Labeling of the PM of protoplasts with lanthanum salts has been reported by Taylor and Hall (1978; 1979). They claim that no redistribution of stain occurs upon homogenization and that marker enzymes such as K^+–ATPase can still be assayed in the presence of lanthanum. Although not yet used to localize the PM in gradients, the application of antimonate ap-

pears to be potentially as good (Quader et al., 1983; Stockwell and Hanckey, 1982).

5. *Callose synthesis.* Segments of higher plant tissue, when incubated *in vivo* with UDP–^{14}C-glucose synthesize radioactive callose (1,3-β-glucan—Anderson and Ray, 1978; Raymond et al., 1978). Since incorporation is restricted to the PM–wall interface of the cut cells at the periphery of the tissue (MacLachlan, 1982; Mueller and MacLachlan, 1980), it is not a method which can be universally applied.

7.3.4. Sidedness of PM Vesicles

Homogenization causes the PM to be fragmented into small pieces that rapidly close into vesicles. Do these vesicles retain a similar orientation with respect to the PM *in situ* [i.e., the cytoplasmic face is now the lumen-facing side [''inside-in'' or ''outside-out'')] or do they show an inversion [i.e., the cytoplasmic face is now the external-facing side [''inside-out'' or ''outside-in'')]? If one assumes that vesicles derived from the PM orient themselves in the same way as do those from the ER or the TP, then they should be outside-out. Speaking for this interpretation would be their labeling *in vitro* with Con A. However, if PM–Con A receptor glycoproteins are cotranslationally synthesized at the ER (see Section 10.1) their oligosaccharide side chains are directed into the lumen of the ER. A specific binding of Con A to PM-destined Con A receptors at the ER surface would therefore not seem possible and yet Con A has been shown to bind to outside-out ER vesicles *in vitro* (see above). If one assumes this latter binding is unspecific, one can also adopt the same argument for the PM vesicles, which would mean that they might be inside out.

Current opinion on extension growth in plant cells centers on a PM-located ATPase which pumps protons into the extracellular space (Rubery, 1981). If the electrogenic ATPase which pumps protons inwardly in PM vesicles is identical with this, then an inversion of the PM has occurred during homogenization. Supporting this contention is the fact that Ca^{2+} is transported out of the cell *in vivo* but is inwardly transported in PM vesicles from both plant (see above) and fungal (Stroobant et al., 1980) cells.

Whereas Con A binding *in vitro* may be subject to some criticism, it is somewhat more difficult to negate the results of (anti) auxin binding and transport. If one assumes that such substances have binding sites exclusively at the external surface of the PM, their binding *in vitro* to putative PM vesicles must indicate an outside-out orientation. This contention finds strong support from experiments with a nonpenetrating albumin–fusicoccin conjugate which not only binds to the surface of protoplasts but competes for fusicoccin binding sites on *Cucurbita* PM vesicles (Aducci et al., 1980; Hertel et al., 1983). Similarly, the kinetics of NPA-binding (Michalke, 1982) as well as *in vitro* auxin uptake (Depta and Hertel, 1982; Hertel et al., 1983) are hard to understand if the PM vesicles do not have an outside-out orientation.

It is difficult to judge which of the two possibilities as to the orientation or sidedness of PM vesicles is the more correct. Having had more experience with glucan synthase, this author tends to believe that PM vesicles are inverted and that this is one of the reasons that the rate of incorporation in this assay is so low. Indeed, the detergent stimulation of sterol-glucosyl transferase activity (Quantin et al., 1980) does suggest a site for this enzyme on the inner surface of the vesicle membrane. On the other hand, it is difficult to give a reason as to why the PM should behave differently to the TP upon homogenization.

7.4. SEPARATION

As with other organelles, attempts at separating the PM from other components in cell homogenates have most often involved density gradient centrifugation, especially with sucrose as the gradient medium. After equilibrium conditions have been achieved, PM vesicles band invariably at sucrose densities somewhere between 1.13 and 1.18 $g \cdot cm^{-3}$. Until recently equilibrium densities for plant PM have been obtained from gradients (and homogenates) which contain low (0.1 m*M*) or no Mg^{2+}. It is now clear that in the presence of high Mg^{2+} concentrations (~3 mM), PM vesicles equilibrate at higher densities (lying roughly 2–5% sucrose deeper in the gradient—Robinson et al., 1982). In terms of a one-step separation from other organelles on sucrose gradients, this property of the PM has only limited usefulness since the presence of Mg^{2+} allows the retention of ribosomes on the ER (see Chapter 2) and shifts the latter to a similar density as that of the PM. A density shift of another sort, and one which is probably of little value as a regular method for PM separation, is that which occurs as a result of tissue incubation in UDPG (see above). The callose formed by the wounded cells apparently becomes trapped within the vesicles resulting in their increased density (>1.19 $g \cdot cm^{-3}$—Anderson and Ray, 1978; Pierce and Hendrix, 1979).

The majority of work involving gradient centrifugation on dicotyledenous plants has been done with members of the Leguminoseae (peas, beans). Typical for these plants is the considerable cross-contamination with the membranes of the GA, as judged by the high degree of overlapping of IDPase/GS I, GS II/K^+–ATPase (Robinson et al., 1982; Samson et al., 1983; Shore and MacLachlan, 1975; Wienecke et al., 1982). In contrast, monocotyledenous plants, especially grasses, seem to have a clear separation of GA and PM marker enzyme activities (Leonard and Van der Woude, 1976; Mandala et al., 1982; Nagahashi and Kane, 1982; Nagahashi et al., 1978; Perlin and Spanswick, 1980; Ray, 1979; Robinson and Glas, 1983).

As with other organelles, the recognition and separation of PM vesicles in gradients can only be as good as the marker employed. Whereas there are several markers available and all show a unimodal distribution, their lack of coincidence in sucrose gradients has occasionally been remarked upon (e.g., Cross and Briggs, 1976; Boss and Ruesink, 1979; Bowles and Kauss, 1976; Hendrix and Kennedy, 1977). Hendriks (1978) has tested for GS II and K^+–ATPase activities on linear and step gradients using 10,000*g* and 10,000–153,000*g* pellets obtained from maize coleoptiles. In the former, a fairly good correlation was observed but this was not so with the lighter differential fraction. A similar lack of coincidence has been noted for PM markers from pea stems when centrifuged rate zonally on renografin (Taiz et al., 1983). As these authors, together with Quail (1979), have suggested, this may well represent a separation of the PM into domains whose sedimentation and/or density characteristics differ slightly.

An alternative method to density gradient centrifugation which apparently produces PM fractions of high purity relatively quickly is the phase-partitioning system developed by Albertsson (Albertsson et al., 1982), which separates organelles according to their surface properties. According to Widell et al. (1982) and Larsson (1983), PM vesicles have a high affinity for the polyethylene glycol-rich phase of a dextran–polyethylene glycol two-phase system and can be thus purified by a repeated mixing and low-speed centrifugation to facilitate phase settling. Using this method purities of 80–90% after only a single partitioning have been claimed (Yoshida et al., 1983).

Finally, another method which should have much promise for the isolation of the PM from protoplasts since it has already been successfully applied at the EM level involved their attachment to poly-L-lysine-coated objects (see also Section 8.3). After cell rupture, relatively pure PM preparations can be "lifted off" coated polyacrylamide or glass beads in the case of erythrocytes, *Dictyostelium*, and HeLa cells (Cohen et al., 1977; Jacobson, 1977; Jacobson et al., 1978; Kalish et al., 1978). In my opinion, it can only be a question of time before plants are added to this list.

REFERENCES

Aducci, P., Federico, R., and Ballio, A. (1980). *Phytopathologia medit.* **19,** 187–188.

Albertsson, P.-A., Albertsson, B., Akerlund, H.-E., and Larsson, C. (1982). In *Methods of Biochemical Analysis* (D. Glick, ed.), Vol. 28, pp. 115–150. Wiley-Interscience, New York.

Anderson, R. M. and Ray, P. M. (1978). *Plant Physiol.* **61,** 723–730.

Baker, D. A. and Hall, J. E. (1973). *New Phytol.* **72,** 1281–1291.

Bal, A. K., Verma, D. P. S., and Byrne, H. (1976). *J. Cell Biol.* **6,** 97–105.

Berg, H. C. and Hirsh, D. (1975). *Anal. Biochem.* **66,** 629–631.

Berkovitz, R. L. and Travis R. L. (1981). *Plant Physiol.* **68,** 1014–1019.

Berkovitz, R. L. and Travis R. L. (1982). *Plant Physiol.* **69,** 379–384.

Boss, W. F. and Ruesink, A. E. (1979). *Plant Physiol.* **64,** 1005–1011.

Bowles, D. J., and Kauss, H. (1976). *Biochem. Biophys. Acta* **443,** 360–374.

Bowman, B. J. and Slayman, C. W. (1979). *J. Biol. Chem.* **254,** 2928–2934.

Briskin, D. P. and Poole, R. J. (1983). *Plant Physiol.* **71,** 969–971.

Buckhout, T. J. (1983). *Plant Physiol.* **72,** 87.

Burgess, J. and Linstead P. J. (1976). *Planta* **130,** 73–79.

Burgess, J. and Linstead P. J. (1977). *Planta* **133,** 267–233.

Cantley, L. C., Josephson, L., Warner, L., Yanagisawa, R., Lechene, M., and Guidotti, G. (1977). *J. Biol. Chem.* **252,** 7421–7423.

Chadwick, C. M. and Northcote, D. H. (1980). *Biochem. J.* **186,** 411–421.

Chrispeels, M. J. (1976). *Ann. Rev. Plant Physiol.* **27,** 19–38.

Cocking, E. C. (1970). *Int. Rev. Cytol.* **28,** 89–124.

Cohen, C. M., Kalish, D. I., Jacobson, B. S., and Branton, D. (1977). *J. Cell Biol.* **75,** 119–134.

Cook, J. A., Fincher, G. B., Keller, F., and Stone, B. A. (1980). In *Mechanisms of Saccharide Polymerization and Depolymerization* (J. J. Marshall, ed.), pp. 301–315. Academic Press, New York.

Cronshaw, J. (1980). *Ber. Dtsch. Bot. Ges.* **93,** 123–139.

Cross, J. W. and Briggs, W. R. (1976). *Carnegie Inst. Washington Yearb.* **75,** 379–383.

Delmer, D. P. (1977). In *Recent Advances in Phytochemistry* (F. Loewus, eds.), Vol. 11, pp. 45–47. Academic Press, New York.

Depta, H. and Hertel, R. (1982). In *Plasmalemma and Tonoplast: Their Functions in the Plant Cell* (D. Marmé, E. Marré and R. Hertel, eds.), pp. 137–145. Elsevier Biomedical Press, Amsterdam.

Dieter, P. and Marmé, D. (1980a). *Proc. Natl. Acad. Sci. USA* **77,** 7311–7314.

Dieter, P. and Marmé, D. (1980b). *Planta* **150,** 1–8.

Dohrmann, U., Hertel, R., Pesci, P., Cocucci, S. M., Marre, E., Randazzo, G., and Ballo, A. (1977). *Plant Sci. Lett.* **9,** 291–299.

Du Pont, F. M., Burke, L. L., and Spanswick, R. M. (1981). *Plant Physiol.* **67,** 59–63.

Edwards, R. M., Kempson, S. A., Carlson, G. L. and Dousa, T. P. (1979). *Biochim. Biophys. Acta* **553,** 54–65.

Ephritikhine, G., Lamant, A., and Heller, R. (1980). *Plant Sci. Lett.* **19,** 55–64.

Esau, K. (1965). *Plant Anatomy*. Wiley, New York.

Essner, E. (1973). In *Electron Microscopy of Enzymes—Principles and Methods* (M. A. Hayat, ed.), Vol. 1, pp. 44–76.

Evert, R. F., Eschrich, W., and Heyser, W. (1977). *Planta* **136,** 77–89.

Fèvre, M. and Rougier, M. (1981). *Planta* **151,** 232–241.

Flagg-Newton, J., Simpson, I., and Loewenstein, W. R. (1979). *Science* **205,** 404–407.

Galbraith, D. W. and Northcote, D. H. (1977). *J. Cell Sci.* **24,** 295–310.

Gallagher, S. R. and Leonard, R. T. (1982). *Plant Physiol.* **70,** 1335–1340.

Goodchild, D. J. and Bergersen, F. J. (1966). *J. Bact.* **92,** 205–213.

Goodwin, P. B. (1983). *Planta* **157,** 124–130.

Gross, J. and Marmé, D. (1978). *Proc. Natl. Acad. Sci. USA* **75,** 1232–1236.

Grove, S. N., Bracker, C. E., and Morré, D. J. (1968). *Science* **161,** 171–173.

Gunning, B. E. S. and Hughes, J. E. (1976). *Austr. J. Plant Physiol.* **3,** 619–637.

Gunning, B. E. S. and Overall, R. L. (1982). *Bioscience* **8,** 262–281.

Gunning, B. E. S. and Pate, J. S. (1974). In *Dynamic Aspects of Plant Ultrastructure* (A. W. Robards, ed.), pp. 441–476. McGraw–Hill, London.

Gunning, B. E. S. and Robards, A. W. (1976a). In *Transport and Transfer Processes in Planta* (I. E. Wardlaw and J. Passioura, eds.), pp. 15–41. Academic Press, New York.

Gunning, B. E. S. and Robards, A. W. (1976b). *Intracellular Communication in Plants: Studies on Plasmodesmata*. pp. 1–387 Springer-Verlag, Berlin.

Hall, J. L. (1983). In *Isolation of Membranes and Organelles from Plant Cells* (J. L. Hall and A. L. Moore, eds.), pp. 55–81. Academic Press, London.

Hall, J. L. and Flowers, T. J. (1976). *J. Exp. Bot.* **27,** 658–671.

Hall, J. L. and Sexton, R. (1972). *Planta* **108,** 103–120.

Hall, J. L., Browning, A. J., and Harvey, D. M. R. (1980). *Protoplasma* **104,** 193–200.

Hall, J. L., Kinney, A. J., Dymott, A., Thorpe, J. R., and Brummell, D. A. (1982). *Histochem. J.* **14,** 323–331.

Hartmann, M. A., Fonteneau, P., and Benveniste, P. (1977). *Plant Sci. Lett.* **8,** 45–51.

Hartmann-Bouillon, M. A., Benveniste, P., and Roland, J.-C. (1979). *Biol. Cellulaire* **35,** 183–194.

Hartmann-Bouillon, M. A., Erhardt, A., and Benveniste, P. (1982). In *Plasmalemma and Tonoplast: Their Functions in the Plant Cell* (D. Marme, E. Marré, and R. Hertel, eds.), pp. 163–169. Elsevier Biomedical Press, Amsterdam.

Heiniger, U. and Delmer, D. P. (1977). *Plant Physiol.* **59,** 719–723.

Hendriks, T. (1978). *Plant Sci. Lett.* **11,** 261–274.

Hendrix, D. L. and Kennedy, R. M. (1977). *Plant Physiol.* **59,** 264–267.

Henry, R. J. and Stone, B. A. (1982). *Plant Physiol.* **69,** 632–636.

Hepler, P. K. (1982). *Protoplasma* **111,** 121–133.

Hertel, R., Lomax, T., and Briggs, W. R. (1983). *Planta* **157,** 193–201.

Jacobs, M. and Hertel, R. (1978). *Planta* **142,** 1–10.

Jacobson, B. S. (1977). *Biochim. Biophys. Acta* **471,** 331–335.

Jacobson, B. S., Cronin, J., and Branton, D. (1978). *Biochim. Biophys. Acta* **506,** 81–96.

Kalish, D. I., Cohen, C. M., Jacobson, B. S., and Branton, D. (1978). *Biochim. Biophys. Acta* **506,** 97–110.

Kemp, J., Loughman, B. C., and Ephhritikhine, G. (1978). In *Cyclotols and Phosphoinositides* (W. W. Wells and F. Eisenberg, eds.), pp. 439–449. Academic Press, New York.

Koehler, D. E., Leonard, R. T., Van Der Woude, W. J., Linkins, A. E., and Lewis, L. N. (1976). *Plant Physiol.* **58,** 324–330.

Larkin, P. J. (1978a). *Plant Physiol.* **61,** 626–629.

Larkin, P. J. (1978b). *Ann. Rev. Plant Physiol.* **61,** 626–629.

Larsson, C. (1983). In *Isolation of Membranes and Organelles from Plant Cells* (J. L. Hall and A. L. Moore, eds.), pp. 277–309. Academic Press, London.

Ledbetter, M. C. and Porter, K. R. (1970). *Introduction to the Fine Structure of Plant Cells.* Springer Verlag, Berlin.

Lembi, C. A., Morre, D. J., Thomson, K. S., and Hertel, R. (1971). *Planta* **99,** 37–45.

Leonard, R. T. (1982). In *Membranes and Transport,* (A. N. Martonosi, ed.), Vol. 2, pp. 633–637. Plenum, New York.

Leonard, R. T. and Hodges, T. K. (1973). *Plant Physiol.* **52,** 6–12.

Leonard, R. T. and Hodges, T. K. (1980). In *The Plant Cell* (N. E. Tolbert, ed.) Vol. 1, The Biochemistry of Plants, pp. 163–208. Academic Press, New York.

Leonard, R. T. and van der Woude, W. J. (1976). *Plant Physiol.* **57,** 105–114.

Levering, C. A. and Thomson, W. W. (1971). *Planta* **97,** 183–196.

Lopez-Saez, J. F., Gimenez-Martin, G., and Risueno, M. C. (1966). *Protoplasma* **61,** 81–84.

Lüttge, U. and Higinbotham, N. (1979). *Transport in Plants,* pp. 145–159. Springer-Verlag, New York.

Maclachlan, G. A. (1982). In *Cellulose and Other Natural Polymer Systems—Biogenesis, Structure, and Degradation* (R. M. Brown, Jr., ed.), pp. 327–339. Plenum Press, New York.

Mandala, S., Mettler, I. J., and Taiz, L. (1982). *Plant Physiol.* **70,** 1743–1747.

Marré, E. (1979). *Plant Physiol.* **30,** 273–288.

Mayo, M. A. and Cocking, E. C. (1969). *Protoplasma* **68,** 231–236.

Michalke, W. (1982). In *Plasmalemma and Tonoplast: Their Functions in the Plant Cell* (D. Marmé, E. Marré, and R. Hertel, eds.), pp. 129–135. Elsevier Biomedical Press, Amsterdam.

Morré, D. J. and Bracker, C. E. (1976). *Plant Physiol.* **58,** 544–547.

Morré, D. J. and Mollenhauer, H. H. (1983). *Eur. J. Cell Biol.* **29,** 126–132.

Mueller, S. C. and Maclachlan, G. A. (1980). *Plant Physiol.* **65,** 106.

Nagahashi, G., Leonard, R. T., and Thomson, W. W. (1978). *Plant Physiol.* **61,** 993–999.

Nagahashi, J. and Kane, A. P. (1982). *Protoplasma* **112,** 167–173.

Oleson, P. (1979). *Planta* **144,** 349–358.

Oross, J. W. and Thomson, W. W. (1982). *Amer. J. Bot.* **69,** 939–949.

Overall, R. L., Wolfe, J., and Gunning, B. E. S. (1982). *Protoplasma* **111,** 134–150.

Perlin, D. S. and Spanswick, R. M. (1980). *Plant Physiol.* **65,** 1053–1057.

Phillips, D. R. and Morrison, M. (1973). *Biochem. Biophys. Acta* **40,** 284–289.

Pierce, W. S. and Hendrix, D. L. (1979). *Planta* **146,** 161–169.

Quader, H., van Kempen, R., Stelzer, R., and Robinson, D. G. (1983). *Eur. J. Cell Biol.* **30,** 283–287.

Quail, P. H. (1979). *Ann. Rev. Plant Physiol.* **30,** 425–484.

Quail, P. H. and Hughes, J. E. (1977). *Planta* **133,** 169–177.

Quantin, E., Hartmann-Bouillon, M. A., Schuber F., and Benveniste, P. (1980). *Plant Sci. Lett.* **17,** 193–199.

Ray, P. M. (1977). *Plant Physiol.* **59,** 594–599.

Ray, P. M. (1979). In *Plant Organelles, Methodological Surveys* (B), Biochemistry 9 (E. Reid, ed.), pp. 135–146. Horwood Publishers, Chichester.

Raymond, Y., Fincher, G. B., and Maclachlan, G. A. (1978). *Plant Physiol.* **61,** 938–942.

Robards, A. W. (1968). *Planta* **82,** 200–210.

Robards, A. W. (1971). *Protoplasma* **72,** 315–323.

Robards, A. W. (1975). *Ann. Rev. Plant Physiol.* **26,** 13–29.

Robinson, D. G. (1977). *Adv. Botanical Res.* **5,** 89–151.

Robinson, D. G. and Glas, R. (1983). *J. Exp. Bot.* **34,** 668–675.

Robinson, D. G., Eberle, M., Hafemann, C., Wienecke, K., and Graebe, J. E. (1982). *Z. Pflanzenphysiol.* **105,** 323–330.

Roland, J.-C. (1978). In *Electron Microscopy of Plant Cells* (J. L. Hall, ed.), pp. 1–320. Elsevier/North-Holland, Amsterdam.

Roland, J.-C., Lembi, C. A., and Morré, D. J. (1972). *Stain Technology* **47,** 195–200.

Rubery, P. H. (1981). *Ann. Rev. Plant Physiol.* **32,** 569–596.

Samson, M., Klis, F. M., Sigon, C. A. M., and Stegwee, D. (1983). *Planta* **159,** 322–328.

Scarborough, G. A. (1975). *J. Biol. Chem.* **250,** 1106–1111.

Schibeci, A., Fincher, G. B., Stone, B. A., and Wardrop, A. B. (1982). *Biochemical J.* **205,** 511–519.

Shore, G. and Maclachlan, G. A. (1975). *J. Cell Biol.* **64,** 557–571.

Smith, M. M. and Stone, B. A. (1973). *Biochim. Biophys. Acta* **313,** 72–94.

Spanswick, R. M., Lucas, W. J., and Dainty, J. (1980). (, ed.), Elsevier/North Holland, Amsterdam.

Stockwell, V. and Hanckey, P. (1982). *Plant Physiol.* **70,** 244–251.

Stroobant, P., Dame, J. B., and Scarborough, G. A. (1980). *Fed. Proc.* **39,** 2437–2441.

Taiz, L., Jacobs, M., Gepstein, A., and Mettler, I. J. (1981). *Plant Physiol.* **67,** 8.

Taiz, L., Murry, M., and Robinson, D. G. (1983). *Planta* **158,** 534–539.

Taylor, A. R. D. and Hall, J. L. (1978). *Protoplasma* **96,** 113–126.

Taylor, A. R. D. and Hall, J. L. (1979). *Plant Sci. Lett.* **14,** 113–126.

Thom, M., Laetsch, W. M., and Maretzki, A. (1975). *Plant Sci. Lett.* **5,** 245–253.

Travis, R. L. and Berkovitz R. L. (1980). *Plant Physiol.* **65,** 871–879.

Tsai, C. M. and Hassid, W. Z. (1971). *Plant Physiol.* **47,** 740–744.

Tsai, C. M. and Hassid, W. Z. (1973). *Plant Physiol.* **51,** 998–1001.

Van Der Woude, W. J., Lembi, C. A., Morre, D. J., Kindinger, J. I., and Ordin, L. (1974). *Plant Physiol.* **54,** 333–340.

Vian, B. and Roland, J.-C. (1972). *J. Microsc.* (*Paris*) **13,** 119–136.

Wallach, D. F. H. (1972). *The Plasma Membrane.* pp. 1–276 English Universities Press, London.

Widell, S., Lundborg, T., and Larsson, C. (1982). *Plant Physiol.* **70,** 1429–1435.

Wienecke, K., Glas, R., and Robinson, D. G. (1982). *Planta* **155,** 58–63.

Williamson, F. A. (1979). *Planta* **144,** 209–215.

Williamson, F. A., Fowke, L. C., Constabel, F. C., and Gamborg, O. L. (1976). *Protoplasma* **89,** 305–316.

Yoshida, S., Uemura, M., Niki, T., Sakai, A., and Gusta, L. V. (1983). *Plant Physiol.* **72,** 105–114.

8

COATED VESICLES

Coated vesicles are, as the name suggests, vesicles with material external to the membrane. In all cases external means directed towards the cytoplasm, thereby differentiating the coating from other surface structures such as the glycocalyx. They were first described in the early 1960s (see Wild, 1980, for historical aspects) under a variety of other names (e.g., fuzzy vesicles, spiny vesicles, bristle-coated vesicles, vesicles-in-a-basket, alveolate vesicles, acanthasomes). It is now recognized that they occur in both animal and plant cells (Nevrotin, 1980; Newcomb, 1980), but while their function(s) is (are) sufficiently well understood in the former, there is no direct experimental evidence concerning their role in the latter.

8.1. STRUCTURE AND CHEMISTRY

Coated vesicles have a diameter varying between 50 and 250 nm. In section they appear to have bristles attached to their surface (Figure 8.1a). When negatively stained (Figure 8.2b) it is apparent that these projections belong to a complex interlocking skeletal covering resembling the stitching in a leather soccer ball. First depicted by Bowers (1964) *in situ*, and confirmed by Kanaseki and Kadota (1969) on crude isolated coated vesicle fractions, the surface of a coated vesicle, in agreement with Euler's theorem, is covered with a network of hexagons and pentagons (Crowther et al., 1976).

The improvements in coated vesicle isolation from brain tissue achieved by Pearse (1975; see also Section 8.3) allowed a biochemical characterization of the coating to be made. Treatment with trypsin or pronase leads to a loss of the coating indicating its proteinaceous nature. SDS–PAGE of purified coated vesicles from brain shows the presence of a major band at 180 kd. This polypeptide was termed *clathrin* (Greek, meaning bar) by Pearse and appears to be characteristic for coated vesicles isolated from a number of other different animal tissues such as liver (Pilch et al., 1983), mammary gland (Kartenbeck et al., 1981), oocytes (Woods et al., 1978), and placentae (Ockleford and Whyte, 1977). The amino acid composition of clathrin has been given (Pearse, 1976; Schjeide et al., 1969) and appears to be more or less the same from a variety of sources. It is rich in glutamic and aspartic acids.

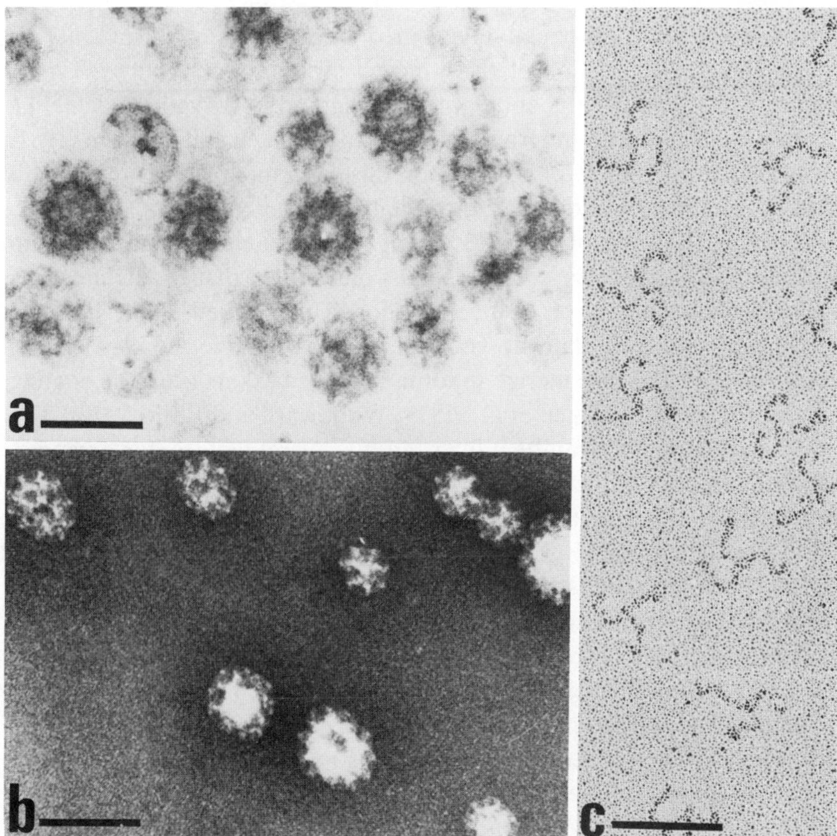

Figure 8.1. (a) Section through a pelleted fraction of bovine-coated vesicles prepared according to the Nandi et al. (1982) method. (b) Negatively stained coated vesicles from a fraction prepared as in a. (c) Rotary shadowed preparation of triskelions prepared from bovine brain-coated vesicles. Bar = 0.1 μm. (a,b) Unpublished micrographs of Resemann and Robinson; (c) unpublished micrograph of Ungewickell.

Coated vesicles can also be denuded of their coating by treatment with 2 *M* urea or 0.5 *M* Tris (Keen et al. 1979). That which is solubilized sediments at 8S and gives rise to three bands upon SDS–PAGE: the 180-kd polypeptide and two other polypeptides in smaller amounts. These are termed *light-chains* (Ungewickell and Branton, 1981) and have MWs of 36 kd (α) and 33 kd (β). Cross-linking experiments (Kirchhausen and Harrison, 1981) have shown that the light chains are indeed associated with clathrin in a molar ratio of 1 clathrin/1 α/2 β.

Rotary-shadowed or negatively stained preparations of 8S subunits reveal a uniform population of three-legged structures termed *triskelions* (Figure 8.1c). Each leg is about 44 nm long and is curved or sicklelike due to a knick about 16 nm from the end. Ungewickell and Branton (1981) have calculated a MW of 630 kd for this structure, suggesting that it consists of three clathrin plus three light chain polypeptides (unclear, however, is the ratio of α and β chains in any one triskelion). Ferritin labeling of elastase-released light chains has shown that the light chains are localized at the vertex of the triskelion (Ungewickell et al., 1982).

8S preparations of clathrin when dialyzed to remove the dissociating factors urea or Tris polymerize to form cages or baskets (i.e., the coating without the vesicle—Keen et al., 1979; Woodward and Roth, 1978). The rate of cage assembly can be influenced by several factors, for example, positively by decreasing the pH to 6 and the salt concentration (Van Jaarsveld et al., 1981). Two types of cages are formed equivalent to sedimentation coefficients of 150S and 300S and it is interesting to note that *in situ* two or more populations of different-sized coated vesicles have occasionally been recorded (e.g., Bowers, 1964; Croze et al., 1983; Friend and Farquhar, 1967; Mollenhauer et al., 1976). The smaller cages have a MW of 25×10^6, which corresponds to more than 100 clathrin molecules. Crowther et al. (1976) have considered the geometrical possibilities for creating a closed polyhedron corresponding to this number and have presented three possibilities all containing 12 pentagons. The most likely structure is one with 6 hexagons. Such a structure would have 36 vertices, suggesting that each vertex is occupied by the center of a triskelion (Ungewickell and Branton, 1981). However, since the distance between vertices is only about half a triskelion leg this means that the legs of triskelion extend along two edges leading to an overlap of four half-legs per edge (Crowther and Pearse, 1982; Ungewickell and Branton, 1982). This overlap does not appear to be obligatory for the formation of a cage since clathrin with short legs prepared by proteolysis is still capable of assembling (Schmid et al., 1982; Ungewickell et al., 1982).

Coated vesicles, upon SDS–PAGE, in addition to the 180-, 36-, and 33-kd polypeptides comprising the clathrin triskelions, are also seen to contain two other polypeptides of MW about 110 and 55 kd (Blitz et al., 1977; Keen et al., 1979). These presumably belong to the vesicle membrane rather than the coat, which is in itself surprising for a structure widely believed to participate in endocytotic and exocytotic transport (see below and chapter 15). There is some evidence that the larger polypeptide has an anchoring function for the clathrin coat. When coated vesicles are denuded and the membrane subjected to a mild proteolysis or to 0.5 *M* Tris, they are no longer capable of binding clathrin. The component which is lost from the membrane after these treatments is the 110-kd polypeptide (Keen et al., 1979; Unanue et al., 1981).

The lipids of the coated vesicle membrane have been investigated by Pearse (1975; 1976) and Schjeide et al. (1969). There is about 80% phospholipid, 10% sterols, and 10% triglycerides present. Phosphatidyl-choline and -ethanolamine represent about 75% of the phospholipids. Cholesterol is low but not entirely absent as shown by the binding of filipin to coated vesicles which have just lost their coats (McGookey et al., 1983). The levels of cholesterol are more characteristic of endo- rather than plasma membranes and it has been suggested that cholesterol-free or -poor areas of the PM are internalized by coated vesicles (Bretscher, 1976).

8.2. COATED PITS AND RECEPTOR-MEDIATED ENDOCYTOSIS

In their 1964 paper on developing mosquito oocytes where they first described coated vesicles, Roth and Porter suggested that they were involved in the uptake of yolk protein. This has now been confirmed for a number of comparable situations in avian and mammalian organisms (Rodewald, 1980; Roth et al., 1976) and is made clear by the presence of localized areas of coating on the cytoplasmic face of the PM in these organisms (e.g., King and Enders, 1970). These observations prompted Kanaseki and Kadota in 1969 to propose that coated vesicles arise from coated regions of the PM which invaginate. Before they become pinched off, these invaginations are called *coated pits* (Figure 8.2d,g).

Yolk proteins are just one of a large number of substances taken up by endo (pino) cytosis via coated pits. These include various nutrients, viruses, toxins, effectors (hormones), and lysosomal enzymes (see, e.g., Table I in Steinman et al., 1983). For each of these substances (ligands) there exists a large number of specific receptors in the PM (Table II in Steinman et al., 1983). With the help of video intensification microscopy it has been possible to show that PM-bound fluorescent-ligand conjugates are at first uniformly distributed over the surface of cells held at 4°C to

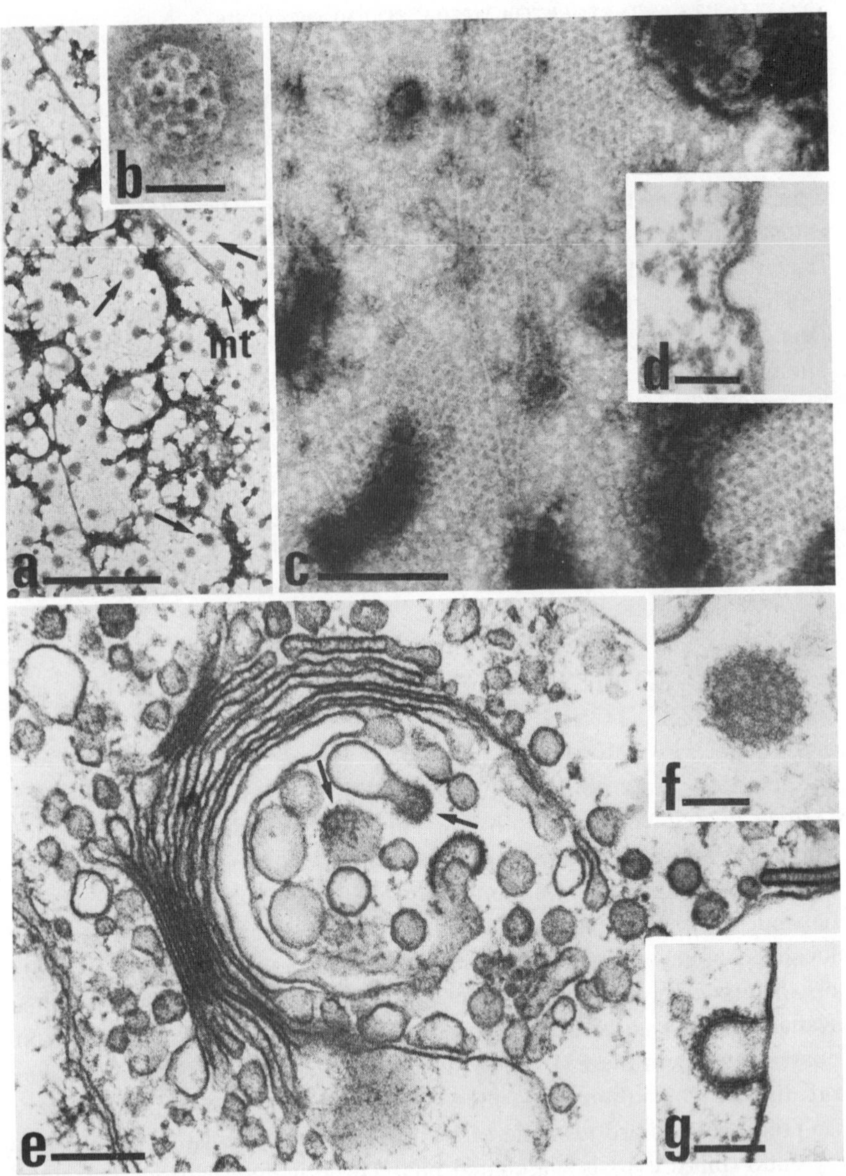

Figure 8.2. (a–d) Coated vesicles (pits?) in protoplast from suspension-cultured tobacco cells. (a) Low-magnification view of negatively stained PM showing microtubules (mt) and numerous coated structures (arrows). Bar = 1 μm. (b) Details of the polyhedral structure of the coated pits seen in a. Bar = 0.1 μm. (c) Patches of coated lying in the plane of the PM. Bar = 0.2 μm. (d) Coated pit in section. Bar = 0.1 μm. (e–g) Coated structures in *Poterioochromonas malhamensis*. (e) The GA showing coated regions on associated vesicles (arrows). Bar = 0.2 μm. (f) Coated "vesicle." Bar = 0.1 μm. (g) Coated pit. Bar = 0.1 μm. Unpublished (a) and published (b–d) micrographs of Van der Falk and Fowke (1981). Unpublished micrographs (e–g) of Van Kempen and Robinson.

prevent endocytosis (Pastan and Willingham, 1981; Schlessinger, 1980). If the temperature is then raised to 37°C, the fluorescence signals are seen clustered at various points on the cell surface. A few minutes later the fluorescent aggregates are seen in the cytoplasm. When instead of a fluorescent label the ligands are tagged with an electron-dense marker such as ferritin or colloidal gold, it can be seen that the loci for clustering and coated pits are one and the same (Anderson et al., 1977; Handley et al., 1981; Helenius et al., 1980; McGookey et al., 1983; Pilch et al., 1983; Wall et al. 1980).

A further feature of these electron microscopical studies is that once in the cell the tagged ligands are surrounded by an uncoated rather than a coated vesicle. In agreement with these observations are older ones showing that coated vesicles shed their coats prior to fusion with other membranes (Friend and Farquhar, 1967; Roth et al., 1976). Furthermore, since receptor-mediated endocytosis can continue for several hours in the absence of protein synthesis (Table II in Steinman et al., 1983), it is necessary to recycle not only the receptors but also the clathrin coat to the PM. According to Anderson et al. (1977). the most efficient way to do this is when the coated vesicles lose their clathrin coating in the near vicinity of the PM. Indeed, it has been occasionally mentioned in the literature (e.g., Ockleford and Whyte, 1977) that coated vesicles are to be found predominantly in the vicinity of the PM. The same would also apply to the coated regions on other membranes (e.g., GA–GERL) (see Figure 8.2e) where receptors carrying, for example, lysosomal proteins (see Chapter 12) might also aggregate.

An alternative and also provocative suggestion is that of Pastan and Willingham (1981) whereby the uncoated vesicle (termed *receptosome*) is never coated and originates as an invagination of the PM in the uncoated neck region adjacent to a coated pit. A consequence of this would be that coated vesicles are artifacts of homogenization generated by cleavage and resealing of coated regions of membranes. These authors have marshalled two lines of evidence in support of this contention. The first is that coated pits may often have very long extenuated necks which can, depending upon the plane of the section, give rise to the illusory appearance of a coated "vesicle." In partial support of this is the demonstration by serial sectioning that in L cells and skin fibroblasts only 10 and 36%, respectively, of putative coated vesicles really are without connection to the PM (Petersen and Van Deurs, 1983). More convincing has been, however, the application of immunological techniques. Free clathrin in the cytosol, which one might expect if coats are shed or if coated vesicles exist, could not be detected either by ultrastructural immunocytochemistry (Willingham, 1981) or by immunofluorescence microscopy after microinjection with antibodies against clathrin (Wehland et al., 1981, 1982). A subsequent injection with triskelions or cages resulted in a marked intracellular immunoprecipitation, indicating that the clathrin antibodies were present in sufficient amounts to recognize and react with free clathrin or coated vesicles. The fact that cells microinjected with clathrin antibodies are still capable of carrying out receptor-mediated endocytosis does not speak in favor of significant populations of coated vesicles. In contrast to the results of Pastan's group are those of Kartenbeck et al. (1981) who have used conventional immunofluorescence microscopy and claim to have visualized, in addition to coated pits and coated regions of the GA–GERL, coated vesicles in the cytoplasm. Clear proof, however, is not given for this.

Coated vesicles have been implicated in exocytotic as well as endocytotic processes [for literature see Franke et al. (1976), Kartenbeck (1980), and Morré et al., 1979]. Invariably the evidence is purely ultrastructural and represents a correlation between their "presence" and the secretory nature of the cell involved. As already mentioned, coated regions of the GA or GERL are well known [see Croze et al. (1982) Kartenbeck et al. (1981) and Morré et al. (1979) for literature] and it would appear logical to assume the pinching-off and transport of coated vesicles from them. This view has been strengthened by a paper by Rothman et al. (1980) claiming the participation of coated vesicles in the intracellular transport of the so-called G protein of chinese hamster ovary cells infected with vesicular stomatitis virus. The results from pulse-chase experiments are quite clear-cut showing that ^{35}S-labeled core- and terminal-glycosylated (see Chapter 10 for an explanation of these terms) forms of G protein are found in coated vesicles extracted from the cells at times corresponding to the loss of radioactive G protein from ER and GA fractions, respectively. Although Rothman et al. (1980) have stated that "coated vesicles budd off from the endoplasmic reticulum and fuse with Golgi apparatus," proof of these events is lacking. Their observations can be subjected to the same rationale as that given above for receptor-mediated endocytosis, whereby the vesicle actually responsible for the transport and fusion is uncoated.

As far as can be judged, there does not appear to be convincing evidence in the literature for the existence of a stable population of coated vesicles. It seems that they are transient in nature, shedding their coats shortly after their formation be that at the PM or GA–GERL.

8.3. COATED VESICLES (PITS) IN PLANT CELLS

Coated vesicles similar in size and appearance to those in animal cells have been described in both lower and higher plants (Newcomb, 1980). In many flagellates they are often seen in the vicinity of, and probably belong to, the contractile vacuole (Leedale et al., 1965; Manton, 1964; Wessel and Robinson, 1979). The presence of a clathrinlike coating on peripheral regions of dictyosomal cisternae and associated large vesicles has also been recorded for both algae (see Figure 8.2e; Manton, 1966) and higher plants (Bonnett and Newcomb, 1966). Particularly because of the large amount of coating on dictyosomal membranes and forming PM in cells during cytokinesis (Cronshaw and Esau, 1968; Fowke et al., 1975; Franke and Herth, 1974), it has become usual (though not without exception—see, e.g., Ryser, 1979) for botanists to ascribe an exocytotic function to coated vesicles. An endocytotic function is, however, just as plausible and indeed finds support in papers published more recently.

In the older literature reference has been made on several occasions to the prevalence of coated vesicles in the vicinity of the PM and to coated invaginations of the PM (Bonnett, 1969; Bonnett and Newcomb, 1966; Hepler and Newcomb, 1967). Taking into consideration the situation in animal cells, these structures should be interpreted as coated pits and endocytotic coated vesicles. This opinion is strengthened by several papers involving the electron microscopy of the PM from protoplasts (Doohan and Palevitz, 1980; Fowke and Gamborg, 1980; Van der Valk and Fowke, 1981; Van der Valk et al., 1980). Protoplasts which are attached to poly-L-lysine-coated grids burst upon immersing in water. The PM remains behind on the grid and can be negatively stained. Besides showing microtubules and microfilaments these "ghosts" show wonderfully coated structures (Figure 8.2a–d). The coating is present on vesicle-type structures as well as flattened patches in the PM. Authors who have published such pictures have maintained that the former are indeed coated vesicles, but in my opinion they more likely represent coated pits (Figure

8.2d,g), since free coated vesicles would probably be washed away during the preparation of the ghosts. Of course, in adopting this line of argument, I am aware that, in comparison to the work on animal cells, there has been next to nothing done on the recognition and characterization of PM-based receptors in plant cells. Furthermore, one must ask oneself "what might be the appropriate ligands: hydrolytic enzymes perhaps?" (See Chapter 12.)

8.4. ISOLATION FROM ANIMAL CELLS

8.4.1. Markers

Coated vesicles can be recognized on the basis of their content. This, however, can vary depending on the nature of the receptor–ligand complex which is collected in them. A general content marker does not yet seem available. With respect to membrane markers a Ca^{2+}–ATPase activity has been described (Blitz et al., 1977), but this now appears to be a component of contaminating PM vesicles present in the coated vesicle preparation (Rubinstein et al., 1981). Forgac et al. (1983) have reported another ATPase activity stimulated by Mg^{2+} but unaffected by vanadate and oligomycin in purified brain-coated vesicles. A Mg^{2+}-dependent protein kinase present in the coat fraction of coated vesicles has recently been described (Kadota et al., 1982; Pauloin et al., 1982), but this has not yet been widely used. At the moment there are only two ways to recognize coated vesicles: by their optical appearance in the ER and by the presence of a major 180-kd polypeptide after SDS–PAGE.

8.4.2. Separation

Coated vesicles have been successfully isolated by several methods. The most-used of these involves centrifugation. Coated vesicles are essentially postmicrosomal particles (Chapter 1), and the first step in their purification by this method is the removal of the endo- and plasma membranes by differential centrifugation (see Figure 8.3). In the method of Pearse (1975, 1976) the postmicrosomal supernatant is then subjected to two rate zonal and one isopycnic centrifugations on sucrose gradients. This method suffers from the disadvantage that sucrose, at the density at which coated

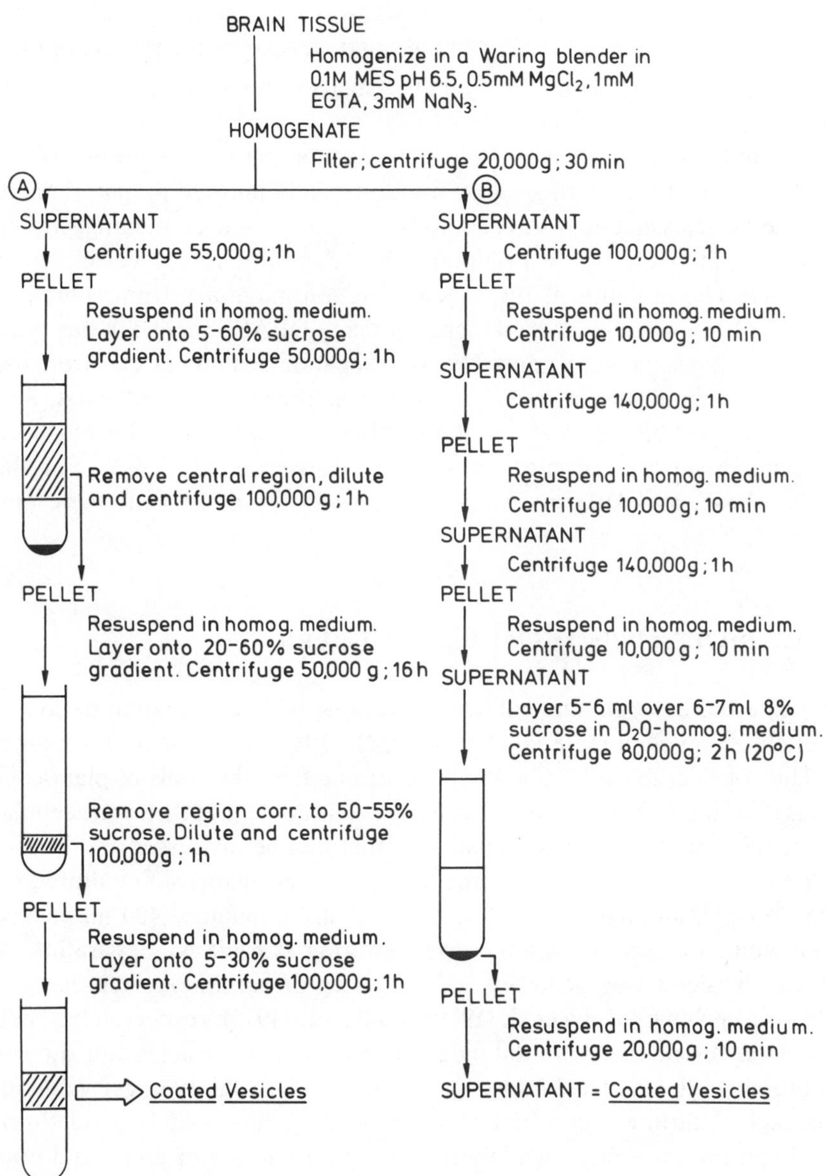

Figure 8.3. Purification schemes for coated vesicles. Route A is that of Pearse (1975); route B, Nandi et al. (1982).

vesicles equilibrate (~ 1.25 g·cm^{-3}), causes a dissociation of clathrin from the vesicle membrane. This feature prompted Nandi et al. (1982) to introduce an alternative method which involves successive high- and low-speed centrifugations and finally a D_2O-sucrose step gradient (Figure 8.3, route B).

Coated vesicles have been isolated by agarose gel electrophoresis (Rubinstein et al., 1981). Using this method, partially purified preparations of coated vesicles can be resolved into two fractions: one containing smooth vesicles putatively of PM origin and the other with coated vesicles plus baskets. The mobility of the second fraction may vary from source to source. Ion exchange chromatography using DEAE–Sephadex has been used by Kanaseki and Kadoto (1969) to separate coated vesicles from the postmicrosomal supernatant, but this has not been widely followed. Perhaps the most elegant method of coated vesicle isolation so far has been by immunoadsorption chromatography (Merisko et al., 1982). For the solid phase, anticlathrin-coated *Staphylococcus aureus* cells were employed.

8.5. ISOLATION FROM PLANT CELLS

To date there is only one publication dealing with the isolation of coated vesicles from plant material (Mersey et al., 1982). As these authors state, "The considerable additional bulk contributed by the walls of plant cells plus dilution of the cytoplasm by the aqueous content of their intracellular vacuoles severely limit the quantities that can be processed by the isolation procedures." To overcome this, they used suspension-cultured tobacco cells, although it must be added that the amount (400 ml packed cell volume) used demands large culturing capacities. Separation of coated vesicles was achieved by using step or continuous sucrose gradients as given by Keen et al. (1979) and Pearse (1975), respectively. Their "coated vesicle fraction" did indeed contain coated vesicles and showed a prominent band at 190 kd by SDS–PAGE. This slightly larger value for tobacco clathrin was confirmed by running it together with brain clathrin.

Experiments in my own laboratory with a number of green and nongreen plants have confirmed the results of Mersey et al. (1982). In addition with bean leaves we have seen a band in SDS–PAGE at 180 kd which in contrast to the gels of Mersey et al. (1982), is clearly recognizable

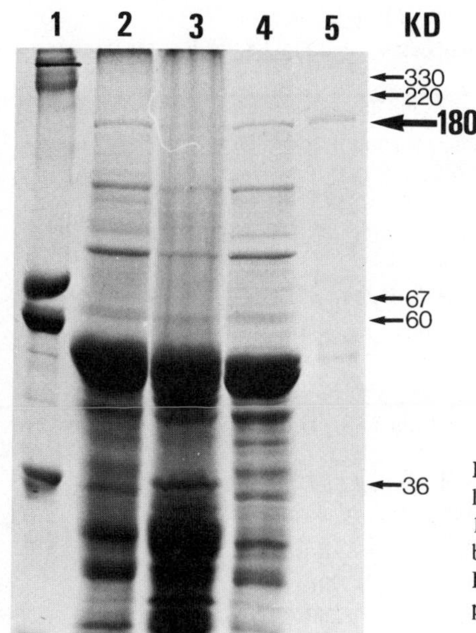

Figure 8.4. Placement of a bean leaf 180-kd polypeptide to a differential gradient fraction. Lane 1—high-molecular-weight marker kit. Lane 2—bean leaf homogenate. Lane 3—100,000*g* pellet. Lane 4—100,000*g* supernatant. Lane 5—purified porcine brain-coated vesicles.

in the homogenate and interestingly in the 100,000*g* supernatant as well (see Figure 8.4). Electron microscopy of this fraction after concentration through lyophilization and purification by gel chromatography shows the presence of what appears to be coated vesicles (Resemann and Robinson, unpublished results). Presumably, these are reassembled clathrin coats; otherwise it would be difficult to give an explanation for their differing sedimentation characteristics.

REFERENCES

Anderson, R. G. W., Brown, M. S. and Goldstein, J. L. (1977). *Cell* **10,** 351–364.

Blitz, A. L., Fine, R. E., and Toselli, P. A. (1977). *J. Cell Biol.* **75,** 135–147.

Bonnett, H. T. (1969). *J. Cell Biol.* **40,** 144–159.

Bonnett, H. T. and Newcomb, E. H. (1966). *Protoplasma* **62,** 59–75.

Bowers, B. (1964). *Protoplasma* **59,** 351–367.

Bretscher, M. S. (1976). *Nature* **260,** 21–23.

Cronshaw, J. and Esau, K. (1968). *Protoplasma* **65,** 1–24.

Crowther R. A. and Pearse, B. M. F. (1982). *J. Cell Biol.* **91,** 790–797.

Crowther, R. A., Finch, J. T., and Pearse, B. M. F. (1976). *J. Molec. Biol.* **103,** 785–798.

Croze, E. M., Morré, D. M., and Morré, D. J. (1983). *Protoplasma* **117,** 45–52.

Croze, E. M., Morré, D. J., Morré, D. M., Kartenbeck, J., and Franke, W. W. (1982). *Eur. J. Cell Biol.* **28,** 130–138.

Doohan, M. E. and Palevitz, B. A. (1980). *Planta* **149,** 389–401.

Forgac, M., Cantley, L., Wiedermann, B., Altstiel, L., and Branton, D. (1983). *Proc. Nat. Acad. Sci.* (*USA*) **80,** 1300–1303.

Fowke, L. C. and Gamborg, O. L. (1980). *Int. Rev. Cytol.* **68,** 9–51.

Fowke, L. C., Bech-Hansen, C. W., Gamborg, O. L., and Constabel, F. (1975). *J. Cell Sci.* **18** 491–507.

Franke, W. W. and Herth, W. (1974). *Exp. Cell Res.* **89,** 447–451.

Franke, W. W., Luder, M. R., Kartenbeck, J., Terban, H., and Keenan, T. W. (1976). *J. Cell Biol.* **69,** 173–195.

Friend, D. S. and Farquhar, M. G. (1967). *J. Cell Biol.* **35,** 357–376.

Handley, D. A., Arbeeny, C. M., Witte, L. D., and Chien, S. (1981). *Proc. Nat. Acad. Sci.* (*USA*) **78,** 368–371.

Helenius, A., Kartenbeck, J., and Fries, E. (1980). *J. Cell Biol.* **84,** 404–420.

Hepler, P. K. and Newcomb, E. H. (1967). *J. Ultr. Res.* **19,** 498–513.

Kadota, K., Usami, M., and Takahashi, A. (1982). *Biomed. Res.* **3,** 575–578.

Kanaseki, T. and Kadota, K. (1969). *J. Cell Biol.* **42,** 202–220.

Kartenbeck, J. (1980). In *Coated Vesicles* (C. D. Ockleford and A. Whyte, eds.), pp. 243–253. Cambridge University Press, Cambridge.

Kartenbeck, J., Schmid, E., Müller, H., and Franke, W. W. (1981). *Exp. Cell Res.* **133,** 191–211.

Keen, J. H., Willingham, M. C., and Paston, I. H. (1979). *Cell* **16,** 303–312.

King, B. F. and Enders, A. C. (1970). *Am. J. Anatomy* **129,** 261–288.

Kirchhausen, T. and Harrison, S. C. (1981). *Cell* **23,** 755–761.

Leedale, G. F., Meeuse, B. J. D., and Pringsheim, E. G. (1965). *Archiv. Mikrobiol.* **50,** 68–102.

Manton, I, (1964). *J. Exp. Bot.* **15,** 399–411.

Manton, I. (1966). *J. Cell Sci.* **1,** 429–438.

McGookey, D. J., Fagerberg, K., Anderson, R. G. W. (1983). *J. Cell Biol.* **96,** 1273–1278.

Merisko, E. M., Farquhar, M. G., and Palade, G. E. (1982). *J. Cell Biol.* **92,** 846–857.

Mersey, B. G., Fowke, L. C., Constabel, F., and Newcomb, E. H. (1982). *Exp. Cell Res.* **141,** 459–463.

Mollenhauer, H. H., Hass, B. S., and Morré, D. J. (1976). *J. Microsc. Biol. Cell.* **27,** 33–36.

Morré, D. J., Kartenbeck, J., and Franke, W. W. (1979). *Biochem. Biophys. Acta* **559,** 72–152.

Nandi, P. K., Irace, G., van Jaarsveld, P. P., Lippoldt, R. E., and Edelhoch, H. (1982). *P.N.A.S.* (*USA*) **79,** 5881–5885.

Nevrotin, A. J. (1980). In *Coated Vesicles* (C. D. Ockleford and A. Whyte, eds.), pp. 25–54. Cambridge University Press, Cambridge.

Newcomb, E. H. (1980). In *Coated Vesicles* (C. D. Ockleford and A. Whyte, eds.), pp. 55–68. Cambridge University Press, Cambridge.

Ockleford, C. D. and Whyte, A. (1977). *J. Cell Sci.* **25,** 293–312.

Pastan, I. H. and Willingham, M. C. (1981). *Science* **214,** 504–509.

Pauloin, A., Bernier, I., and Jolles, P. (1982). *Nature* **298** 574–575.

Pearse, B. M. F. (1975). *J. Molec. Biol.* **97,** 93–98.

Pearse, B. M. F. (1976). *Proc. Nat. Acad. Sci. (USA)* **73,** 1255–1259.

Petersen, O. W. and van Deurs, B. (1983). *J. Cell Biol.* **96,** 277–281.

Pilch, P. F., Silva, M. A., Benson, R. J. J., and Fine, R. E. (1983). *J. Cell Biol.* **96,** 133–138.

Rodewald, R. (1980). In *Coated Vesicles* (C. D. Ockelford and A. Whyte, eds.), pp. 69–101. Cambridge University Press, Cambridge.

Roth, T. F. and Porter, K. R. (1964). *J. Cell Biol.* **20,** 313–332.

Roth, T. F., Cutting, J. A., and Atlas, S. B. (1976). *J. Supramolec. Struct.* **4,** 527–548.

Rothman, J. E., Brusztyn-Pettegrew, H., and Fine, R. E. (1980). *J. Cell Biol.* **86,** 162–171.

Rubinstein, J. L. R., Fine, R. E., Luskey, B. D., and Rothman, J. R. (1981). *J. Cell Biol.* **89,** 357–361.

Ryser, U. (1979). *Protoplasma* **98,** 223–239.

Schjeide, O. A., San Lin, R. T., Grellert, E. A., Galey, F. R., and Mead, J F. (1969). *Physiological Chem. Physics* **1,** 141–163.

Schlessinger, J. (1980). *TIBS* **5,** 210–214.

Schmid, S. L., Matsumoto, A. K., and Rothman, J. E. (1982). *Proc. Nat. Acad. Sci. (USA)* **79,** 91–95.

Steinman, R. M., Mellman, I. S., Mueller, W. A., and Cohn, Z A. (1983) *J. Cell Biol.* **96,** 1–27.

Unanue, E. R., Ungewickell, E. and Branton, D. (1981). *Cell.* **26,** 439–446.

Ungewickell, E. and Branton, D. (1981). *Nature* **289,** 420–422.

Ungewickell, E. and Branton, D. (1982). *TIBS* **7,** 358–361.

Ungewickell, E., Unanue, E. R., and Branton, D. (1982). *Cold Spring Harbor Symp.* **46,** 723–731.

van der Valk, P. and Fowke, L. C. (1981) *Can. J. Bot.* **59,** 1307–1313.

van der Valk, P., Rennie, P. J., Connolly, J. A. and Fowke, L. C. (1980). *Protoplasma* **105,** 27–43.

Van Jaarsveld, P. J., Nandi, P. K., Lippoldt, R. E., Saroff, H., and Edelhoch, H. (1981). *Biochem.* **20,** 4129–4135.

Wall, D. A., Wilson, G., Hubbard, A. L. (1980). *Cell* **21,** 79–93.

Wehland, J., Willingham, M. C., Dickson, R., and Pastan, I. H. (1981). *Cell* **25,** 105–121.

Wehland, J., Willingham, M. C., Gallo, M. G., Rutherford, A. V., Rudick, J., Dickson, R. B., and Pastan, I. (1982). *Cold Spring Harbor Symp.* **46,** 743–753.

Wessel, D. and Robinson, D. G. (1979). *Eur. J. Cell Biol.* **19,** 60–66.

Wild, A. E. (1980). In *Coated Vesicles* (C. D. Ockleford and A. Whyte, eds.), pp. 1–24. Cambridge University Press, Cambridge.

Willingham, M. C., Keen, J. H., and Pastan, I. H. (1981). *Exp. Cell Res.* **132,** 329–338.

Woods, J. W., Woodward, M. P., and Roth, T. F. (1978). *J. Cell Sci.* **30,** 87–97.

Woodward, M. and Roth, T. (1978). *Proc. Natl. Acad. Sci. (USA)* **75,** 4394–4399.

Part 2

Membranes and Particular Functions of Plant Cells

9

MICROFIBRIL SYNTHESIS AND ORIENTATION

Plant cells are surrounded by a cell wall for most of their lives. With the exception of some unusual algae (see Preston, 1974, for further details) cell walls contain appreciable amounts of cellulose. Cellulose is important for the plant; one has only to think of the various industrial cellulose products (wood, paper, cotton, etc.) to appreciate its economic value. The synthesis of this most important biopolymer has occupied the attention of scientists for several decades, but success in terms of understanding this process has come slowly and mainly in the last 7 years.

9.1. WHAT IS CELLULOSE?

Cellulose is our most abundant biopolymer and one would think that everyone is in agreement about its structure. Unfortunately, this has not always been the case, particularly when claims as to its synthesis *in vitro* have been made. The difficulty has been one of differentiating the molecular form from the physical entity. Chemically speaking, cellulose is an unbranched, linear 1,4-linked β-D-glucan having a substantial degree of polymerization. In cell walls and in the cell, individual chains of this type do not exist; instead many chains fasciate laterally to produce crystalline microfibrils with sizes ranging from 2 nm (slime microfibrils—Franke and Ermen, 1969) to almost 20 nm (Valonia—Harada, and Goto, 1982) in diameter. In any one cell and at any one particular point during the differentiation of a plant cell, the diameter of the microfibrils synthesized is constant, indicating the cellular control over this event.

Although it is now possible to experimentally prevent crystallization of microfibrils without inhibiting glucan polymerization (see Robinson, 1981, for references), in nature the two processes are so tightly coupled that it is convenient to think of them as occurring almost simultaneously. Thus, when considering claims for the *in vitro* synthesis of cellulose, the difference between a polymer and a microfibril must be kept in mind. The synthesis of the former may satisfy the biochemist, but "the synthesis of cellulose *in vitro* can only be regarded as being biologically relevant when the product synthesized is in microfibrillar form" (Robinson and Quader, 1981b). Since it has been possible to synthesize chitin microfibrils *in vitro* (Ruiz-Herrera et al., 1975), it is not unreasonable to impose this criterion of crystallinity (Carpita, 1982) for cellulose too.

9.2. SITES OF CELLULOSE SYNTHESIS

The title of this section is somewhat a misnomer. With the exception of a group of planktonic algae, cellulose synthesis is a property of the plant PM. In those members of the Chrysophyceae, Haptophyceae, and Prasinophyceae which possess a pseudowall consisting of cellulosic scales, it is clear that the GA is involved. Much work has been done by Brown's group (Brown, 1969; Brown et al., 1970, 1973; Brown and Romanovicz, 1976; Romanovicz, 1981) on scale production in the GA of *Pleurochrysis scherffelii*. Sections through this flagellate reveal most beautifully the development of "bicycle-wheel"-type scales in the GA, as well as their release to the cell exterior by fusion of GA cisternae with the PM. Such pictures present both beautiful and convincing evidence for the role of the GA in cellulose biogenesis, but despite having been at times overgeneralized (Brown et al., 1970; Brown, 1973), it must be emphasized that scale-producing flagellates are an exception to the rule.

Until freeze-fracturing became applied to problems of cell wall formation, there was very little direct evidence available which implicated the PM in cellulose biosynthesis. Preston (Frei and Preston, 1961; Preston and Goodman, 1968) had published micrographs of inner wall layers of plasmolyzed *Chaetomorpha* cells which revealed patches of granular material with microfibrils emanating from them. Observations of this type, together with the fact that no evidence could be found for an intracellular synthesis of cellulose, led to the view that the PM had to be the site of microfibril deposition. This has now been fully substantiated on the basis of freeze-fracturing data obtained with a number of different organisms.

9.3. IMPS AND CELLULOSE SYNTHESIS

In one of the first papers including freeze-fracture micrographs Moor and Mühlethaler (1963) showed paracrystalline arrays of particles associated with the PM of yeast cells. Projecting from these particles, particularly at the edges where the cell wall was exposed, were microfibrils. Attempts to obtain similar pictures for plants were carried out in the ensuing years but were not entirely convincing (Barnett and Preston, 1970; Peng and Jaffe, 1976; Robinson and Preston, 1971; Robinson et al., 1972; Sassen et al., 1970; Staehelin, 1966; Willison and Cocking, 1972, 1975). Highly

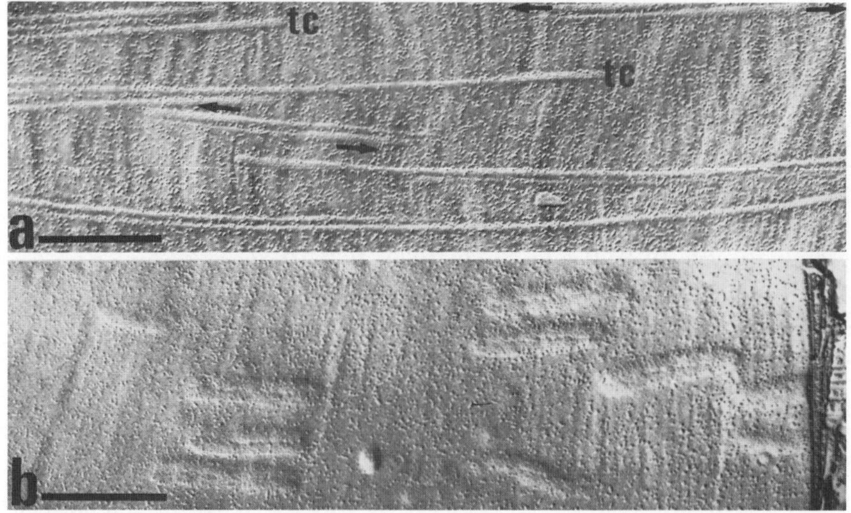

Figure 9.1. E-face freeze-fracture replicas of *Oocystis solitaria*. (a) Normal cells frozen during microfibril deposition. Terminal complexes (tc) are at the ends of microfibril imprints (mi). The tc's are paired and separate in an antiparallel manner (see arrows). (b) Cells treated with the crystallization inhibitor congo red (0.5 $mgml^{-1}$; 2 hr). Microfibril imprints are lost. Terminal complexes are exclusively paired and thicker. Bar = 0.5 μm. Published micrographs of Robinson and Quader (1981).

organized arrays of particles had occasionally been reported (Robinson and Preston, 1972; Robenek and Preveling, 1977), but their connection to or relationship with cell wall microfibrils was assumed rather than proved.

The breakthrough came in 1976 with a paper by Brown and Montezinos which, while still not showing a direct attachment of particle to microfibril end, did provide suggestive evidence for it. The organism employed in this study, the alga *Oocystis solitaria*, has very thick cellulose microfibrils in its wall which in any one layer are arranged strictly parallel to one another. E-face replicas of the PM (i.e., those revealing the outer bimolecular leaflet viewed from inside the cell—see Branton et al., 1975, for nomenclature) show the imprints of newly synthesized microfibrils which terminate in particular complexes (“terminal complexes”—see Figure 9.1a). Confirmation that these structures are indeed the structures responsible for cellulose synthesis has come from studies using the crys-

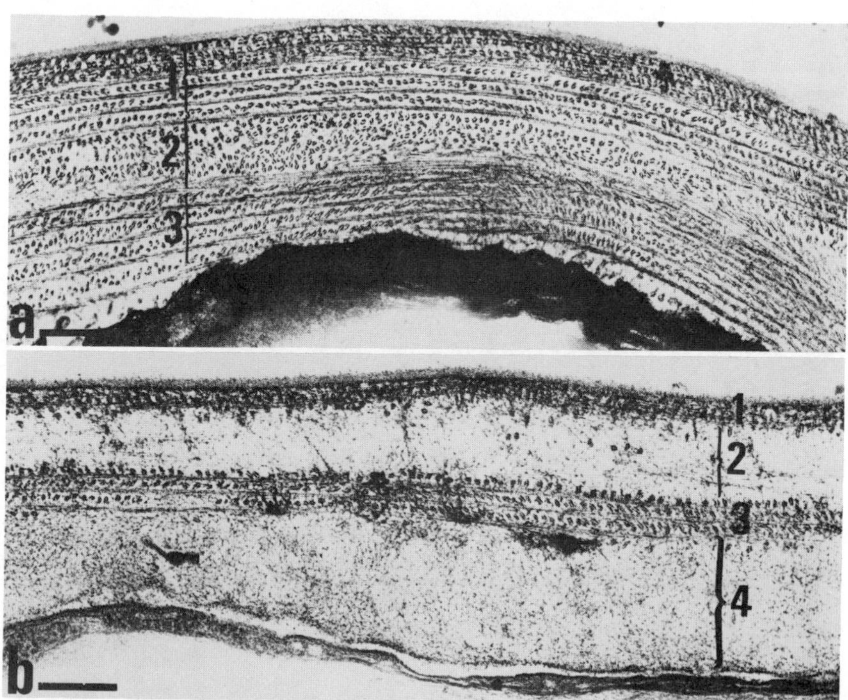

Figure 9.2. "Sandwich-walls" produced in *Oocystis solitaria* through the action and recovery from inhibitors of microtubules and cellulose crystallization. (a) Parallel microfibrils (zone 2) produced in the presence of colchicine (10^{-2} *M*; 6 hr). Zones 1 and 3 are normal microfibril layers deposited before and after the colchicine treatment. (b) Noncrystalline glucan (zones 2 and 4) produced in the presence of congo red (0.5 mg·ml^{-1}; 6 hr. Zones 1 and 3 are normal microfibril layers deposited before and between congo red treatments. Bar = 0.25 μm. Published micrographs of Quader et al. (1978, 1983).

tallization inhibitors congo red and calcofluor white (Quader, 1983; Quader et al., 1983; Robinson and Quader, 1981a). Application of these substances to cells undergoing wall formation results in the production of nonmicrofibrillar glucan (see Figure 9.2b; also Herth, 1980; Quader, 1981) and in the loss of microfibril imprints and the prevention of terminal complex separation (see Figure 9.1b). The terminal complexes become thicker and, when the treatment is long enough, gradually disappear, be-

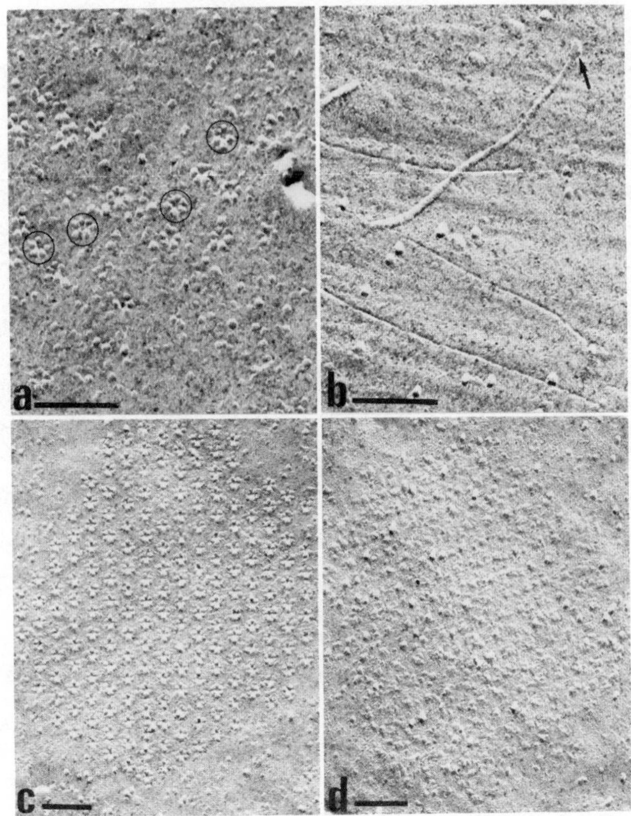

Figure 9.3. Rosette complexes and terminal globules involved in microfibril synthesis. (a) P-face freeze-fracture replica of rye (*Secale cereale* L.) root tip cell: Rosettes are circled. (b) E-face freeze-fracture replica as for a. Terminal globule (arrow) at the end of a microfibril which has been ripped through the membrane due to the fracturing process. (c,d) Aggregates of rosette complexes and terminal globules in complimentary P- and E-face freeze-fracture replicas from *Microsterias denticulata*. Bar = 0.1 μm. Unpublished (a,b) micrographs of Willison and Pearce and published micrographs (c,d) of Giddings et al. (1980).

coming replaced by a new set oriented at 90° to the original ones. Since the layers of microfibrils in the wall also alternate regularly through 90° (Quader and Robinson, 1981), this is in good correlation with wall development.

At this time no other organism appears to have terminal complexes of the type visible in *Oocystis*. Instead "rosette" complexes appear to be the rule. These have been detected in a number of algae (Giddings et al.,

1980; Herth, 1983; Kiermayer and Sleytr, 1979; Noguchi et al., 1981; Staehelin and Giddings, 1982), mosses (Wada and Staehelin, 1981), and higher plant cells (Mueller and Brown, 1981; Volkmann, 1983; Willison, 1983). The rosette complex, which is typical of P-face fractures, is about 25 nm in diameter and consists of six particles (see Figure 9.3a,c). On the E-face, complementary to the depression between rosette particles, is a single particle. Occasionally one sees microfibrils attached to such particles (see Figure 9.3b) and in other cases microfibril imprints (see Giddings et al., 1980).

Although there is a problem about their frequency in higher plant cells (Willison, 1983), the rosette and central particle appears to be a general feature of the PM of most cells involved in microfibril synthesis. Indeed it is interesting to note that the original demonstration of PM-associated intramembrane particles (IMPs) in yeast has now been improved to reveal rosettelike tubes on the E-face and complementary single particles on the P-face (Sleytr and Messner, 1978; Steere et al., 1980). There can, therefore, be no longer any doubt as to the role of such intramembrane particles in microfibril synthesis.

9.4. MICROTUBULES AND MICROFIBRIL ORIENTATION

Cells which are spherical usually have randomly oriented microfibrils. Higher plant cells are rarely spherical and only some algae are so. As cells differentiate and elongate, the microfibrils in their walls are deposited with an increasing parallelity of orientation. Clearly the orientation is under cellular control and must be exerted across the PM. Because of their length, their frequency, and more particularly their parallelity to microfibrils (Figure 9.4), microtubules have been held to be involved in this orientation mechanism ever since their discovery in plant cells in 1963 by Ledbetter and Porter. This opinion has, however, not always found universal support. Thus, one searches in vain for a mention of microtubules in Preston's classic textbook on cell walls (Preston, 1974).

The published micrographs showing a coparallelity between microfibrils and microtubules on opposing faces of the PM are legion [see, e.g., the review articles of Gunning and Hardham (1983); Hepler and Palevitz (1974), Newcomb (1969), and Robinson and Quader (1982)] but in themselves afford no proof for a causal relationship in terms of orientation

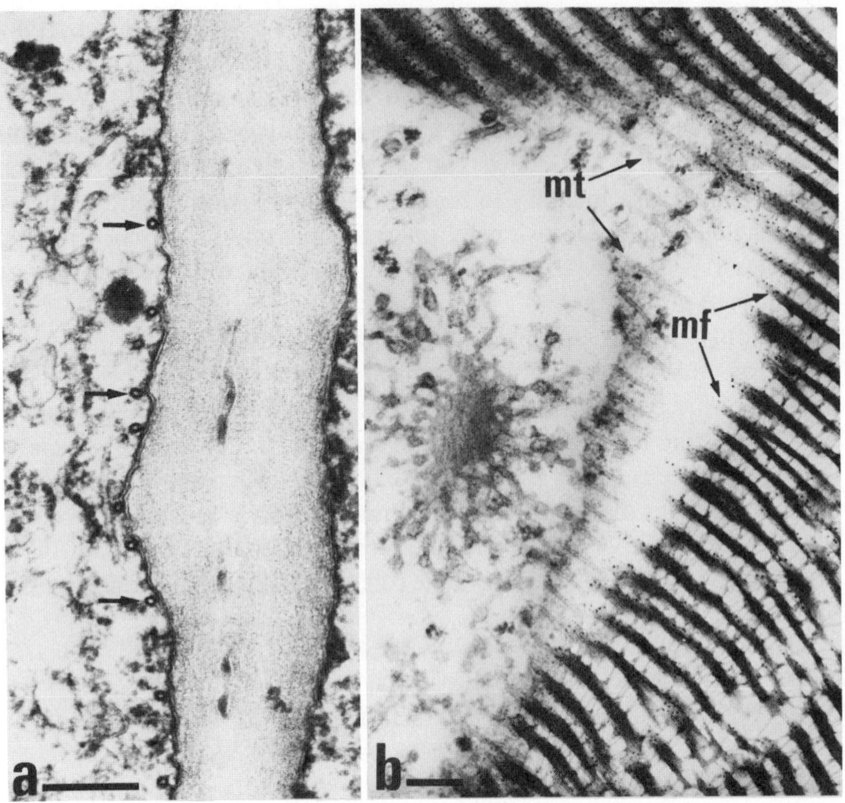

Figure 9.4. (a) Cortical microtubules (arrows) subjacent to the PM in suspension-cultured cells of sugar cane. (b) Tangential section through a seed epidermis cell of *Ruellia* showing the parallelity of cortical microtubules (mt) and bundles of mucilage fibrils (mf). Bar = 0.25 μm. Unpublished (a) micrograph of the author and published (b) micrograph of Schnepf and Deichgräber (1983).

between the two. More conclusive evidence for the participation of microtubules in microfibril orientation has been provided by experiments using substances which selectively interfere with either microtubule or microfibril assembly. A prerequisite for investigations of this type has been the choice of cell types in which microfibrils as well as microtubules could be easily and individually identified. Only few organisms fulfill this criterion and the best ones are algae such as *Oocystis solitaria* and the flagellate *Poterioochromonas malhamensis*.

Microtubule inhibitors of two types have been employed: those which bind to the microtubule protein, tubulin (e.g., colchicine) and those which raise cytosolic Ca^{2+} concentrations (e.g., dinitroaniline herbicides). Both lead to a shift in the microtubule–tubulin equilibrium toward tubulin, resulting in a disassembly of microtubules. Both result in a loss of control over microfibril orientation (Hertel et al., 1980; Robinson and Herzog, 1977; Schnepf et al., 1975). In *Oocystis* the effect is quite spectacular, (Figure 9.2a), and is reversible by washing out the inhibitor involved. The importance of these studies is the fact that microtubules, at least in this organism, appear to be more involved in redirecting microfibril synthesis than in maintaining the orientation of microfibrils in a given direction. This is apparent from the more or less parallel orientation of microfibrils synthesized in the presence of colchicine. Although these effects are best seen in *Oocystis,* there are indications of the same in cells of higher plants, particularly those with thickened cell walls (e.g., Brower and Hepler, 1976; Hardham and Gunning, 1980; Palevitz and Hepler, 1976).

The regular (hourly) change in the direction of microfibril synthesis in *Oocystis* presupposes a similar change in microtubule orientation if the latter are indeed acting as guide elements. Of the two possibilities accommodating such a feat—microtubules regularly switching directions or a disassembly–reassembly through 90°—it would appear that the latter is physiologically more appropriate and is to a certain extent supported by experiments on the recovery of *Oocystis* cells from cold treatments (Robinson and Quader, 1980) which also cause a disassembly of microtubules.

New protein synthesis is not necessary for these changes in microtubule–tubulin equilibrium (Quader et al., 1978), but just how the directional effect on the reassembly of microtubules is involved is not known. Organizing centers for cortical microtubules (MTOCs) have been described for higher plants (Gunning et al., 1978; Galatis et al., 1983) and consist of an aggregation of small vesicles. Such structures have not yet been found in *Oocystis* but might be expected to exist here too.

Uncertainty also exists as to how microtubules exert their influence on the terminal complexes responsible for the synthesis of the microfibrils. In *Poterioochromonas* there appears to be one microtubule per microfibril (Schnepf et al., 1975) and the same is true for the slime fibril bundles of *Ruellia* (Schnepf and Deichgräber, 1983). *Oocystis*, however, has a ratio of 1:2.5. Two possibilities therefore exist: in the one (model I, Figure 9.5) the terminal complexes ride along the microtubule; in the other (model II, Figure 9.5) the microtubules create channels between them through which the terminal complexes may move (Heath and Seagull, 1982; Staehelin and Giddings, 1982). Both models have in common

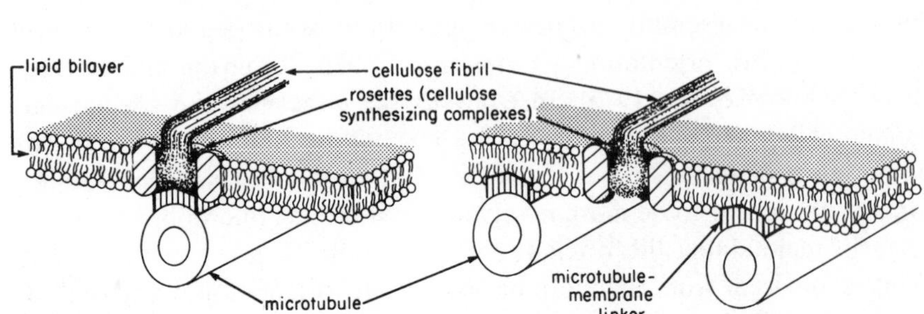

Figure 9.5. Two alternative models for the synthesis and orientation of cellulose microfibrils achieved through the participation of intermembrane particle complexes (rosettes and terminal globules) and cortical microtubules. Published drawing of Staehelin and Giddings (1982).

the movement of the terminal complex in the plane of the membrane, and this would appear to be generally the case for eukaryotic cells. As far as the motive force for this movement is concerned, there are several mechanisms under discussion (see Heath and Seagull, 1982). One of the most attractive, in my opinion, is also the simplest and is based on the crystallization process itself (Herth, 1980; Quader, 1983). This entails a shoving of the terminal complexes rather than their being pulled. One might describe this effect as "cellulose motive force" and its expression in the case of the bacterium *Acetobacter xylinum* results literally in a propulsion of the organism (Brown, 1978).

One exception to the rule appears to exist, namely *Glaucocystis nostochinearum*. This alga, which is superficially similar to *Oocystis* has microtubules parallel to the wall microfibrils but are separated by cisternae (ER origin?) from the PM (Robinson, 1977). According to Brown (1978), microfibril synthesis in this organism probably occurs via a "fixed site" in the PM, with the orientation achieved by the rotation of the protoplast in a particular direction. A "fixed site" for microfibril synthesis in the PM and wall of *Acetobacter* has been clearly demonstrated (Brown et al., 1976; Haigler and Benziman, 1982; Delmer et al., 1982) and it may well be that *Glaucocystis* is one of the few eukaryotes which have retained this prokaryotic feature. On the other hand, claims have been made for the existence of terminal complexes in *Glaucocystis* (Willison and Brown,

1978), although the quality of the preservation of these structures is not comparable to that of *Oocystis*.

9.5. CELLULOSE SYNTHESIS: SOME CHEMICAL ASPECTS

Nucleotide sugars are the substrates for the synthesis of complex oligosaccharides and polysaccharides (Hassid, 1970; Hubbard and Ivatt, 1981). It is also well known that they serve this function in the case of the microfibrillar polysaccharides chitin (McMurrough et al., 1971; Cabib et al., 1982) and cellulose (Carpita and Delmer, 1980; MacLachlan and Fèvre, 1982). Under the right conditions the provision of the appropriate nucleotide sugar to a subcellular fraction can result in the formation of the desired product. This is the case with chitin which has been synthesized in microfibrillar form (Ruiz-Herrera et al., 1975) but has hitherto not been possible in the case of cellulose with extracts from plant cells.

Compared to the rates of synthesis *in vivo* the rates of glucan synthesis *in vitro* with plant extracts are pitifully low and the products have not always contained exclusively 1,4-β-D-linkages but sometimes 1,3-β-D as well (Delmer, 1977; Franz and Heiniger, 1981). In contrast, very high incorporation rates into 1,4-β-D-glucan from UDP–^{14}C-glucose have been achieved with membrane preparations derived from the bacterium *Acetobacter* (Aloni et al., 1982). Several requirements have to be met to enable such high amounts of incorporation in the presence of guanosine triphosphate (GTP), polyethylene glycol 4000 (PEG 4000) and a protein factor present in the 100,000*g* supernatant. Individually these factors have little effect on incorporation but together can give rise to rates of glucan synthase activity up to 60% of those *in vivo*. The activation with GTP is specific for this nucleotide and occurs only in the presence of PEG 4000. It resembles the requirement for GTP in 1,3-β-D-glucan synthesis in yeast (Notario et al., 1982; Shematek and Cabib, 1980) and 1,4-β-D-glucan synthesis in *Saprolegnia* (Fèvre and Rougier, 1981). The role of PEG 4000 is unclear, its promotory effect can be replaced by Ca^{2+} ions (Aloni et al., 1983). It is possible that both of these factors may interact with the membrane environment of the synthase in a manner analogous to that required for some protein kinases whose function are dependent on the phospholipid environment (e.g., Mori et al., 1980). In support of this contention is the fact that solubilized preparations of *Acetobacter* glucan synthase are much less responsive toward PEG 4000 or Ca^{2+} than the membrane-bound form. At this time all that is known about the protein factor is that it is a high-molecular-weight, heat-labile protein. To date, the product has not yet been identified as a crystalline microfibril but its linear 1,4-β-D-nature is certain (Aloni et al., 1982).

It is interesting to note that PEG 4000 and Ca^{2+} also have a stimulatory effect on cellulose synthesis *in vivo* in detached cotton fibers (Carpita and Delmer, 1980). The importance of Ca^{2+} in microfibril synthesis is also inherent in the results of Quader and Robinson (1979) and Quader (1982) on *Oocystis*. In addition, a protein synthesis requirement for microfibril deposition in this organism has been concluded on the basis of experiments with cycloheximide (Quader et al., 1978; Robinson and Quader, 1981b). It is difficult to say how far one can stretch the comparison between bacterial and plant systems. Certainly, for both, the maintenance of a *trans* membrane electrical potential ($\Delta\psi$) has been shown to be of great importance for glucan synthesis (Bacic and Delmer, 1981; Delmer, Benziman, and Padan 1982) but cytologically, in terms of the mobility of the synthesizing complexes and the maintenance of PM integrity in general, there are great differences between the two systems (Delmer, Benziman, Klein, Bacic, Mitchell, Weinhouse, Aloni, and Callaghan 1982; Robinson and Quader, 1981b).

Nucleotide sugars are usually considered to be the substrates for glucan synthesis, but a direct transfer of glucose from them to the growing glucan chain is no longer accepted (MacLachlan and Fèvre, 1982). Instead glycosylated lipid or protein intermediates have been considered for some years. The participation of lipidlike intermediates in the synthesis of 1,4-β-glucan synthesis was suggested in the early papers of Kahn and Colvin (1961), Forsee and Elbein (1973), and Brett and Northcote (1975), but has not been substantiated in the hands of Morohashi and Bandurski (1976) or Helsper (1979). Lipid intermediates of the dolichol (polypropyl) phosphate with either one or two phosphate groups before the terminal glucose residue(s) have been identified in both algal (Hopp et al., 1978; Pont Lezica et al., 1978) and higher plants (Dürr et al., 1979; Bailey et al., 1979) cells, but there is no evidence that these molecules are associated with PM preparations; indeed their presence instead in ER and GA preparations suggests their participation in the synthesis of complex oligosaccharides (see Chapter 10).

A protein which accepts glucose residues from dol-PP-(Glc) has been detected by Pont-Lezica et al. (1978), but it is not clear whether this is an intermediate in cellulose synthesis or a membrane-bound lectin. Whelan (1976) suggested that a protein intermediate is a universal requirement for the synthesis of all polysaccharides, and there is evidence for the participation of such molecules in 1,3-β-glucan synthesis in *Euglena* (Tomos and Northcote, 1978). As far as 1,4-β-glucan synthesis is concerned, Franz (1976) has shown that a glucoprotein (as well as other products) is produced by membrane fractions of *Phaseolus* with UDP–Glc. Analysis of this product indicated long oligosaccharide chains and pulse-chase experiments were consistent with its potential role as an intermediate in 1,4-β-glucan synthesis. Once again, however, the localization of these molecules/events at the PM is questionable.

Before closing this section a few words should be spent on chitin synthesis simply as a comparison to that of cellulose. Microfibrillar chitin has been synthesized *in vitro* with extracts from a number of fungi (Bartnicki-Garcia et al., 1978; Mills and Cantino, 1978). *In vitro* synthesis with a solubilized enzyme has also been achieved (Gooday, 1979). Although a synthesis with isolated PM from *Saccharomyces* has been shown (Cabib et al., 1979), it appears that the majority of the recoverable chitin synthase is in an inactive ("zymogenic") form and is found in the postmicrosomal fraction. Analysis of this fraction has given rise to the recognition of so-called chitosomes, which are essentially microvesicles about 40–70 nm in diameter. Proteolytic treatment of these particles and then their incubation in the presence of UDP-GlcNAc and Mg^{2+} ions gives rise to microfibrillar chitin (Bracker et al., 1976). Although it is still not certain whether or not chitosomes are true vesicles or homogenization artifacts (see also Chapter 8), there have been as yet no reports of a significant amount of glucan synthetase activity in the postmicrosomal supernatant.

REFERENCES

Aloni, Y., Cohen, R., Benziman, M., and Delmer, D. (1983). *J. Biol. Chem.* **258,** 4419–4423.

Aloni, Y., Delmer, D. P., and Benziman, M. (1982). *P.N.A.S. (USA)* **79,** 6448–6452.

Bacic, A. and Delmer, D. P. (1981). *Planta* **152,** 346–351.

Bailey, D. S., Durr, M., Burke, J., and MacLachlan, G. A. (1979). *J. Supramol. Struc.* **11,** 123–138.

Barnett, J. R. and Preston, R. D. (1970). *Ann. Bot. N. S.* **34,** 1011–1017.

Bartnicki-Garcia, S., Bracker, C. E., Reyes, E., and Ruiz-Herrera, J. (1978). *Exp. Mycol.* **2,** 173–192.

Bracker C. E., Ruiz-Herrera, J., and Bartnicki-Garcia, S. (1976). *Proc. Nat. Acad. Sci.* (*USA*) **73,** 4570–4574.

Branton, D., Bullivant, S., Gilula, N. B., Karnovsky, M. J., Moor, H., Mühlethaler, K., Northcote, D. H., Packer, L., Satir, B., Satir, P., Speth, V., Staehelin, L. S., Steere, R. L., and Weinstein, R. S. (1975). *Science* **190,** 54–56.

Brett, C. T., and Northcote, D. H. (1975). *Biochem. J.* **148,** 107–117.

Brower, D. L. and Hepler, P. K. (1976). *Protoplasma* **87,** 91–111.

Brown, R. M. (1969). *J. Cell Biol.* **41,** 109–123.

Brown, R. M., (1973). In *Plant Cell Wall Biogenesis* (C. J. Arceneaux, ed.), 31st Ann. Proc. Electron Microscopy Soc. Amer.

Brown, R. M. (1978). In The Third Philip Morris Science Symposium, pp. 52–123.

Brown, R. M. and Montezinos, D. (1976). *Proc. Nat. Acad. Sci.* (*USA*) **73,** 143–147.

Brown, R. M. and Romanovicz, D. K. (1976). *Appl. Polym. Symp.* **28,** 537–585.

Brown, R. M., Franke, W. W., Kleinig, H., Falk, H., and Sitte, P. (1970). *J. Cell Biol.* **45,** 246–271.

Brown, R. M., Herth, W., Franke, W. W., and Romanovicz, D. (1973). In *Biogenesis of Plant Cell Wall Polysaccharides* (F. Loewus, ed.), pp. 207–257, Academic Press, New York.

Brown, R. M., Willison, J. H. M., and Richardson, C. L., (1976). *Proc. Nat. Acad. Sci.* (*USA*) **73,** 4564–4569.

Cabib, E., Duran, A., and Bowers, B. (1979). In *Fungal Walls and Hyphal Growth* (J. H. Burnett and A. P. J. Trinci, eds.), pp. 180–201, Cambridge University Press, Cambridge.

Cabib, E., Roberts, R., and Bowers, B. (1982). *Ann. Rev. Biochem.* **51,** 763–793.

Carpita, N. C. (1982). In *Cellulose and Other Natural Polymer Systems* (R. M. Brown, Jr., ed.), pp. 225–242, Plenum Press, New York.

Carpita, N. C. and Delmer, D. P. (1980). *Plant Physiol.* **66,** 911–916.

Delmer, D. P. (1977). In *Recent Advances in Phytochemistry* (F. Loewus, ed.), Vol. 11, pp. 45–77, Academic Press, New York.

Delmer, D. P., Benziman, M., and Padan, E. (1982). *Proc. Natl. Acad. Sci.* (*USA*) **79,** 5282–5286.

Delmer, D. P., Benziman, M., Klein, A. S., Bacic, A., Mitchell, B., Weinhouse, H., Aloni, Y., and Callaghan, T. (1982). *J. Appl. Polym. Sci.* (*Appl. Polym. Symp.*)

Dürr, M., Bailey, D. S., and MacLachlan, G. (1979). *Eur. J. Biochem.* **97,** 445–453.

Fèvre, M., and Rougier, M. (1981). *Planta* **151,** 232–241.

Forsee, W. T. and Elbein, A. O. (1973). *T. Biol. Chem.* **248,** 2858–2867.

Franke, W. W. and Ermin, B. (1969). *Z. Naturforsch.* **24b,** 918–922.

Franz, G. (1976). *J. Polym. Sci. Part.* **C28,** 611–621.

Franz, G. and Heiniger, U. (1981). In *Encyclopedia Plant Physiology* Vol. 13B, pp. 47–67, Springer Verlag, Berlin.

Frei, E. and Preston, R. D. (1961). *Proc. Roy. Soc. B* **154,** 70–94.

Galatis, B., Apostolakos, P., and Katsaros, C. (1983). *Protoplasma* **115,** 176–192.

Giddings, T. H., Brower, D. L., and Staehelin, L. A. (1980). *J. Cell Biol.* **84,**327–339.

Gooday, G. W. (1979). In *Fungal Walls Hyphal Growth* (J. H Burnett and A. P. J. Trinci, eds.), pp. 203–233, Cambridge University Press, Cambridge.

Gunning, B. E. S. and Hardham, A. R. (1983). *Ann. Rev. Plant Physiol.* **33,** 651–698.

Gunning, B. E. S., Hardham, A. R., and Hughes J. E. (1978). *Planta* **143,** 161–179.

Haigler, C. H. and Benziman, M. (1982). In *Cellulose and Other Natural Polymer Systems* (R. M. Brown, ed.), pp. 273–297, Plenum Press, New York.

Harada, H. and Goto, T. (1982). In *Cellulose and Other Natural Polymer Systems* (R. M. Brown, ed.), pp. 383–401, Plenum Press, *New York.*

Hardham, A. R. and Gunning, B. E. S. (1980). *Protoplasma* **102,** 31–51.

Hassid, W. Z. (1970). In *The Carbohydrates* (W. Pigman and D. Horton, eds.), Vol. 11A, pp. 301–373, Academic Press, New York.

Heath, I. B. and Seagull, W. (1982). In *The Cytoskeleton in Plant Growth and Development* (C. W. Lloyd, ed.), pp. 163–182, Academic Press, London.

Helsper, J. P. F. G. (1979). *Planta* **144,** 443–450.

Hepler, P. K. and Palevitz, B. A. (1974). *Ann. Rev. Plant Physiol.* **25,** 309–362.

Hertel, C., Quader, H., Robinson, D. G., and Marmé, D. (1980). *Planta* **149,** 336–340.

Herth, W. (1980). *J. Cell Biol.* **87,** 442–450.

Herth, W. (1983). *Planta* **159,** 347–356.

Hopp, E. H., Romero, P. A., Daleo, G. R., and Pont Lezica, R. (1978). *Eur. J. Biochem.* **84,** 561–571.

Hubbard, S. C. and Ivatt, R. J. (1981). *Ann. Rev. Biochem.* **50,** 555–584.

Kahn, A. W. and Colvin, J. R. (1961). *Science* **133,** 2014–2015.

Kiermayer, O. and Sleytr, U. B. (1979). *Protoplasma* **101,** 133–138.

Ledbetter, M. and Porter, K. (1963). *J. Cell Biol* **19,** 239–250.

MacLachlan, G. A. and Fèvre, M. (1982). In *The Cytoskeleton in Plant Growth and Development* (C. W. Lloyd, ed.), pp. 127–146, Academic Press, London.

McMurrough, I., Flores-Carreon, A., and Bartnicki-Garcia, S. (1971). *J. Biol. Chem.* **246,** 3999–4007.

Mills, G. L. and Cantino, E. C. (1978). *Exp. Mycol.* **2,** 99–109.

Moor, H. and Mühlethaler, K. (1963). *J. Cell Biol.* **17,** 609–628.

Mori, T., Takai, Y., Minakuchi, R., Yu, B., and Nishizuka, Y. (1980). *J. Biol. Chem.* **255,** 8378–8390.

Morohashi, Y. and Bandurski, R. S. (1976). *Plant Physiol.* **57,** 846–849.

Mueller, S. C. and Brown, R. M. (1980). *J. Cell Biol.* **84,** 315–326.

Newcomb, E. H. (1969). *Ann. Rev. Plant Physiol.* **20,** 253–288.

Noguchi, T., Tanaka, K., and Ueda, K. (1981). *Cell Struct. Funct.* **6,** 217–229.

Notario, V., Kawai, H., and Cabib, E. (1982). *J. Biol. Chem.* **257,**1902–1905.

Palevitz, B. A. and Hepler, P. K. (1976). *Planta* **132,** 71–93.

Peng, H. B. and Jaffe, L. F. (1976). *Planta* **133,** 57–71.

Preston, R. D. (1974). *The Physical Biology of Plant Cell Walls.* Chapman and Hall, London.

Preston, R. D. and Goodman, G. N. (1968). *Proc. Roy. Microsc. Soc.* **88,** 513–527.

Pont Lezica, R., Romero, P. A., and Hopp, H. E. (1978). *Planta* **140,** 177–183.

Quader, H. (1981). *Die Naturwiss* **76,** 428.

Quader, H. (1982). In *Microtubules in Microorganisms* (P. Cappuccinelli and N. R. Morris, eds.), pp. 313–324, Marcel Dekker, New York.

Quader, H. (1983). *Eur. J. Cell Biol.* **32,** 174–177.

Quader, H, and Robinson, D. G. (1979). *Eur. J. Cell Biol.* **20,** 51–56.

Quader, H. and Robinson D. G. (1981). *Ber, Dtsch. Bot. Ges.* **94,** 75–84.

Quader, H., Robinson, D. G., and van Kempen, R. (1983). *Planta* **157,** 317–323.

Quader, H., Wagenbretly, I., and Robinson, D. G. (1978). *Cytobiol.* **18,** 39–51.

Robenek, H. and Peveling, E. (1977). *Planta* **136,** 135–145.

Robinson, D. G. (1977). *Adv. Bot. Res.* **5,** 89–154.

Robinson, D. G. (1981). *Engyl. Plant Physiology* **13B,** 25–28.

Robinson, D. G. and Herzog, W. (1977). *Cytobiologie* **15,** 463–474.

Robinson, D. G., and Preston, R. D. (1971). *J. Cell Sci.* **9,** 581–601.

Robinson D. G. and Preston, R. D. (1972). *Planta* **104,** 234–246.

Robinson, D. G. and Quader, H. (1980). *Eur. J. Cell Biol.* **21,** 229–230.

Robinson, D. G. and Quader, H. (1981a). *Eur. J. Cell Biol.* **25,** 278–288.

Robinson, D. G. and Quader, H. (1981b). *J. Theor. Biol.* **92,** 483–495.

Robinson, D. G., and Quader, H. (1982). In *The Cytoskeleton in Plant Growth and Development* (C. Lloyd, ed.), pp. 109–126, Academic Press, London.

Robinson, D. G., White, R. K., and Preston, R. D. (1972). *Planta* **107,** 131–144.

Romanovicz, D., (1981). In *Cytomorphogenesis in Plants* (O. Kiermayer, ed.), pp. 27–62. Springer Verlag, Vienna.

Ruiz-Herrera, J., Sing, V. O., van der Woude, W. J., and Bartnicki-Garcia, S. (1975). *Proc. Nat. Acad. Sci. (USA)* **72,**2706–2710.

Sassen, A., van Eyden-Emons, S., Lamers, A., and Wanka, F. (1970). *Cytobiol.* **1,**373–382.

Schnepf, E and Deichgräber, G. (1983). *Protoplasma* **114,** 222–234.

Schnepf, E., Röderer, G., and Herth, W. (1975). *Planta* **125,** 45–62.

Shematek, E. M. and Cabib, E. (1980). *J. Biol. Chem.* **255,** 895–902.

Sleytr, U. B. and Messner, P. (1978). *J. Cell Biol.* **79,** 276–280.

Staehelin, L. A. and Giddings, T. H. (1982). In *Developmental Order: Its Origin and Regulation* (S. Subtelny, ed.), pp. 133–147. A. R. Liss, New York.

Staehelin, L. A. (1966). *Z. Zellforsch.* **74,** 325–350.

Steere, R. L., Erbe, E. F. and Moseley, J. M. (1980). *J. Cell Biol.* **86,** 113–122.

Tomos, A. D. and Northcote, D. H. (1978). *Biochem. J.* **174,** 283–290.

Volkmann, D. (1983). *Eur. J. Cell Biol.* **30,** 258–265.

Wada, M. and Staehelin, L. A. (1981). *Planta* **151,** 462–468.

Whelan, W. H. (1976). *Trends Biochem. Sci.* **1,** 13–15.

Willison, J. H. M. (1978). *J. Cell Biol.* **77,** 103–119.

Willison, J. H. M. (1983). *J. Appl. Polym. Sci.* **37,** 91–105.

Willison, J. H. M. *Int. Rev. Cytol.* (in press)

Willison, J. H. M. and Cocking, E. C. (1972). *Protoplasma* **75,** 397–403.

Willison, J. H. M. and Cocking, E. C. (1975). *Protoplasma* **84,**147–159.

Willison, J. H. M. and Brown, R. M. (1978). *J. Cell Biol.* **77,** 103–119.

10

SYNTHESIS AND SECRETION OF EXTRACELLULAR MACROMOLECULES

In the previous chapter we considered the production of microfibrils, one component of the cell wall of plants and fungi. Other cell wall components may be conveniently termed *matrix* substances in the sense that they fill in the spaces between the microfibrils. Although the macromolecules which comprise this group of substances are quite variable in structure, they share one thing in common: they reach the cell wall through fusion of a vesicle with the PM and expulsion of the contents. This is the ultimate act of what is termed *granulocrine secretion*, and it is the purpose of this chapter to trace the origin of such vesicles and their contents.

I use the term *matrix substance* in the widest possible sense, including macromolecules covalently bound to one another and constituting the cell wall *sensu stricto* and extracellular enzymes which happen to be located in the acqueous environment of the wall whether they have something to do with cell wall function or not. An example of the latter is invertase, whose role in the improved unloading of photosynthetic assimilate in root cells is not to be underestimated (Eschrich, 1980). Therefore, both enzymic and structural proteins as well as polysaccharides are secreted by plant cells, their relative amounts in the cell wall depending upon cell type and developmental stage. However, with the exception of special glandular cells (e.g., in insectivorous plants), plant cell walls contain relatively little protein (usually less than 10% of the total fresh weight), a feature of plant cells that must not be forgotten when comparing them with animal cells in terms of secretion.

The proteins secreted by plant cells are also qualitatively different from those released from animal cells. In animals cells a large number of secretory proteins are actually glycoproteins possessing oligosaccharide side chains which are usually covalently linked through the NH_2 group of asparagine (Asn) residues. *N*-glycosidic linkages to Asn are typical of reserve proteins in seeds (see Chapter 11) and some enzymes and lectins, but the principal extracellular plant glycoproteins, together with some other lectins, are characterized by *O*-glycosidic linkages to hydroxyproline (Hyp) or serine (Ser) (Selvendran and O'Neill, 1982).

10.1 THE SITUATION IN ANIMAL CELLS: A SUMMARY

As a rule animal cells do not secrete polysaccharides. There may be cases where appreciable amounts of uronic acids (e.g., hyaluronic acid in connective tissue) are released, but these are never the only or predominant secretory product. Instead pure or glycosylated proteins are synthesized and exported. It is well known that ribosomes are the sites of protein synthesis (see Siekevitz and Zamecnik, 1981, for a concise historical review), and thanks in particular to the now-classical research of Palade and co-workers (Jamieson, 1978; Jamieson and Palade, 1977; Palade, 1975), we recognize that those ribosomes that become associated with the ER are responsible for the synthesis of secretory proteins.

The rER synthesizes not only *bona fide* secretory proteins but also a number of other proteins whose destination are an intracellular rather than an extracellular site (Jamieson, 1981). Some remain in the ER lumen, some are transported to lysosomes, and some remain embedded in (integral) or attached to (associated with) the ER membrane and later find their way into other membranes (Lodish et al., 1981). Since their syntheses are not temporally separated from one another, some kind of sorting mechanism(s) must be involved to avoid their chaotic distribution within the cell or their uncontrolled loss from the cell. We will consider elsewhere the redistribution of ER-synthesized membrane proteins (Chapter 15) and the problem of lysosomal enzymes (Chapter 12) and will deal here only with the case of *bona fide* secretory proteins.

Historically one can divide the work carried out in this field into two phases. In the first phase, which essentially spans the period from 1965 to 1975, the export route for secretory proteins was mapped out. The second phase, from 1975 until the present, has seen the application of molecular biological rather than cytological or cell physiological techniques.

Numerous studies with a variety of different cell types involving the secretion of enzymes, hormones, and extracellular structures (e.g., collagen) have established the secretory route as beginning at the ER and passing through the GA before ending with the fusion of a secretory vesicle at the PM. In elucidating this pathway the approach has been that of the pulse-chase experiment whereby the location of the radioactivity in the cell in question is determined either by autoradiography or by fractionation (see Whaley, 1975, for a summary of the various experiments of this nature carried out on animal cells). Invariably, when radioactive amino acids are used, the first labeled organelle is the rER; with chase-out the radioactivity is then seen to leave this organelle and pass through smooth membrane fractions (gER $\rightarrow$ GA $\rightarrow$ SV) before appearing in the extracellular milieu (Bergeron et al., 1978; Jamieson and Palade, 1977; Palade, 1975). When, instead, incubations with radioactive sugars are carried out, the first labeled organelle can be either the ER of the GA, depending on the sugar employed. If galactose, glucose, or fucose is used, the GA is labeled first (Haddad et al., 1971; Halbhuber et al., 1972; Neutra and Leblond, 1966; Whur et al., 1969). In contrast, radioactivity from mannose and glucosamine is first detectable in the ER (Whur et al., 1969; Zagury et al., 1970). These early papers thus indicate the existence of two sites of glycosylation in the cell, one at the site of polypeptide synthesis (i.e., the rER) and the other in the GA.

The localization of radiolabel by autoradiography does not allow for a differentiation between membrane and content. Separation of content from membrane by salt or mild detergent treatment indicates that the major portion of the radiolabel is associated with the former (Franke et al., 1971; Wallach et al., 1975). In contrast, membrane proteins are labeled at a much slower rate than secretory proteins (Castle and Palade, 1978; Meldolesi, 1974). Estimations of the displacement times for secretory proteins lie in the order of approximately 10 min and 5–20 min for ER and GA fractions, respectively (see Table II in Robinson and Kristen, 1982). The displacement times for secretory vesicles are quite variable and one must be careful to distinguish between glands with a regulated or cyclic discharge from those with a continuous discharge (Amsterdam et al., 1969; Tartakoff and Vassalli, 1978).

Inhibitor studies have confirmed the participation of the stations along the route for secretory proteins. The inclusion of KCN, antimycin A, or DNP at various times during the chase-out results in a rapid cessation of secretion, indicating the energy dependence of this process. Such blockages, which are reversible, are possible at each vesiculation and vesicle fusion stage—that is, between ER and GA, between GA and SV, and between SV and PM (Jamieson and Palade, 1968, 1971; Tartakoff and Vassalli, 1977, 1978).

In the early seventies Blobel and Sabatini (1971) suggested that polypeptides destined for secretion must have something in common to enable them to be recognized and segregated by the membranes of the ER. Shortly afterwards Milstein et al. (1972) found that the cell-free translation of an immunoglobulin light chain mRNA from a mouse myeloma resulted in a polypeptide some 15 kd larger than the native species. This increase in size was due to an additional 15 amino acids situated at the *N*-terminus end of the molecule. Based on this observation and corroborated by their own work, Blobel and Dobberstein (1975a,b) formulated the so-called Signal hypothesis according to which the additional, or "leader," peptide acts as a recognition signal for a receptor at the surface of the ER. A number of secretory and membrane proteins are now known to be synthesized with an N-terminus leader sequence (Blobel et al., 1979; Lodish et al., 1981) and the process by which the vectorial segregation of the polypeptide occurs simultaneously with its translation is termed *cotranslation*. In contrast, those events which occur after the polypeptide has been released from the polysome are collectively designated *posttranslational*. The latter events are also sometimes termed *processing* and occur mainly but not exclusively in the GA, confirming the earlier observations that the GA is a major site of glycosylation reactions. The major changes which are known to occur to secretory polypeptides in the ER and GA are presented in Table 10.1. (Additional useful reviews on these features are Bowles, 1982; Hubbard and Ivatt, 1981; Sabatini et al., 1982; Sharon and Lis, 1981; and Snider and Robbins, 1981).

Blobel's signal hypothesis is widely accepted, but there are exceptions (see Waksman et al., 1980, for examples) to the rule that an N-terminus sequence is obligatory. Probably the most well-known is that of chicken oviduct ovalbumin (Palmiter et al., 1978) which *in vitro* is synthesized and inserted into microsomal vesicles without undergoing a proteolytic cleavage. More recent work by Blobel's group (Lingappa et al., 1979) has suggested that ovalbumin does have a peptide sequence which is recognized by the ER membrane, but this is not terminally located in the molecule. As a result, Kreil (1981) concludes it is hard to envisage the same tunnel or pore being capable of allowing the passage of both linear as well as folded, hairpinlike polypeptides. However, the insertion into and passage through the ER membrane of secretory proteins like ovalbumin may not require a tunnel and might be achieved spontaneously based on the physical conformation of the molecule (see Engelman and Steitz, 1981; and Von Heijne and Blomberg, 1979, for theoretical consideration of this possibility).

Although some secretory proteins are not glycosylated, one must ask whether or not glycosylation is obligatory for secretion since, as Lehle (1981) states, "one should, however, keep in mind the possibility that nonglycosylated and yet secreted proteins were synthesized as glycoproteins and have lost their carbohydrate part at some stage during their

TABLE 10.1. THE ROLES OF THE ER AND THE GA IN SECRETION IN ANIMAL CELLS: A SYNOPSIS

Step	Location	Remarks
A. Cotranslational events		
1. Initiation of translation	Free polysomes	N-terminus leader sequence of 15–30 AAs; variable length and structure; evolutionary unconservative; central region of hydrophobicity which is α-helical (Kreil, 1981)
2. Recognition of leader peptide (signal sequence)	Cytoplasmic surface of the ER membrane	Peptide–phospholipid (Chan et al., 1979) or peptide–protein (Prehn et al., 1980) interactions? Participation of 11S protein complex (Walter and Blobel, 1981a,b)?
3. Polysome–membrane interaction	Cytoplasmic surface of the ER membrane	Ribosome (60S subunit) binding (glyco)proteins (Ribophorins—Kreibich et al., 1978; see however Bielinska, 1979). Ionic interaction?
4. Polypeptide chain elongation	Within-ER membrane	Interaction of ribosomes and membrane proteins leads to the formation of a "tunnel or pore" (Blobel and Dobberstein, 1975a,b)
5. Removal of leader peptide	Luminal surface of the ER membrane	Signal (endo)peptidase; integral membrane protein (Kreil et al., 1980). Cleaved leader sequence is short-lived (Habener et al., 1979)
6. "Core glycosylation" of nascent polypeptide	Luminal surface of the ER membrane	Transfer of Glc_3-Man_9-$GlcNAc_2$ oligosaccharide from dolichol phosphate to some, but not all, Asn residues (Kronquist and Lennarz, 1978). Conformation restriction?

TABLE 10.1. (*Continued*)

Step	Location	Remarks
B. Posttransational events		
7. Removal of Glc residues from Asn-linked oligosaccharide	ER lumen	α-glucosidase I and II/III; integral membrane proteins (Grinna and Robbins, 1979; Elting et al., 1980)
8. Removal of outer four Man residues	*Cis* GA cisternae	Mannosidase I and II; specific for α-1,2 linkages (Tabas and Kornfeld, 1979; Dunphy et al., 1981)
9. Addition of GlcNAc to core Man residue	*Cis* GA cisternae	GlcNAc Transferase I (Harpaz and Schachter, 1980)
10. Removal of two (none-core) terminal Man residues	*Cis* GA cisternae	"Late mannosidase" (Tabas and Kornfeld, 1978)
11. Addition of GlcNAc to second core Man residue	*Cis* (?) GA cisternae	GlcNAc Transferase II (Harpaz and Schachter, 1980)
12. Addition of Gal, and sialic acid "terminal glycosylatin"	*Trans* GA cisternae	Galactosyl, fucosyl, and sialyl transferases (Dunphy et al., 1981; Tabas and Kornfeld, 1978; Griffiths et al., 1982; Bretz et al., 1980). See however (Bergeron et al., 1982) for possible location in cis-cisternae.

formation." Of use in determining this is the antibiotic tunicamycin. This substance prevents the formation of dolichol pyrophosphoryl G1cNAc and therefore inhibits the synthesis of lipid-linked oligosaccharides (see Elbein, 1981, for a short review). At appropriate concentrations it can be extremely selective, effectively inhibiting core glycosylation, while being without effect on translation. Currently we can point to cases where glycosylation clearly is necessary for secretion (Duksin and Bornstein 1977; Hickman et al., 1977; Tanzer et al., 1977), but one is also able to cite examples where it is apparently not (Keller and Swank, 1978; Loh and Grainer, 1978; Olden et al., 1978; Struck et al., 1978).

10.2 HYDROXYPROLINE-RICH GLYCOPROTEINS (HRGP)

A number of plant glycoproteins contain hydroxyproline residues (Lamport, 1980; Selvendran and O'Neill, 1982). These are usually structural components of the cell wall (Lamport and Catt, 1981) but, in the case of the arabinogalactan proteins (AGPs—Clarke et al., 1979) and potato lectins (Allen and Neuberger, 1973; Leach et al., 1982), may also be intracellularly localized. The mole percent of hydroxyproline in a HRGP ranges between 9 and 60% with the second most frequent amino acid usually being serine (see Tables 6 and 7 of Selvendran and O'Neill, 1982). In contrast to the collagens of animal connective tissue, which have unconjugated Hyp and glycosylated hydroxylysine residues (Kornfeld and Kornfeld, 1976), HRGPs are characterized by *O*-glycosidic linkages to Hyp and Ser.

Depending on the type of attached carbohydrate moiety we can differentiate HRGPs of the extensin type from the AGPs. The former, characteristic of the structural protein of plant cell walls, has between one and four β-linked arabinofuranosyl residues attached to the Hyp residue and single α-linked galactosyl residues on the Ser residue. AGPs, on the other hand, have a much larger carbohydrate moiety (~90% w/w) which in addition to the Hyp arabinosides consists of a branched galactan (main chain β, 1 → 3; side chain α, 1 → 6; terminal arabinofuranosyl residues), also attached to Hyp residues. *N*-acetylglucosamine also appears to be attached to the Ser residues. Moreover, a significant amount of uronic acids are present (10–20% of total carbohydrate) which are the cause of the very low isoelectric point (between pH 2 and 3). In contrast to extensin, AGPs are freely soluble, which in the case of suspension-cultured cells leads to their accumulation in the growth medium (Anderson et al., 1977; Hori and Sato, 1977; Pope, 1977; Van Holst et al., 1981), and in the case of tissue cells to their predominantly extracellular location (Clarke et al., 1975; Gleeson and Clarke, 1979). There is no evidence that

AGPs become structurally bound to the cell wall. Whether the different patterns of glycosylation for HRGPs is of importance for determining the speed and route of their export is at this time unclear.

The synthesis of collagen is well known, both biochemically (Bornstein, 1974; Fessler and Fessler, 1978) and cytologically (Ehrlich et al., 1974; Weinstock and Leblond, 1974). For this analog to the HRGPs we know the sites of polypeptide synthesis and the posttranslational events such as hydroxylation and glycosylation (Harwood et al., 1974; Olsen et al., 1975; Peterkovsky and Assad, 1976), and these confirm the ER → GA → PM pathway for secretory proteins. In marked contrast, due to contradictory observations, there is still uncertainty as to the secretory route for HRGPs. A critical appraisal of the current situation is given in the following three sections, with specific references to published work.

10.2.1. Sites of Translation and Transport

Short-term labeling studies have indicated both in the case of collagen (Vuust and Piez, 1972) and HRGPs (Sadava and Chrispeels, 1971a) that the hydroxylation of proline-containing polypeptides occurs after their translation is completed. Although it has been known for some time that HRGPs become associated with membranous organelles before their transfer to the cell wall (see, e.g., Brysk and Chrispeels, 1972; Chrispeels, 1969; Cleland, 1968; Dashek, 1970; and more recently Kawasaki, 1982b; Van Holst et al., 1981), it has not been clear which organelles are involved.

The major peak of Hyp-containing macromolecules in isopycnic sucrose density gradients coincides with GA marker enzymes. This is true for homogenates from carrot roots (Gardiner and Chrispeels, 1975; Wienecke et al., 1982) and from suspension-cultured tobacco cells (Kawasaki, 1982a). In the latter case a considerable amount of ER contaminates the GA fraction (see Figure 1 in Kawasaki, 1981) as documented by the more or less confluent profiles of CCR and IDPase activities. The reason for this is unclear but Kawasaki (personal communication) believes that this might reflect the vacuolar storage of large amounts of Mg^{2+} ions. These, upon release through homogenization, prevent the loss of ribosomes from the ER and result in a CCR profile approaching that seen in a high Mg^{2+} sucrose gradient (see Chapter 2).

The GA fractions mentioned above for the carrot system are also not pure. Just like the legumes (Chapter 7) markers for the PM in low Mg^{2+}

gradients of carrot root homogenates overlap with those for the GA (Robinson et al., 1982). Certainly some of the HRGPs appear to be PM associated in this system since the broad peak for ^{14}C-Hyp macromolecules does shift somewhat to a higher density in high Mg^{2+} gradients (Wienecke et al., 1982). These authors suggested that the PM-associated ^{14}C-Hyp was in the form of an AGP rather than extensin and showed later (Robinson and Glas, 1982) that the more dense portion of this broad peak did not chase out as a true secretory glycoprotein in an organelle fraction might be expected to do.

The problem of PM contamination of GA fractions is largely overcome by employing monocotolydenous cells. As previously mentioned (see Chapters 5 and 7, also Robinson and Glas, 1982) the PM has an appreciably higher density in low Mg^{2+} sucrose gradients than is the case with legumes/carrots. For this reason we recently examined homogenates from suspension-cultured sugar cane cells (Robinson et al., 1984). Figure 10.1 shows the relative amounts of Hyp-containing macromolecules present in the lumen of the various organelles as separated on a low Mg^{2+} sucrose gradient. There is no doubt that such molecules are associated with both GA and PM fractions.

Although there is considerable evidence for the presence of Hyp-rich macromolecules in the GA, the situation is not quite so clear with respect to the ER. In their 1975 paper Gardiner and Chrispeels were unable to detect radioactive Hyp polypeptides in ER fractions. Wienecke et al. (1982) were however able to demonstrate small, but significant amounts of Hyp in fractions bearing CCR activity. These fractions appeared to shift in density in response to Mg^{2+} and became labeled with ^{14}C-Hyp before the GA fractions in pulse-chase experiments performed by Robinson and Glas (1982). The presence of Hyp polypeptides in ER fractions has been confirmed by Samson et al. (1983) but these authors have claimed, in contrast to Wienecke et al. (1982), that the GA contains negligible amounts of the same.

10.2.2. Site of Hydroxylation

As already mentioned, the hydroxylation of peptidyl proline occurs posttranslationally. This fact has recently been confirmed by *in vitro* translation experiments which have shown that in the absence of microsomal

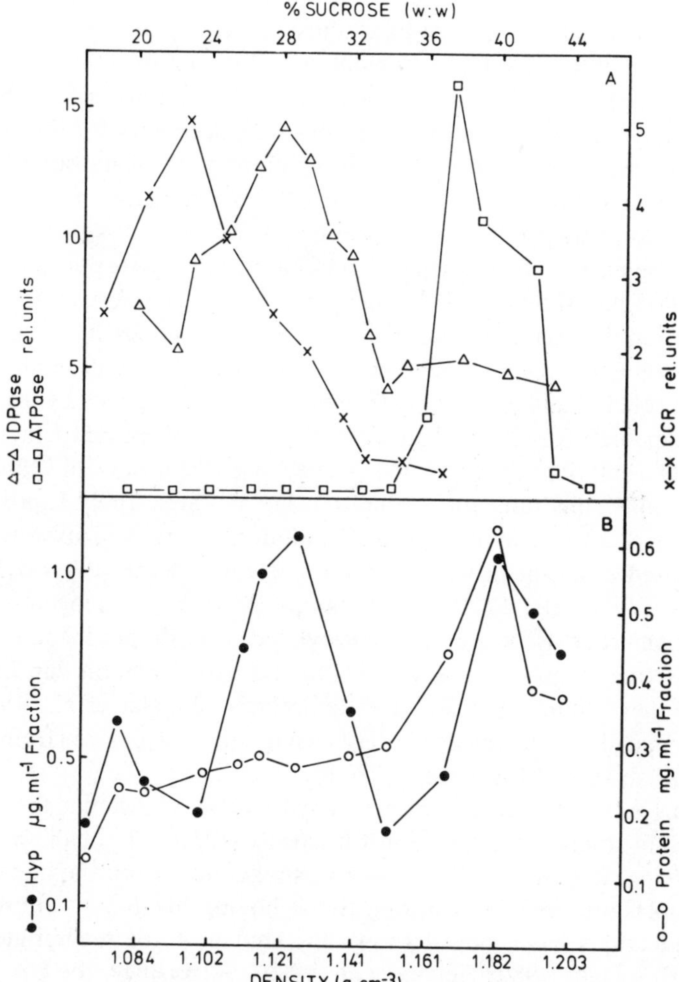

Figure 10.1. The distribution of HRGPs (b) in isopycnic sucrose gradients (low Mg^{2+} conditions) prepared from suspension-cultured sugar cane cells. HRGPs were released from the membranes in each fraction by sonification with 0.1% Triton X-100. For comparison marker enzymes for the ER (CCR), GA (IDPase), and PM (K^+-ATPase) are also given (a). Unpublished data of Glas and Robinson.

vesicles a nonhydroxylated, unglycosylated polypeptide is synthesized (Smith, 1981a).

In contrast to collagen, where hydroxylation leads to helical folding and is a prerequisite for intracellular transport (Grant and Jackson, 1976), the inhibition of hydroxylation in plant cells does not result in a blockage of secretion. This inhibition is brought about by incubation in solutions of α, α′-dipyridyl which binds the ferrous ions necessary for the activity of the enzyme prolyl hydroxylase. In common with many other dioxygenases this enzyme also requires ascorbate as a cofactor and α-ketoglutarate as cosubstrate (Fessler and Fessler, 1978). Recent work on prolyl hydroxylase isolated from suspension cultured *Vinca rosea* cells (Tanaka et al., 1981) has shown that this enzyme, like its counterpart in animal systems, requires as substrate at least five Pro residues in a sequence. This indicates that the enzyme recognizes a secondary conformation (polyproline II helix) and is probably the reason why plant prolyl hydroxylase is also capable of hydroxylating protocollagen (Sadava and Chrispeels, 1971b). The activity of the enzyme can be measured in several ways: with an exogenous substrate, for example, using synthetic (poly-L-proline—Tanaka et al., 1980) or natural (cell-wall-derived nonhydroxylated–deglycosylated extensin produced in the presence of α,α′-dipyridyl—Sadava and Chrispeels, 1971b) polypeptides, or with an endogenous substrate (membrane associated nonhydroylated extensin produced by α,α′-dipyridyl treatment—Kawasaki, 1980). The hydroxylation can be estimated either photometrically after hydrolysis (Tanaka et al., 1981) or radiochemically by the release of ^3H_2O using substrate derived from 3,4^{3}H proline (Sadava and Chrispeels, 1971b).

An early attempt at localizing prolyl hydroxylase placed it in the soluble supernatant fraction (Sadava and Chrispeels, 1971b) although, as Chrispeels (1976) later stated, "little case was taken to ensure the integrity of the cytoplasmic organelles during tissue homogenization." Subsequent communications have indeed shown the enzyme to be membrane-associated (Kawasaki, 1980; Tanaka et al., 1980). Surprisingly the GA, rather than the ER, appears to be the site for the hydroxylation of peptidyl proline. This has been demonstrated by Cohen et al. (1983) and Robinson et al. (1984) for ryegrass endosperm cells, sugar cane cells, and maize roots, respectively. These are all monocotyledenous types but there is no reason for thinking that dicotyledenous plants will be different. At the time of writing the localization has been determined on pooled gradient fractions and, although Mg^{2+}-shifting experiments have not yet been carried out, it is clear that the bulk of the activity is not associated with the ER. These results are in contradiction to those that show the presence of Hyp macromolecules in the ER, and the reason for this is, at the moment, unclear.

10.2.3. Site of Glycosylation

Since the majority of glycosylating reactions, both in animal and in plant (see Sections 10.4. and 10.5.) take place in the GA it might be expected that the glycosylation of Hyp polypeptides occurs there too. One must however be aware of the fact that the glycosylation of hydroxylysyl residues in collagen is a posttranslational event associated with ER rather than the GA (Fessler and Fessler, 1978).

In vivo labeling experiments with ^{14}C-arabinose by Kawasaki (1982a) have shown that radioactive HRGPs are associated with GA fractions from suspension-cultured tobacco cells. Unfortunately, because of the heavy ER contamination in these fractions (see Section 10.2.1.), the results are of an equivocal nature. Samson et al. (1983) have recently investigated the subcellular distribution of AGPs in etiolated bean hypocotyls by running isoelectric focussing gels of the various fractions from low and high Mg^{2+} sucrose gradients and staining the with the Yariv reagent (Figure 10.2). [This is an artificial lectin prepared by coupling diazotized 4-aminophenyl glycosides to phoroglucinol (Yariv et al., 1967). It apparently recognizes the sugar moieties of the AGPs (presumably arabinose) and can be used cytochemically as a stain for AGPs (Jermyn and Yeow, 1975).] In contrast to the PM and soluble overlay fractions neither ER nor GA fractions stained appreciably. The staining of the PM shifted in response to Mg^{2+} and also became more intense, indicating an increased binding of AGPs to the cell surface during homogenization. That no evidence could be obtained as to the intracellular site of glycosylation probably lies in the relative insensitivity of the Yariv staining method coupled with a relatively high turnover and secretion of AGPs in these organelles.

Attempts at localizing hydroxyproline-arabinosyl transferase activity have been undertaken but the results are somewhat contradictory. Using an assay identical to one employed in a study on the synthesis of neutral arabinans or arabinogalactans (Johnson and Chrispeels, 1973). Gardiner and Chrispeels (1975) have shown that arabinosyl transferase activity lies

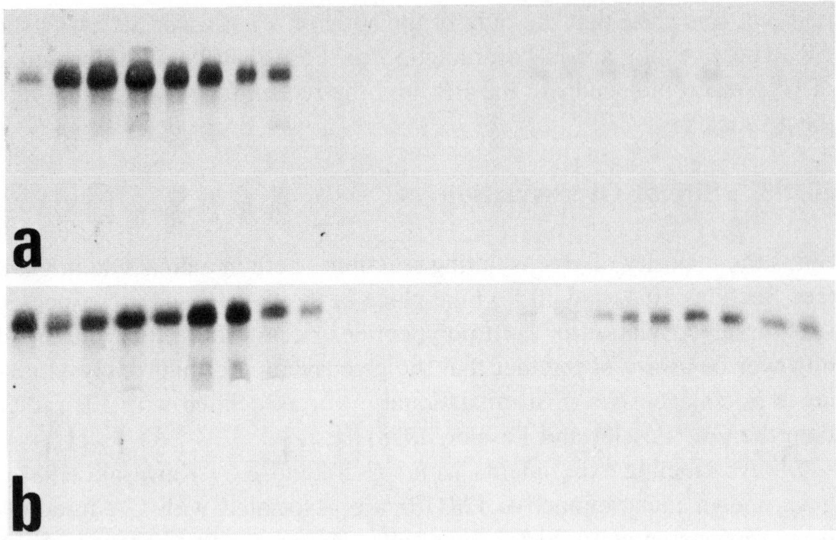

Figure 10.2. Isoelectric focusing of the matrix of various membrane fractions collected from isopycnic sucrose gradients of bean hypocotyl homogenates. In (a) the homogenates and gradients were prepared under low Mg^{2+} conditions; in (b) under high Mg^{2+} conditions. AGPs, which have a characteristically low isoelectric point, are recognized by staining with the Yariv reagent. GA-containing fractions (d = 1.12–1.15 g·cm^{-3}) do not appear to react positively. Published results of Samson et al. (1983).

primarily in GA fractions from carrot tissue. In a similar assay carried out with mixed membrane preparations from suspension-cultured sycamore cells Karr (1972) has shown that at least two thirds of the label enters products which are not HRGPs. Particularly in view of the high glycosylation of HRGPs much higher concentrations of trichloracetic acid are required for effective precipitation of the product from such *in vitro* nucleotide transfer reactions than were used in these earlier studies. The more recent studies of Owens and Northcote (1981) and Robinson et al. (1984) have taken this into consideration and have provided the corresponding product characterization. Incorporation is increased when the assay is carried out in the presence of detergent (Triton-X-100) and when an exogenous acceptor is provided. Poly-L-hydroxyproline, in contrast to deglycosylated extensin, does not fulfill this latter role (Robinson, unpublished observations) which is in agreement with the emerging significance of Ser-Hyp_4 sequences as the potential sequon for HRGP glycosylation (Lamport, 1980; Smith, 1981b).

Owens and Northcote (1981) have localized the majority of their arabinosyl transferase activity in the GA of suspension-cultured potato cells. Their use of step gradients, however, led to appreciable amounts of activity being present in fractions rich in ER and PM marker enzymes. In contrast, Robinson et al. (1984) have not been able to provide evidence for a clear allocation of the arabinosyl transferase activity to the GA. Instead, the peak of activity in low Mg^{2+} sucrose gradients lies more in the ER region than in the GA region. More recent experiments (Andreae and Robinson, unpublished results) have shown that this peak of arabinosyl transferase activity is indeed influenced by Mg^{2+} concentration and behaves in a manner similar to that of the ER marker enzyme CCR. This result would make the synthesis of HRGPs somewhat more similar to that of collagen as had previously been thought, but brings with it the problem of explaining the relatively larger amounts of peptidyl hydroxylase activity in GA rather than ER fractions.

With respect to galactosyl transferase activities Mascara and Fincher (1982) have characterized the *in vitro* products obtained with a mixed membrane preparation from suspension-cultured ryegrass cells. Not only did the label enter galactose-containing glycoproteins of the AGP type but it also went into high MW 1,6 galactans. Localization of this activity has not yet been accomplished.

10.3. ROOT CAP SLIME

The production of slime as a lubricant is a general property of angiosperm root tips (Rougier, 1981). Particularly in grasses such as maize (Morré et al., 1967) and wheat (Northcote and Pickett-Heaps, 1966), considerable quantities are exuded in droplet form at the root surface. In the well-studied maize root tip it has been possible to show that slime secretion follows a 3-hr cycle. According to Paull and Jones (1976b), this characteristic may well result from a rhythmic mitosis in the root cap initials and relates also to a corresponding regular cell sloughing at the root surface.

The secretory activity at the root tip is restricted to the so-called secretory or mantle cells which lie, several layers deep, outside of the core cells of the root cap (Juniper, 1972). Characteristic of the secretory cells are (a) numerous (about 200 per cell—Juniper and Clowes, 1965) hypertrophied dictyosomes (see Chapter 5 for details of their ultrastructure), (b) abundant rER, (c) numerous smaller vacuoles, rather than a large central vacuole, and (d) a withdrawn plasma membrane, always at the morphological "outer" surface of the cell (Juniper and Pask, 1973). Amyloplasts are also present and have been suggested as being the suppliers of the carbohydrate for slime production (Juniper and Roberts, 1966), although in my opinion this cannot be the sole source.

10.3.1. Slime Composition and Structure

Maize root cap slime has been reported on several occasions (Floyd and Ohlrogge, 1970; Jones and Morré, 1967; Wright, 1975) to contain up to 10% protein. Some of this certainly represents contamination from degenerating cells or bacteria. CsCl centrifugation experiments (Paull et al., 1975a) with purified slime have demonstrated the presence of only one component with a density of 1.63 $g \cdot cm^{-3}$, indicative of a polysaccharide. The molecule is large (MW 18×10^7—Floyd and Ohlrogge; 2×10^6—Paull et al., 1975a) and characterized by a high proportion of fucose (about 30% of the sugars), glucose and galactose (at least 20%), as well as uronic acids (Harris and Northcote, 1970; Paull et al., 1975). The molecule appears to bear a certain resemblance to the pectin fraction of the cell wall (Wright and Northcote, 1974) and is made up of a core of short $\beta,1 \rightarrow 4$ glucan chains to which fuco- and glucouronans are attached (Wright, 1975; Wright and Northcote, 1976). At least some of the fucose residues are terminally linked, as judged by their removal through mild acid hydrolysis and exofucosidase treatments, but there are different opinions as to how much (cf. James and Jones, 1979a, with Wright et al., 1976).

10.3.2. Slime Biosynthesis: *In Vivo* Labeling Experiments

A variety of cytochemical staining procedures (see Rougier, 1981, for a summary) indicate the presence of polysaccharides in both attached and released dictyosome vesicles, which, because of their hypertrophied nature, can easily be recognized. Autoradiographic investigations using ^{3}H-glucose (Northcote and Pickett-Heaps, 1966), ^{3}H-galactose (Dauwalder and Whaley, 1974), and ^{3}H-fucose (Kirby and Roberts, 1971; Paull and Jones, 1975b; Rougier, 1976) confirm this. Because of the poor resolution with autoradiography, a participation of the ER could not be ascertained in these studies.

The first fractionation experiments employed ^{14}C-glucose (Bowles and Northcote, 1972, 1974, 1976). Although the separation of organelles was carried out with step rather than continuous sucrose density gradients, leading to fractions of questionable purity (Robinson, 1977), the first labeled organelle appeared to be the ER. After about 30 min of incubation the membrane compartments reached saturation with respect to polymer-based radioactivity. During this period all membrane fractions gained in radioactivity with the GA, on a protein basis, incorporating relatively more than the other organelles. Radiochromatographic analysis of hydroylysates of both GA and ER fractions revealed the presence of arabinose, galactose, xylose, fucose, and galacturonic acids. Smaller amounts of glucose were also detected. The polymers synthesized in the two fractions have been fractionated and subjected to a size determination by gel-filtration. Water-soluble polymers are present in both (two-thirds of the GA; one-third of the ER) fractions; in contrast, radioactivity recoverable as insoluble polymers was much greater for the ER fractions but became soluble after proteolysis. Whereas the water-soluble polymers were mainly comprised of components of MW of at least 40 kd and contained fucose, the insoluble ones were usually less than 4 kd in size and had no fucose.

Although there are minor disagreements as to the exclusive localization of fucose-containing polymers (cf. Harris and Northcote, 1970, with Paull and Jones, 1975a), radioactive fucose is taken up, not metabolized, and finds its way primarily into the slime polysaccharide. Paull and Jones (1976a) have analyzed the distribution of radioactivity on linear isopycnic sucrose gradients from homogenates of 1-mm root tips incubated with ^{3}H-fucose for 2 hr. A broad band of radioactivity between 1.12 and 1.18 $g{\cdot}cm^{-3}$ was obtained. The peak at 1.15 $g{\cdot}cm^{-3}$ coincided with the presence of dictyosomes as judged by electron microscopy of the fraction. Unfortunately, due to the equally broad distribution of the so-called ER marker enzyme CDP–choline transferase (see Chapters 2 and 15) together with the lack of other marker enzymes suggest that the participation of the ER in slime biosynthesis cannot, with this study, be ruled out.

Using a mixed-membrane preparation from ^{3}H-fucose prelabeled maize root tips, Green and Northcote (1978) have been able to demonstrate that two glycoproteins are present in this fraction, as well as the slime polysaccharide. When the roots were incubated with radioactive leucine and fucose, both glycoproteins became labeled. Pulse-chase experiments indicated that the glycoproteins were precursors of the slime polysaccharide, indeed one of them could be β-eliminated to give a carbohydrate portion similar to that of the slime polysaccharide. The carbohydrate appears to be linked to the protein through an *O*-glycosidic bond between xylose and threonine. Green and Northcote have followed up their interesting observation that slime biosynthesis involves glycoprotein precursors with attempts at isolating dolichol phosphate–fucose derivatives from their membrane preparations (Green and Northcote, 1979a). They have succeeded in showing that fucosylated polar lipids were predominantly found in a 20,000–100,000*g* fraction which contained most of the ER and have interpreted their results in the same way as glycoprotein biosynthesis in animal cells (see Table 10.1).

10.3.3. Slime Biosynthesis: *In Vitro* Transferase Studies

The presence of fucosyl and glycosyl transferase activities in membranous organelles from maize root tips has been shown. James and Jones (1979a) have improved the fucosyl transferase assay which required GDP–fucose as donor by giving defucosylated slime as exogenous acceptor and by adding small amounts of detergent, two peaks of fucosyl transferase activity coinciding with ER and GA markers, respectively (see Fig. 10.2 and James and Jones, 1979b). Green and Northcote (1979b) have confirmed the existence of ER- and GA-associated fucosyl transferase activities and have shown in addition that the ER-based activity *in vivo* probably transfers fucose to dolichol phosphate, whereas the GA activity appears to be involved in the transfer to the polysaccharide portion of the glycoprotein slime precursor. Glycosylation and therefore presumably the synthesis of the $\beta,1 \rightarrow 4$ chains which are present in the slime (Wright and Northcote, 1976) occurs only in the GA, as signified by the absence of this transferase activity in the ER (Figure 10.3).

10.4. CELL WALL MATRIX POLYSACCHARIDES

The matrix (noncellulosic) polysaccharides of the plant cell wall are chemically heterogeneous but, based on solubility properties, are conveniently

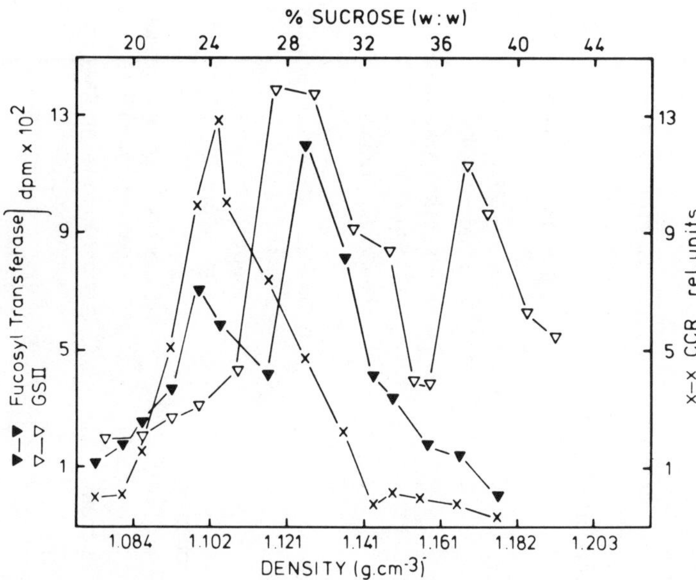

Figure 10.3. The distribution of fucosyl- and glucosyl (GS II) transferase activities in isopycnic sucrose gradients (low Mg^{2+} conditions) prepared from maize root tip homogenates. For comparison marker enzymes for the ER (CCR); GA (IDPase) are also given. The denser peak of GS II activity (between 35–40% sucrose) corresponds to the peak for K^+-ATPase activity (not shown here). Unpublished data of Robinson and Freundt.

divided into pectins and hemicellulose. The polymers which comprise these classes of polysaccharides have been extensively reviewed (see, e.g., Aspinall, 1970, 1980; Darvill et al., 1980; Kato, 1981; McNeil et al., 1979; Whistler and Richards, 1970) so that a cursory description will suffice here. This is done in table form (Table 10.2). It is immediately apparent that monocotyledonous and dicotyldenous angiosperms and gymnosperms are quite different in their polysaccharide composition; so much so that it is impossible to make a generalization. Compared to dicots, for example, the monocots have very small amounts of pectins and the structure of these is quite different. There are also significant qualitative and quantitative differences in the hemicellulose fraction of these groups. Furthermore, the type of wall polymer synthesized and secreted by a plant cell is dependent on its developmental stage: pectins being more characteristic of growing cells (primary walls) and hemicelluloses predomi-

TABLE 10.2. THE MATRIX POLYSACCHARIDES OF THE PLANT CELL WALL

Polymer and category	Sugars; linkages; conformation[a]; size[b]	Occurrence; location
A. Pectins		
Rhamnogalacturonan I (II)	-[α(1,4)DGalpA]$_n$-β(1,2)LRhap-; MW 5 × 10^{-6}; DP ~ 2000 (25–50)	Dicots, primary wall
Homogalacturonan	-[α(1,4)DGalpA]$_n$-; DP ⩾ 25	Dicots, primary wall
Araban	-[α(1,5)LAraf]$_n$-main chain -α(1,3)LAraf-side terminal monomers; DP 34–100	Dicots, primary wall
Galactan	-[β(1,4)DGalp]$_n$-; DP 33–50 -[β(1,6)DGalp-β(1,4)DGalp]$_n$-	Gymnosperms, secondary wall Dicots, primary and secondary walls
Arabinogalactan	-[β(1,4)DGalp]$_n$-main chain -[α(1,5)LAraf]$_2$-side terminal dimers	Dicots, primary/secondary (?) walls
Glucogalacturonan	-GalpA; Glc	Monocots, primary wall
B. Hemicelluloses		
Xyloglucan	-[β(1,4)DGlcp]$_n$-main chain; MW 7600; DP > 50 -α(1,6)DXylp-side terminal monomers	Dicots ⩾ Monocots, primary wall
Xylan	-[β(1,4)DXylp]$_n$-main chain -α(1,3)LAraf-side terminal monomers -α(1,2)D4-MeGlcpA-side terminal monomers	Monocots ⩾ Dicots, secondary > primary walls
Mixed glucan	-[β(1,4)DGlcp]$_{2or3}$-[β(1,3)DGlcp]$_1$-; MW 10^5–10^7	Monocots ⩾ Dicots, primary > secondary walls
Galacto(gluco)mannan	-[β(1,4)DManp]$_{3orn}$-β(1,4)DGlcp-main chain -α(1,6)DGalp-side terminal monomers	Gymnosperms, primary and secondary walls Reserve carbohydrates in legumes

[a] β—configuration; D,L—anomer; p,f—pyranose, furanose; Ara—arabinose; Gal—galactose; GalA—galacturonic acid; Glc—glucose; GlcA—glucuronic acid; Man—mannose; Me—methyl; Rha—rhamnose; Xyl—xylose.

[b] MW—molecular weight; DP—degree of polymerization.

nantly deposited by cells no longer growing (secondary walls). Finally, it must be stressed that the matrix polysaccharides of plant cells, in contrast to the connective tissue polysaccharides of animal cells and many capsular polysaccharides of bacteria, rarely contain long stretches of repeating monomers or regularly placed side chains and reveal a molecular weight polydispersity which is indicative of enzyme- rather than template-directed polymerization (see Fincher and Stone, 1981, for a discussion).

Different polymers usually mean different enzymes so that a comparison between the organelles responsible for their synthesis must be undertaken with care. In addition, since there have as yet been no studies documenting the synthesis and intracellular transport of a single wall polysaccharide the information available remains empirical in nature. Nevertheless, this is sufficient to determine which organelles are involved.

10.4.1. Matrix Polysaccharide Biosynthesis: *In Vivo* Studies

As in the case of root cap slime (see Section 10.3.2) a number of cytochemical investigations have established the presence of carbohydrate in dictyosomes (e.g., Amelunxen et al., 1976; Pickett-Heaps, 1968; Roland and Sandoz, 1969; Ryser, 1979; Thiéry, 1967) and in a variety of smooth-surfaced, presumably secretory vesicles (e.g., Albersheim et al., 1960; Conrad et al., 1982; Dashek and Rosen, 1966; Van der Woude et al., 1971). Similarly, pulse-chase autoradiography has generally confirmed the participation of the GA and GA-derived vesicles in matrix polysaccharide production (e.g., Coulomb and Coulon, 1971; Fowke and Pickett-Heaps, 1972; Pickett-Heaps, 1966). With the exception of the curious case of galactomannan synthesis in the developing endosperm of Fenugreek seeds, which occurs in dilated ER cisternae, there is no clear evidence from cytological studies for participation of the ER in this process (see O'Brien, 1972, for a review of the older literature of this type).

Fractions of varying degrees of purity (see Robinson, 1977, for a discussion) from a number of different pulse-chased objects have also been analyzed (Harris and Northcote, 1971; Jilka et al., 1972; Mertz and Nordin, 1971; Ray et al., 1976; Robinson et al., 1976). As Fincher and Stone (1981) state, "in general, the monosaccharide composition of dictyosome material is qualitatively similar to pectic and other noncellulosic polysaccharides of the wall, that is xylose, galactose, arabinose, fucose, glu-

TABLE 10.3. REVERSIBILITY OF CYANIDE INHIBITION IN RELATION TO TRANSFER BETWEEN ENDOMEMBRANES AND CELL WALL FRACTIONS (FROM ROBINSON AND RAY, 1977)

Fraction	Radioactivity (dpm) in contained polymers after KCN chase[a]	Radioactivity (dpm) in contained polymers after KCN removal[b]	Change in radioactivity as a result of KCN removal
Endomembranes	36,000	20,777	−15,223
Cell wall			
Matrix polysaccharides	4,644	21,000	+16,356
Cellulose	8,655	10,155	+ 1,500

[a] Pea stem segments were pretreated for 2 hr with 50 m*M* 2-deoxyglucose, incubated for 5 min in ^{14}C-glucose and then chased for 5 min with 50 m*M* glucose containing 1 m*M* KCN.
[b] Same as *a* but with an extra chase-out period of 15 min without KCN.

cose, and uronic acids are found." The majority of the radiolabeled material associated with such fractions is hot-water soluble, of high MW, and not degradable through proteolysis (Jilka et al., 1972; Mertz and Nordin, 1971; Robinson, 1977). This polymeric material gives the same oligosaccharide products upon treatment with endoglucanases as the matrix fractions of the cell wall do (Ray et al., 1976; Robinson, 1977).

A clear relationship between glucose-derived radiolabel in the endomembrane system and the matrix polysaccharides in the cell wall has been provided by Robinson and Ray (1977). By pretreating pea stem segments with 2-deoxyglucose, which reduces the pool size of nucleotide sugar precursors available for polysaccharide synthesis (Datema et al., 1983), and by including KCN in the chase-out medium, it was possible to "load" the endomembranes with radioactive polysaccharides destined for export. After release of the energy blockage on secretion, the labeled polysaccharides were "unloaded" into the cell wall in almost equivalent amounts (see Table 10.3).

In contrast to the situation with root caps of maize (see Section 10.3), fractionation of *in vivo* labeled elongating plant tissue has failed to implicate the participation of the ER in matrix polysaccharide biosynthesis. Careful experiments carried out with pea stem segments have shown that the dictyosomes are the first labeled membrane fraction when incubations are carried out with ^{14}C-glucose. The label contained in the ER fraction actually resides in secretory vesicles that have similar sedimentation rates to ER vesicles in rate zonal sucrose gradients (Ray et al., 1976; Robinson et al., 1976; Robinson, 1980).

10.4.2. Matrix Polysaccharide Biosynthesis: *In Vitro* Transferase Studies

A number of glycosyl transferases have been shown to be present in the endomembranes of plant cells involved in synthesizing matrix polysaccharides. Mixed-membrane preparations capable of transferring arabinose (Odzuk and Kauss, 1972), galactose (Panayotatos and Villemez, 1973), galacturonic acid (Villemez et al., 1965), glucose (Clark and Villemez, 1972), mannose (Villemez, 1971), and xylose (Bailey and Hassid, 1966) from nucleotide sugars to polysaccharides have been recorded (see Table 1, of Fincher and Stone, 1981). Some of these activities (e.g., arabinosyl—Gardiner and Chrispeels, 1975; galactosyl, xylosyl—Ray, 1980; and glycosyl—Ray et al., 1969; Robinson and Glas, 1983) transferases have been shown to reside in the GA.

In addition to the transfer of such sugars to polysaccharides, there are a number of communications documenting their transfer to lipidlike compounds (see Elbein, 1979). Thus, the glycosylation of sterols tends to be a feature of the PM (Chadwick and Northcote, 1980; Hartmann-Bouillon et al., 1979) or GA (Lercher and Wojciechowski, 1976; Bowles et al., 1977) fractions. In contrast, dolichol (polyprenol) phosphate glycosyl transferases appear to be localized either in the GA (Brett and Northcote, 1975; Dürr et al., 1979) or ER (Hopp et al., 1979; Dürr et al., 1979; Lehle et al., 1978). Although in analogy to the results obtained with root cap slime one might expect that glycoprotein intermediates are also involved (with sugar donated from lipid carriers), there is as yet no definite proof of this for cell wall matrix polysaccharides. Nor can it be assumed that the synthesis of all matrix polysaccharides begins in the ER with core glycosylation. There is at the moment simply no evidence for this.

10.5. SECRETION BY-PASSING THE GA?

In Sections 10.1–10.4 various examples of secretion which involve the GA either with or without a contribution from the ER were considered.

To make this survey complete, we will not consider some cases where ER-derived secretory products apparently reach the PM without proceeding through the GA.

10.5.1. Glandular Exudates

Glandular exudates include the products of secretory trichomes, ovarial and stigmatic papillae, nectaries, and the trapping and digestive juices from the glands of insectivorous plants. High-molecular-weight polymers are released in all cases accompanied, especially in the case of nectaries, by varying amounts of monosaccharides. With few exceptions nearly all of the work in this field is purely ultrastructural in nature. There are no publications involving fractionation studies witnessing the fact that the collection of material for this type of work presents almost unsurmountable problems.

Although many glands produce lipidlike or oillike secretes, and the ER (mostly smooth) is extremely well developed in the cells involved in their formation (Schnepf, 1969a–d), their way to the cell exterior is either unclear or is achieved through cell lysis ("holocrine secretion"). For our discussion we will therefore exclude these types, concentrating instead on polysaccharide–protein secretion.

In many plant glandular cells both ER and GA are present in great amounts, and from the structure of the GA (hypertrophied vesicles) it is almost impossible to deny a role for the latter organelle in the secretion of at least some molecules. This is certainly the case in many nectaries (e.g., Benner and Schnepf, 1975; Figier, 1971; Rachmilevitz and Fahn, 1975; Schnepf, 1977). In some other nectarial types a purely eccrine secretion process appears to take place on the basis of the relative lack of endomembranes present (Eriksson, 1977; Meybert and Kristen, 1981). However, some authors have claimed that nectar is synthesized exclusively in the ER and reaches the cell exterior by fusion of ER vesicles with the PM (Bellini-Depoux and Clair-Maczulajtys, 1975; Rachmilevitz and Fahn, 1973).

The participation of both ER and GA has been involved in a number of gland cells whose secrete, unlike that of nectaries, is predominantly polymeric. This is the case for example in the secretory trichomes of *Psychotria bacteriophila* (Horner and Lersten, 1968) and in the ovary glands of *Aptenia cordifolia* and *Platythyra haeckeliana* (Kristen, 1976,

1977). Sometimes the exudate becomes stored in vacuoles before being released through exocytosis (Figier, 1968, 1969) or through cell lysis (Kristen et al., 1980). The secreted substances, in those cases where it has been examined, contain both polysaccharide and protein with the former predominating (e.g., Kristen et al., 1980).

Claims for glandular exudates originating in the ER but not being exported via or modified by the GA are to be found in the literature. Thus, in the stigmatic papillae of *Aptenia cordifolia* (Kristen et al., 1980), *Crocus* (Heslop-Harrison and Heslop-Harrison, 1975), *Forsythia* (Dumas, 1973-74), and *Petunia hybrida* (Kroh, 1967) there are few dictyosomes. Here the ER appears to pinch off vesicles, and similar vesicles are also occasionally seen fusing with the PM. This is also said to be the case in the leaf glands of the aquatic *Nomaphila stricta* (Kristen, 1975) and *Iosetes lucustris* (Kristen et al., 1982).

Large amounts of secreted enzymic protein are characteristic of the digestive glands of carnivorous plants (Lloyd, 1942). In nearly all of the cases studied (*Dionaea*—Robins and Juniper, 1980a; Scala et al., 1968; Schwab et al., 1969; *Drosera*—Ragetli et al., 1972; *Drosophyllum*—Schnepf, 1961, 1963, 1964; and *Pinguicula*—Vögel, 1960) the digestive enzymes are stored in vesicles, vacuoles, or vacuolar aggregates and sometimes in the periplasmic space between adjacent gland cells (Heslop-Harrison and Knox, 1971). Because of their content the vacuoles are intensely osmiophilic. When the gland cells are triggered, the vacuoles discharge their content through the cell walls, which tend to be very labyrinthine (see also Chapter 7), and the cells begin to synthesize the digestive enzyme once again. The cyclic nature of production, storage, and release make these cells analogous to those regulated glands of animals (e.g., pancreas exocrine gland), a feature which has not escaped the attention of those who have worked in this area (e.g., Schwab et al., 1969).

Although both GA and ER are present in these cells, there are doubts as to the role for the former in the secretion of digestive enzymes. According to Schnepf (1974), "the GA shows such a diversity of structure that it has, as yet, been impossible to draw reasonable conclusions about its role." As a result of their investigations on the Venus flytrap, *Dionaea*, Robins and Juniper (1980c) state, "Dictyosomes do not appear to play any role in the secretion of hydrolase activity in *Dionaea*. Not only is there apparently a decrease in the total number of dictyosomes but no significant stimulus-induced change occurs in either structure of the number of associated vesicles. Dictyosomal vesicles are rarely seen fusing to the plasmalemma and where this does occur it does not appear to be involved with the secretion of hydrolases." Robins and Juniper have based this statement on autoradiographic observations and have been able to show that the synthesis of new enzyme protein already occurs during the secretory phase of previously synthesized and stored enzyme (Robins and Juniper, 1980b). They suggest that digestive enzyme protein synthesized on the rER is transferred to ribosome-free regions of ER which by fusing with one another give rise to small vacuoles. A vesiculation of the ER may also be involved. In addition, the ER membranes are often seen to fuse directly with the PM so that the periplasmic space seems to act as a storage site for the digestive enzymes as well. Robins and Juniper (1980c) draw attention to similar ER–PM fusion profiles in the secretory trichomes of *Pharbitis nil* which are also in contact with large storage vacuoles, although the authors of this cited work (Unzelman and Healey, 1974) do not themselves entirely rule out the participation of the GA in secretion in this case. According to Robins and Juniper (1980a), the GA in *Dionaea* gland cells is involved in the production of cell wall material and in the supply of vacuolar membrane rather than in the modification of secretory enzyme protein.

10.5.2. α-Amylase Secretion

The endosperm in cereals is rich in starch whose breakdown is effected by amylases (Dunn, 1964). These amylases, especially α-amylase, are synthesized by the cells of the aleurone layer, which lie immediately underneath the testa, as well as the scutellum and are released into the endosperm. By excising the embryo and testing the resultant "half-seeds" after water inbibition with various substances, it was established almost a quarter of a century ago (Paleg, 1960; Yomo, 1960) that the signal the growing embryo sends to the aleurone layers to initiate the production of hydrolytic enzymes is gibberellic acid.

Intact barley seeds start the production of α-amylase after 2 days of inbibition but stop after a further 2 days (Jones and Armstrong, 1971). Although the aleurone layers when dissected from the grain are still capable of responding to exogenous gibberellin at this stage, the addition of gibberellin to intact seeds does not result in more or prolonged α-

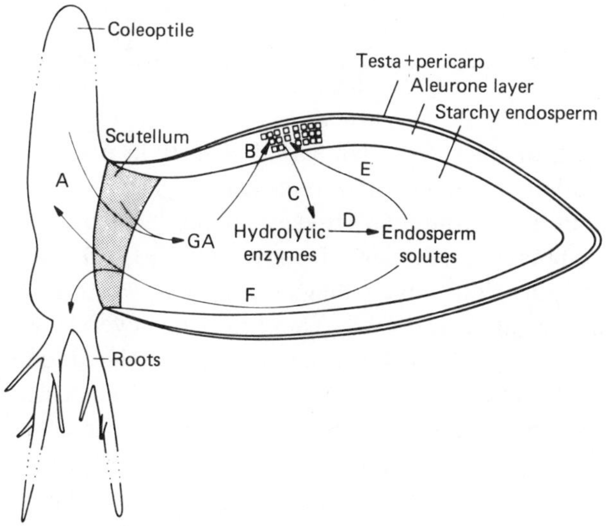

Figure 10.4. Diagrammatic representation of the events surrounding α-amylase induction and release in germinating barley seeds. GA—gibberellic acid; B—aleurone cells; C—release of hydrolytic enzymes; D—production of sugar, etc.; E—osmotic suppression of α-amylase secretion; F—transport of sugars, etc., into the growing embryo (A). Published drawing of Jones and Armstrong (1971).

amylase synthesis. As demonstrated by Jones (1969c) and Jones and Armstrong (1971), the cells of the aleurone layer respond negatively to increased external osmotic pressures which also occur naturally. Thus, in the words of Armstrong and Jones (1973), "although gibberellins act as the initiators of hydrolase synthesis, enzyme production is also regulated by the osmotic pressure exerted by the end products of enzyme action." The cogent anatomical and physiological features of these events are depicted in Figure 10.4.

Although other carbohydrases (Bennett and Chrispeels, 1972) and proteolytic enzymes (Jacobsen and Varner, 1967) are also released by the aleurone layers, it is the synthesis and secretion of α-amylase to which most attention has been devoted and reflects the fact that it is indeed the principal protein whose synthesis is induced as a result of gibberellin application (Mozer, 1980a). Gibberellin-induced α-amylase production occurs through a *de novo* enzyme protein synthesis as judged by density-labeling experiments (Filner and Varner, 1967; Hardie, 1975) and is preceeded by the preferential coding of increased amounts of α-amylase mRNA (Bernal-Lugo et al., 1981; Higgins et al., 1976; Mozer, 1980b). This is supported by experiments with inhibitors of transcription and translation (Chrispeels and Varner, 1967; Ho and Varner, 1974) and is consistent with the fact that a lag period of 8–10 hr exists between hormone application and detection of released enzyme.

At least four isoenzymes of α-amylase are produced by barley aleurone layers after gibberellin treatment (Jacobsen et al., 1970). They comigrate electrophoretically, indicating a similar molecular weight. This has been estimated in the range 42–44 kd. There may be several subunits in the active molecule each of which is irregularly glycosylated (on the average one residue each of glucose, mannose, and *N*-acetylglucosamine per polypeptide–Rodaway, 1978). *In vitro* translation experiments with barley (Mozer, 1980a) and wheat (Okita et al., 1979) α-amylase mRNA have resulted in the the synthesis of a species 1–2 kd larger than the secreted native form. In accordance with Blobel's signal hypothesis (see Section 10.1) when the translations are carried out in the presence of microsomal vesicles, the smaller native form is obtained (Boston et al., 1982). The assumed cleavage of the leader sequence has been achieved only with the inclusion of foreign (canine) microsomal vesicles; experiments with vesicles prepared from aleurone tissue have been as yet unsuccessful (Jones and Jacobsen, 1982; Okita et al., 1979). These results suggest that, in common with secreted (glyco)proteins in animal cells (Section 10.1) and with storage (glyco)proteins in developing seeds (Sections 11.2 and 11.3), the ER is also involved in the sequestration of α-amylase.

A role for the ER in the synthesis and secretion of α-amylase is implicated in a number of ultrastructural studies and is borne out by fractionation studies. Jones (1969a–c) has shown that a number of cytological changes occur during both the lag phase following gibberellin application and the subsequent secretory phase in barley aleurone cells (see Figures 10.5 and 10.6). During the 8- to 10-hr-long lag phase an increase in cell volume together with a gradual swelling of the aleurone grains (protein bodies, see also Section 11.2) is started which ends ultimately (after about 2 days) with the conversion of the latter into the central vacuole through fusion with one another (Jones and Price, 1970). Significant changes in the amount and morphology of the ER have also been registered. At the beginning of the lag phase the ER is present in small amounts and usually as single cisternae. After 8 hr of gibberellin treatment there is clearly

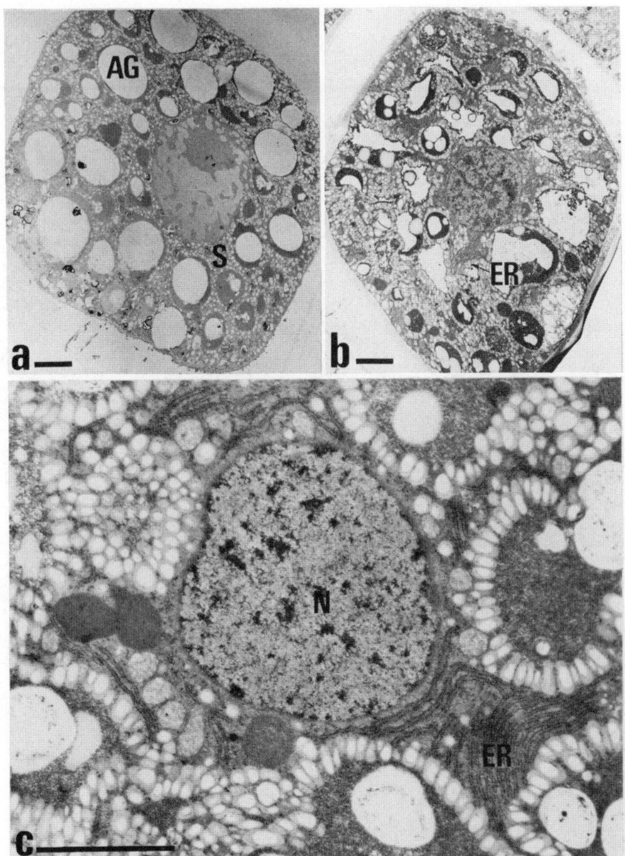

Figure 10.5. (a) Aleurone cell from a 3-day water-inhibited barley grain. The cytoplasm is packed with numerous protein bodies (aleurone grains, AG) and lipid bodies (spherosomes, S). (b) Aleurone cell as in a but after a subsequent 9–10-hr treatment with 0.5 $\mu g \cdot ml^{-1}$ gibberellic acid. The globoid contents of the aleurone grains is less evident but the presence of stacked ER is clear. (c) As for b but at higher magnification. Bar = 3 μm. Published micrographs of Jones (1980, 1969b).

much more and it is stacked. During the first hours of the secretory phase (i.e., about 10–14 hr after gibberellin addition), the ER dilates (see Figure 10.6) and subsequently (2–6 hr later) fragments into small, nondilated pieces.

The visual increase in ER in electron micrographs has been substantiated by measurements of the ER marker enzyme CCR and in the amount

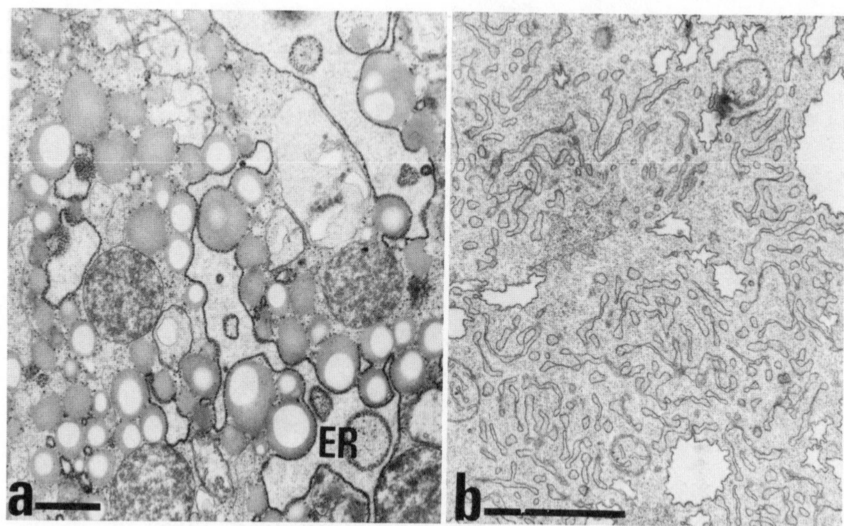

Figure 10.6. Changes in rER morphology in aleurone cells during the secretory phase in germinating barley grain. (a) 12–14 hr after gibberellic acid application. (b) 16 hr after gibberellic acid application. The ER has regained a normal luminal width but is now beginning to fragment. Bar = 1 μm. Published micrographs of Jones (1969c).

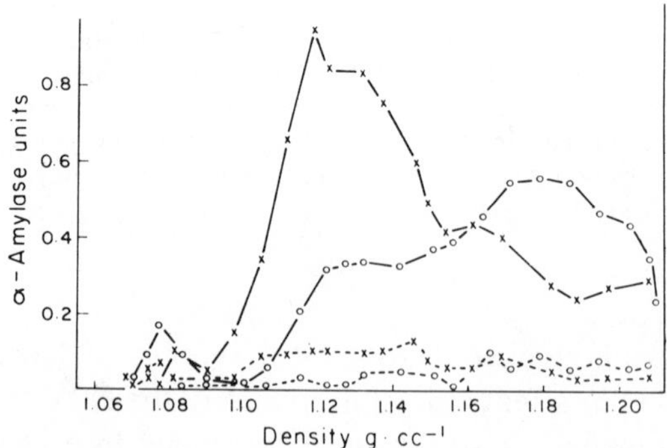

Figure 10.7. The distribution of membrane-bound (detergent-released) α-amylase activity in isopycnic sucrose density gradients prepared from barley aleurone homogenates under low (×) and high (○) Mg^{2+} conditions. Germination in the presence of gibberellic acid is indicated (——) and in water alone (---). Published results of Jones (1980).

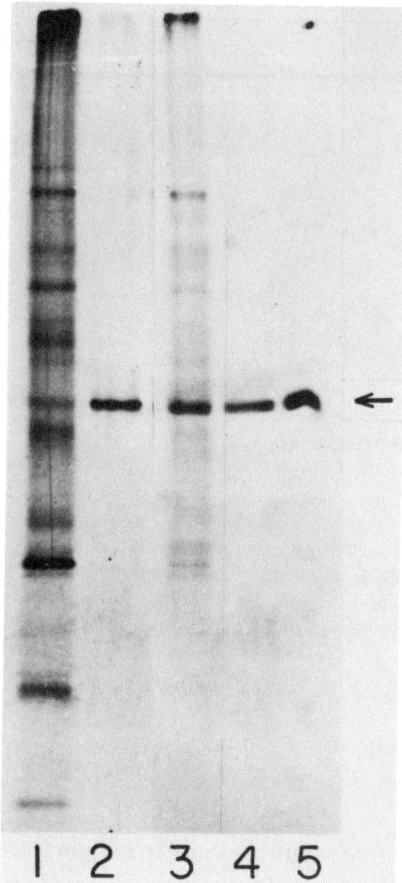

Figure 10.8. Identification of α-amylase in barley aleurone ER fractions by SDS–PAGE fluorography. Lanes 1 and 2 are from H_2O controls; lanes 3 and 4 from gibberellic-acid-treated grains. Lanes 1 and 3 show total radioactivity in fractions; lanes 2 and 4 reveal that which has bound to an α-amylase immunoaffinity column. Lane 5 is marker α-amylase. Published photograph of Jones and Jacobsen (1982).

of protein in this gradient fraction over similar time periods (Jones, 1980). α-amylase activity has also been detected in fractions rich in ER (Firn, 1975), and the enhancement of its measurement by detergent treatment (Gibson and Paleg, 1976; Jones and Jacobsen, 1982; Locy and Kende, 1978) further suggests its luminal location. Furthermore, a Mg^{2+}-shifting effect (Figure 10.7) typical for ER has also been demonstrated for this activity (Jones, 1980; Locy and Kende, 1978). Incubation *in vivo* with radioactive amino acids followed by autoradiography (Chen and Jones, 1974b) or fractionation (Locy and Kende, 1978), as well as the identification of membrane-associated activity as authentic α-amylase by im-

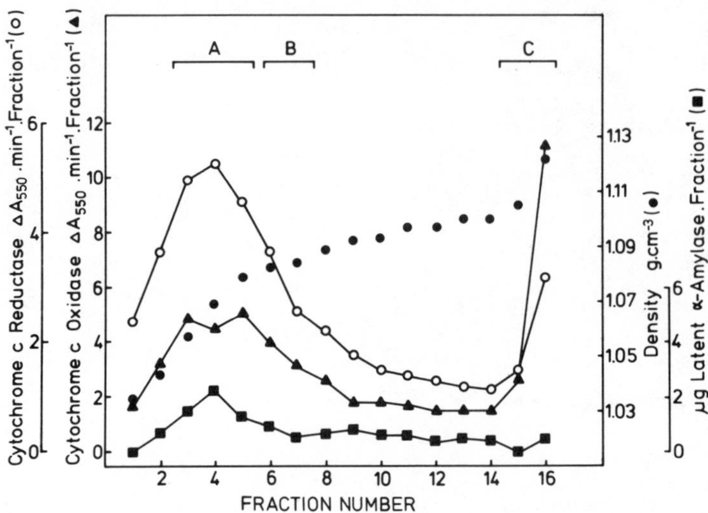

Figure 10.9. Marker enzyme distribution for organelles from gibberellic-acid-treated barley aleurone cells separated on rate zonal Metrizamide gradients. The ER marker, CCR, is enriched in two zones, A (mainly vesicular ER) and C (mainly cisternal ER). Zone B represents a fraction held to be rich in PM vesicles. Results of Gregerson and Taiz (1984a).

munoabsorbent chromatography (Figure 10.8); Jones and Jacobsen, 1982), all point to the ER as the site of α-amylase synthesis in barley aleurone layers.

Although the release of α-amylase from aleurone cells is temperature (Fadeel et al., 1980) and energy (Varner and Mense, 1972) dependent, both of which are characteristics typical of granulocrine secretion, a case has been made (Chen and Jones, 1974a; Jones, 1972) for a soluble (i.e., eccrine) mode for the passage of the enzyme through the PM. Clearly, in the light of recent evidence that secretion-destined α-amylase is sequestered within the lumen of the ER, this hypothesis is no longer tenable. Still uncertain, however, is the route from the ER to the PM and whether or not the GA is involved. Dictyosomes are present in barley aleurone cells, particularly during the phase when the ER fragments and secretion is most intense (Jones, 1969c). Unfortunately, attempts at measuring the GA marker enzymes, IDPase, and GS I (see Section 5.5) in sucrose density gradients of aleurone homogenates have as yet been unsuccessful (Gregerson and Taiz, in press; Locy and Kende, 1978). Because of this

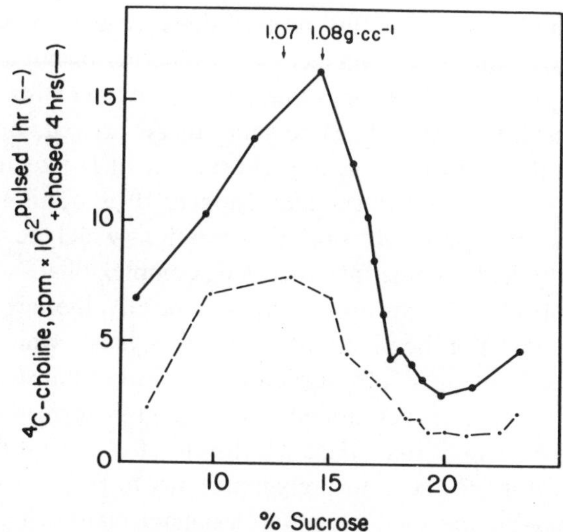

Figure 10.10. Pulse-chase labeling of organelles from 14–16-hr gibberellic-acid-treated barley aleurone cells isolated on rate zonal Metrizamide gradients (compare Figure 10.9). Radioactivity from ^{14}C-choline which is incorporated into membrane phospholipids is first seen associated with fractions rich in ER vesicles. After the chase-out radioactivity is found in PM-rich fraction. Published results of Gregerson and Taiz (1983b).

these authors have refrained from giving a role to the GA in α-amylase secretion.

Gregerson and Taiz (1984a) have been able to separate two classes of CCR-bearing membranes from barley aleurone homogenates by rate-zonal centrifugation in metrizamide gradients (see Figure 10.9). One class travels faster and pellets together with mitochondria and corresponds to cisternal ER. The other (zone A in Figure 10.9) bears α-amylase activity and is composed of small vesicles. By pulse-chasing with ^{14}C-choline these authors have also been able to demonstrate a shift in the distribution of radioactivity from the light ER fraction to a fraction richer in putative PM vesicles (Figure 10.10), which they interpret as representing secretion-associated membrane flow from ER to PM (Gregerson and Taiz, 1984b).

In contrast to the supposition of Akazawa and Miyata (1982), this author believes that the evidence available at the moment points to a secretory route for α-amylase which by-passes the GA, although until the GA from aleurone cells is identified in gradients, isolated, and shown not to contain α-amylase, one cannot be certain of this fact. The very low amount of glycosylation of this protein does at least make a journey through the GA superficially unnecessary. Whether or not a complicated processing of sugar residues takes place as we have seen for many animal glycoproteins (see Section 10.1) remains to be proved, but it may be inferred from the fact that tunicamycin (section 10.1) effectively inhibits α-amylase secretion (Schwaiger and Tanner, 1979). This inhibition, of course, refers only to core glycosylation reactions which occur in the ER. Whereas the typical scheme for processing complex oligosaccharide proteins in animal cells (Section 10.1.) involves the GA, there is considerable evidence in support of the reactions involved in yeasts which do not possess a GA (see Lehle, 1981, for a review). Moreover the successful demonstration of α-amylase secretion in yeast cells transformed with a cDNA clone coding for wheat α-amylase (Rothstein et al., 1984) indicates that it is not obligatory for secretory glycoproteins to pass through the GA. α-amylase may be one such exception in higher plant cells.

REFERENCES

Akazawa, T. and Miyata, S. (1982). *Essays in Biochem.* **18,** 40–78.

Albersheim, P., Mühlethaler, K., and Frey-Wyssling, A. (1960). *J. Cell Biol.* **8,** 501–506.

Allen, A. K. and Neuberger, A. (1973). *Biochem. J.* **135,** 307–314.

Amelunxen, F., Thio Tiang Nio, E., and Spiess, E. (1976). *Cytobiologie* **13,** 233–250.

Amsterdam, A., Ohad, I., and Schramm, M. (1969). *J. Cell Biol.* **41,** 753–773.

Anderson, R. L., Clarke, A. E., Jermyn, M. A., Knox, R. B., and Stone, B. A. (1977). *Austr. J. Plant Physiol.* **4,** 143–158.

Amstrong, J. E. and Jones, R. L. (1973) *J. Cell Biol.* **59,** 444–455.

Aspinall, G. O. (1979). In *The Carbohydrates*, (W. Pigman and D. Horton, eds.), Vol. IIB, pp. 515–536. Academic Press, New York.

Aspinall, G. O. (1980). *Biochemistry of Plants* **3,** 473–500.

Bailey, R. W. and Hassid, W. Z. (1966). *P.N.A.S.* (*USA*) **56,** 1586–1593.

Bellini-Depoux, M. and Clair-Maczulajtys, D. (1975). *Rev. Gen. Bot.* **82,** 119–155.

Benner, U. and Schnepf, E. (1975). *Protoplasma* **85,** 337–349.

Bennett, P. A. and Chrispeels, M. J. (1972). *Plant Physiol.* **49,** 445–447.

Bergeron, J. J. M., Borts, D., and Cruz, J. (1978). *J. Cell Biol.* **76,** 87–97.

Bergeron, J. J. M., Rachubinski, R. A., Sikstrom, R. A., Posner, B. I., and Paiement, J. (1982). *J. Cell Biol.* **92,** 139–146.

Bernal-Lugo, I., Beachy, R. N., and Varner, J. E. (1981). *Biochem. Biophys. Res. Comm.* **102,** 617–623.

Beyer, T. A., Rearick, J. I., Paulson J. C., Prieels, J.-P., Sadler, J. E., and Hill, R. L. (1978). *J. Biol Chem.* **253,** 5617–5624.

Bielinska, M., Rogers, G., Rucinsky, T., and Boime, I. (1979). *Proc. Nat. Acad. Sci (USA)* **76,** 6152–6156.

Blobel, G. and Dobberstein, B. (1975a). *J. Cell Biol.* **67,** 835–851.

Blobel, G. and Dobberstein, B. (1975b). *J. Cell Biol.* **67,**852–862.

Blobel, G. and Sabatini, D. D. (1971). *Biomembranes* **2,** 193–195.

Blobel, G., Walter, P., Chang, C. N., Goldman, B. M., Frickson, A. H., and Lingappa, V. R. (1979). *Symp. Soc. Biol.* **33,** 9–36.

Bornstein, P. (1974). *Ann. Rev. Biochem.* **43,** 567–603.

Boston, R. S., Miller, T. J., Mertz, J. E., and Burgess, R. R. (1982). *Plant Physiol.* **69,** 150–154.

Bowles, D. (1982). *Encyclopedia of Plant Physiology*, Vol. 13A, pp. 584–600.

Bowles, D. J. and Northcote, D. H. (1972). *Biochem. J.* **130,** 1133–1145.

Bowles, D. J. and Northcote, D. H. (1974). *Biochem. J.* **142,** 139–144.

Bowles, D. J. and Northcote, D. H. (1976). *Planta* **128,** 101–106.

Bowles, D. J., Lehle, L., and Kauss, H. (1977). *Planta* **134,** 177–181.

Brett, C. T. and Northcote, D. H. (1975). *Biochem. J.* **148,** 107–117.

Bretz, R., Bretz, H. and Palade, G. E. (1980). *J. Cell Biol.* 87–101.

Brysk, M. M. and Chrispeels, M. J. (1972). *Biochim. Biophys. Acta* **257,** 421–432.

Castle, J. D. and Palade, G. E. (1978). *J. Cell Biol.* **76,** 323–340.

Chadwick, C. M. and Northcote, D. H. (1980). *Biochem. J.* **186,** 411–421.

Chan, S. J., Patzelt, C., Duguid, J. R., Quinn, P., Labrecque, A., Noyes, B., Keim, P., Henrikson, R. L., and Steiner, D. F. (1979). *Miami Winter Symp.* **16,** 361–378.

Chen, R. and Jones, R. L. (1974a). *Planta* **119,** 193–206.

Chen, R. and Jones R. L. (1974b). *Planta* **119,** 207–220.

Chrispeels, M. J. (1969). *Plant Physiol.* **44,** 1187–1193.

Chrispeels, M. J. (1976). *Ann. Rev. Plant Physiol.* **27,** 19–38.

Chrispeels, M. J. and Varner, J. E. (1967) *Plant Physiol.* **42,** 1008–1016.

Clark, A. F. and Villemez, C. L. (1972). *Plant Physiol.* **50,** 371–374.

Clarke, A. E., Anderson, R. L., and Stone, B. A. (1979). *Phytochemistry* **18,** 521–540.

Clarke, A. E., Knox, R. B., and Jermyn, M. A. (1975). *J. Cell Sci.* **19,** 157–161.

Cleland, R. (1968). *Plant Physiol.* **43,** 865–870.

Cohen, P. B., Schibeci, A., and Fincher, G. B. (1983). *Plant Physiol.* **72,** 754–758.

Conrad, P. A., Binari, L. L. W., and Racusen, R. H. (1982). *Protoplasma* **112,** 196–204.

Coulomb, P. and Coulon, J. (1971). *J. Microsc.* **10,** 203–214.

Darvill, A., McNeil, M., Albersheim, P., and Delmer, D. P. (1980). *Biochemistry of Plants* **1,** 91–162.

Dashek, W. V. (1970). *Plant Physiol.* **46,** 831–838.

Dashek, W. V. and Rosen W. G. (1966). *Protoplasma* **61,** 192–204.

Datema, R., Schwartz, R. T., Rivas, L. A., and Pont Lezica, R. (1983). *Plant Physiol.* **71,** 76–81.

Dauwalder, M. and Whaley, W. G. (1974). *J. Cell Sci.* **14,** 11–27.

Duksin, D. and Bornstein, P. (1977). *P.N.A.S. (USA)* **74,** 3433–3437.

Dumas, C. (1973–74). *Le Botaniste* **56,** 59–80.

Dunn, G. (1964). *Phytochem.* **13,** 1341–1346.

Dunphy, W. G., Fries, E., Urbani, L. J., and Rothman, J. E. (1981) *P.N.A.S.* (*USA*) **78,** 7453–7457.

Dürr, M., Bailey, D. S., and MacLachlan, G. (1979). *Eur. J. Biochem.* **97,** 445–453.

Ehrlich, H. P., Ross, R., and Bornstein, P. (1974). *J. Cell Biol.* **62,** 390–405.

Elbein, A. D. (1979). *Ann. Rev. Plant Physiol* **30,** 239–272.

Elbein, A. D. (1981). *Encyclopedia of Plant Physiology*, Vol. 13B, pp. 166–193.

Elting, J. J., Chen, W. W., and Lennarz, W. J. (1980). *J. Biol. Chem.* **255,** 2325–2331.

Engelman, D. M. and Steitz, T. A. (1981) *Cell* **23,** 411–422.

Eriksson, M. (1977). *J. Apic. Res.* **16,** 184–193.

Eschrich, W. (1980). *Ber. Dt. Bot. Ges.* **93,** 363–378.

Fadeel, A., Moll, B. A. and Jones, R. J. (1980). *Plant Physiol.* **66,** 466–470.

Fessler, J. H. and Fessler, L. I. (1978). *Ann. Rev. Biochem.* **47,** 129–162.

Figier, J. (1968). *Comp. Rendus, Acad. Sci. D* **267,** 491–494.

Figier, J. (1969). *Planta* **87,** 275–289.

Figier, J. (1971). *Planta* **98,** 31–49.

Filner, P. and Varner, J. E. (1967). *P.N.A.S.* (*USA*) **58,**, 1520–1526.

Fincher, G. B. and Stone, B. A. (1981). *Encycl. Plant. Physiol.* **133,** 68–132.

Firn, R. D. (1975). *Planta* **125,** 227–233.

Floyd, R. A. and Ohlrogge, A. J. (1970). *Plant Soil.* **33,** 331–343.

Fowke, L. C. and Pickett-Heaps, J. D. (1972). *Protoplasma* **74,** 19–32.

Franke, W. W., Morre, D. J., Deumling, B., Cheetham, R. D., Kartenbeck, J., Jarasch, E. D., and Zentgraf, H. W. (1971). *Z. Naturforsch.* **26B,** 1031–1039.

Gardiner, M. and Chrispeels, M. J. (1975). *Plant Physiol.* **55,** 536–541.

Gibson, R. A. and Paleg, L. G. (1976). *J. Cell Sci.* **22,** 413–426.

Gleeson, P. A. and Clarke, A. E. (1979). *Biochem. J.* **181,** 607–621.

Grant, M. E. and Jackson, D. S. (1976). *Essays in Biochem.* **12,** 77–113.

Green, J. R. and Northcote, D. H. (1978). *Biochem. J.* **170,** 599–608.

Green, J. R. and Northcote, D. H. (1979a). *Biochem. J.* **178,** 661–671.

Green, J. R. and Northcote, D. H. (1979b). *J. Cell. Sci.* **40,** 235–244.

Gregerson, E. L. and Taiz, L. (1984a). In preparation.

Gregerson, E. L. and Taiz, L. (1984b). In preparation.

Griffiths, G., Brandi, R., Burke, B. Louvard, D., and Warren G. (1982). *J. Cell Biol.* **15,** 781–792.

Grinna, L. S. and Robbins, P. W. (1979). *J. Biol. Chem.* **254,** 8814–8818.

Habener, J. F., Rosenblatt, M., Dee, P. C., and Potts, J. T. (1979). *J. Biol Chem.* **254,** 10596–10599.

Haddad, A., Smith, M. D., Herscovics, A., Stadler, N. J. and Leblond, C. P. (1971). *J. Cell Biol.* **49,** 856–882.

Halbhuber, K.-J., Christner, A., and Schirrmeister, W. (1972). *Acta Histochem.* **42,** 157–161.

Hardie, D. G. (1975). *Phytochem.* **14,** 1719–1722.

Harpaz. N. and Schacter, H. (1980). *J. Biol. Chem.* **255,** 4885–4893.

Harris, P. J. and Northcote, D. H. (1970). *Biochem. J.* **120,** 479–491.

Harris, P. J. and Northcote, D. H. (1971). *Biochim. Biophys. Acta* **237,** 56–64.

Hartmann-Bouillon, M. A., Benveniste, P., and Roland, J.-C. (1979). *Biol. Cellulaire* **35,** 183–194.

Harwood, R., Grant, M. E., and Jackson, D. S. (1974). *Biochem. J.* **144,** 123–130.

Heslop-Harrison, J. and Heslop-Harrison, Y. (1975). *Micron* **6,** 45–52.

Heslop-Harrison, Y. and Knox, R. B. (1971). *Planta* **96,** 183–211.

Hickman, S., Kulczycki, A., Lynch, R. G., and Kornfeld, S. (1977). *J. Biol. Chem.* **252,** 4402–4408.

Higgins, T. J. V., Zwar, J. A., and Jacobsen, J. V. (1976). *Nature* **260,** 166–169.

Ho, D. T. H. and Varner, J. E. (1974). *P.N.A.S.* (*USA*) **71,** 4793–4786.

Hopp, E., Romera, P., and Pont Lezica, R. (1979). *Plant & Cell Physiol.* **20,**1063–1069.

Hori, H. and Sato, S. (1977). *Phytochemistry* **16,** 1485–1487.

Horner, H. T. and Lersten, N. R. (1968). *Amer. J. Bot.* **55,** 1089–1099.

Hubbard, S. C. and Ivatt, R. J. (1981). *Ann. Rev. Biochem.* **50,** 555–584.

Jacobsen, J. V. and Varner, J. E. (1967). *Plant Physiol.* **42,** 1596–1600.

Jacobsen, J. V., Scandalios, J. G., and Varner, J. E. (1970).*Plant Physiol.* **45,** 367–371.

James, D. W. and Jones. R. L. (1979a). *Plant Physiol.* **64,** 909–913.

James, D. W. and Jones, R. L. (1979b). *Plant Physiol.* **64,** 914–918.

Jamieson, J. D. (1978). In *Transport of Macromolecules in Cellular Systems* (S. C. Silverstein, ed.), pp. 273–288. Abakon Verlagsgesellschaft, Berlin.

Jamieson, J. D. (1981). *Meth. Cell Biol.* **23,** 547–558.

Jamieson, J. D. and Palade, G. E. (1968). *J. Cell Biol.* **39,** 589–603.

Jamieson, J. D. and Palade, G. E. (1971). *J. Cell Biol.* **48,** 503–522.

Jamieson, J. D. and Palade, G. E. (1977). In *International Cell Biology* (B. R. Brinkley and K. R. Porter, eds.), pp. 308–317. Rockefeller University Press, New York.

Jermyn, M. A. and Yeow, Y. M. (1975). *Austr. J. Plant Physiol.* **2,** 501–531.

Jilka, R., Brmwn O., and Nordin, P. (1972). *Arch. Biochem* Biophys.* **152,** 702–711.

Johnson, K. D. and Chrispeels, M. J. (1973). *Planta* **111,** 353–364.

Jones, D. D. and Morré, D. J. (1967). *Z. Pflanzenphysi/l.* **56,** 166–169.

Jones, R. L. (1969a). *Planta* **85,** 359–375.

Jones, R. L. (1969b). *Planta* **87,** 119–133.

Jones, R. L. (1969c). *Planta* **88,** 73–86.

Jones, R. L. (1972). *Planta* **103,** 95–109.

Jones, R. L. (1980). *Planta* **150,** 70–81.

Jones, R. L. and Armstrong, J. E. (1971). *Plant Physiol.* **48,** 137–142.

Jones, R. L. and Jacobsen, J. V. (1982) *Planta* **156,** 421–432.

Jones, R. L. and Price, J. M. (1970). *Planta* **94,** 191–202.

Juniper, B. E. (1972). In *The Dynamics of Meristem Cell Populations* (M. W. Miller and C. C. Kuehnert, eds.), pp. 119–131. Plenum Press, New York.

Juniper, B. E. and Clowes, F. A. L. (1965). *Nature* **208,** 864–865.

Juniper, B. E. and Pask, G. (1973). *Planta* **109,** 225–231.

Juniper, B. E. and Roberts, R. M. (1966). *J. Roy. Micr. Soc.* **85,** 63–72.

Karr, A. L. (1972). *Plant Physiol.* **50,** 275–282.

Kato, K. (1981). *Encyclopedia of Plant Physiology*, Vol 13B, pp. 29–46.

Kawasaki, S. (1980). Ph.D. thesis. University of Tokyo.

Kawasaki, S. (1981). *Plant & Cell Physiol.* **22,** 431–442.

Kawasaki, S. (1982a). *Planta & Cell Physiol.* **23,** 1443–1452.

Kawasaki, S. (1982b). Proc. 5th Int. Congr. Plant Tissue and Cell Culture, 63–64.

Keller, R. K. and Swank, G. D. (1978). *Biochem. Biophys. Res. Comm.* **85,** 762–768.

Kirby, E. G. and Roberts, R. M. (1971). *Planta* **99,** 211–221.

Kornfeld, R. and Kornfeld, S. (1976). *Ann. Rev. Biochem.* **45,** 217–237.

Kreibich, G., Ulrich, B., and Sabatini, D. D. (1978). *J. Cell Biol.* **77,** 464–487.

Kreil, G. (1981). *Ann. Rev. Biochem.* **50,** 317–348.

Kreil, G., Mollay, G., Kaschnitz, R., Haiml, L. and Vilas, U. (1980). *Ann. NY Acad. Sci.* **343,** 338–346.

Kristen, U. (1975). *Cytobiologie* **11,** 438–447.

Kristen, U. (1976). *Protoplasma* **89,** 221–233.

Kristen, U. (1977). *Protoplasma* **92,** 243–251.

Kristen, U., Biedermann, M., and Liebezeit, G. (1980). *Z. Pflanzenphysiol.* **96,** 239–249.

Kristen, U., Liebezeit, G., and Biedermann, M. (1982). *Ann. Bot.* **49,** 569–584.

Kroh, M. (1967). *Planta* **77,** 250–260.

Kronquist, K. E. and Lennarz, W. J. (1978). *J. Supramol. Struct.* **8,** 51–65.

Lamport, D. T. A. (1980). *The Biochemistry of Plants* **3,** 501–541.

Lamport, D. T. A. and Catt, J. W. (1981). *Encyclopedia of Plant Physiology* Vol. 13B, 133–165.

Leach, J. E., Cantrell, M. A., and Sequeira, L. (1982). *Plant Physiology*, **70,** 1353–1358.

Lercher, M. and Wojciechowski, Z. A. (1976). *Plant Sci. Lett* **7,** 337–340.

Lehle, L. (1981). *Encyclopedia of Plant Physiology*, Vol. 13B, pp. 459–483.

Lehle, L., Bowles, D. J. and Tanner, W. (1978). *Plant Sci. Lett.* **11,** 27–34.

Lingappa, V. R., Lingappa, J. R., and Blobel, G. (1979). *Nature* **281,** 117–121.

Lloyd, F. E. (1942). *The Carnivorous Plants*, Chronica Botanica Co., Waltham MA.

Locy, R. and Kende, H. (1978). *Planta* **143,** 89–99.

Lodish, H. F., Braell, W. A., Schwartz, A. L. Strous, G. J. A. M. and Zilberstein, A. (1981). *Int. Rev. Cytol. Suppl.* **12,** 247–307.

Loh, J. P. and Grainer, H. (1978). *FEBS Lett.* **96,** 262–272.

Mascara, T. and Fincher, G. B. (1981). *Austr. J. Plant Physiol.* **9,** 31–45.

McNeil, M., Darvill, A. G. and Albersheim, P. (1979). *Prog. Chem. Org. Natl. Prod.* **37,** 191–249.

Meldolesi, J. (1974). *J. Cell. Biol.* **61,** 1–13.

Mertz, J. and Nordin, P. (1971). *Phytochem.* **10,** 1223–1227.

Meybert, M. and Kristen, U. (1981). *Z. Pflanzenphysiol.* **104,** 139–147.

Milstein, C., Brownlee, G. G., Harrison, T. M., and Mathews, M. B. (1972). *Nature (New Biol.)* **239,** 117–120.

Morré, D. J., Jones D. D. and Mollenhauer, H. H. (1967). *Planta* **74,** 286–301.

Mozer, T. J. (1980a). *Plant Physiol.* **65,** 834–837.

Mozer, T. J. (1980b). *Cell* **20,** 479–485.

Neutra, M. and Leblond, C. P. (1966). *J. Cell Biol.* **30,** 137–150.

Northcote, D. H. and Pickett-Heaps, J. D. (1966). *Biochem. J.* **98,**159–167.

O'Brien, T. P. (1972). *Bot. Rev.* **38,** 87–118.

Odzuk, W. and Kauss, H. (1972). *Phytochemistry* **11,** 2489–2494.

Okita, T. W., DeCaleys, R., and Rappaport, L. (1979). *Plant Physiol.* **63,** 195–200.

Olden, K., Pratt, R. M., and Yamada, K. M. (1978) *Cell* **13,** 461–473.

Olsen, B. R., Berg, R. A., Kischida, Y., and Prockop, D. J. (1975). *J. Cell Biol.* **64,** 340–355.

Owens, R. J. and Northcote, D. H. (1981). *Biochem. J.* **195,** 661–667.

Palade, G. E. (1975). *Science* **189,** 347–358.

Paleg, L. (1960). *Plant Physiol.* **35,** 902–906.

Palmiter, R. D., Gagnon, J., and Walsh, K. A. (1978). *Proc. Nat. Acad. Sci. USA* **76,** 94–98.

Panayotatos, N. and Villemez, C. L. (1973). *Biochem. J.* **133,** 263–271.

Paull, R. E. and Jones, R. L. (1975a). *Plant Physiol.* **56,** 307–312.

Paull, R. E. and Jones. R. L. (1975b). *Planta* **127,** 97–110.

Paull, R. E. and Jones, R. L. (1976a). *Plant Physiol.* **57,** 249–256.

Paull, R. E. and Jones. R. L. (1976b). *Z. Pflanzenphysiol.* **79,** 154–164.

Paull, R. E., Johnson, C. M., and Jones. R. L. (1975). *Plant Physiol.* **56,**300–306.

Peterkofsky, B. and Assad, R. (1976). *J. Biol. Chem.* **251,** 4770–4777.

Pickett-Heaps, J. D. (1966). *Planta* **71,** 1–14.

Pickett-Heaps, J. D. (1968). *J. Cell Sci.* **3,** 55–63.

Pope, D. G. (1977). *Plant Physiol.* **59,** 894–900.

Prehn, S., Tsamaloukas, A., Rapoport, T. A. (1980). *Eur. J. Biochem.* **107,** 185–195.

Rachmilevitz, T. and Fahn, A. (1973). *Ann. Bot.* **37,** 1–9.

Rachmilevitz, T. and Fahn, A. (1975). *Ann. Bot.* **39,** 721–728.

Ragetli, H. W. J., Weintraub, M., and Lo, E. (1972). *Can. J. Bot.* **50,** 159–168.

Ray, P. M. (1980). *Biochim. Biophys. Acta* **629,** 431–444.

Ray, P. M., Shininger, T. L., and Ray, M. M. (1969). *P.N.A.S. (USA)* **64,** 605–612.

Ray, P. M., Eisinger, W. R. and Robinson, D. G. (1976). *Ber. Dtsch. Bot. Ges.* **89** 121–146.

Robins, R. J. and Juniper, B. E. (1980a). *New Phytol.* **86,** 279–296.

Robins, R. J. and Juniper, B. E. (1980b). *New Phytol.* **86,** 297–311.

Robins, R. J. and Juniper, B. E. (1980c). *New Phytol.* **86,** 313–327.

Robinson, D. G. (1977). *Adv. Botanical Res.* **5,** 89–151.

Robinson, D. G. (1980). In *Cell Compartmentation and Metabolic Channeling* (L. Nover, F. Lynen, and K. Mothes, eds.), pp. 485–493. Gustav Fischer Verlag, Jena and Elsevier/North-Holland Biomedical Press, Amsterdam.

Robinson, D. G. and Glas, R. (1982). *Plant Cell Reports* **1,** 197–198.

Robinson, D. G. and Glas, R. (1983). *J. Exp. Bot.* **34,** 668–675.

Robinson, D. G. and Kristen, U. (1982). *Int. Rev. Cytol.* **77,** 89–127.

Robinson, D. G., and Ray, P. M. (1977). *Cytobiologie* **15,** 65–77.

Robinson, D. G., Andreae, M., Glas, R., and Saner, A. (1984). In Proceedings 7th Annual Winter Symposium in Plant Physiology, Riverside (S. Bartnicki-Garcia and W. M. Dugger, eds.) Amer. Soc. Plant Physiologists, Rockville, Maryland.

Robinson, D. G., Eberle, M., Hafemann, C., Wienecke, K., and Graebe, J. E. (1982). *Z. Pflanzenphysiol.* **105,** 323–330.

Robinson, D. G., Eisinger, W. R., and Ray, P. M. (1976). *Ber. Dtsch. Bot. Ges.* **89,** 147–161.

Rodaway, S. J. (1978). *Phytochemistry* **17,** 385–389.

Roland, J.-C., and Sandoz, D. (1969) *J. Microsc.* **8,** 263–268.

Rothstein, S. J., Lazarus, C. M., Smith, W. E., Baulcombe, D. C. and Gatenby, A. A. (1984), *Nature* **308**, 662–667.

Rougier, M. (1976). *J. Microsc.* **26,** 161–166.

Rougier, M. (1981). *Encyclopedia of Plant Physiology*, Vol. 13B, 542–574.

Ryser, U. (1979). *Protoplasma* **98,** 223–239.

Sabatini, D. D., Kreibich, G., Morimoto, T., and Adesnik, M. (1982). *J. Cell Biol.* **92,** 1–22.

Sadava, D. and Chrispeels, M. J. (1971a). *Biochem.* **10,**4290–4294.

Sadava, D. and Chrispeels, M. J. (1971b). *Biochem. Biophys. Acta* **227,** 278–287.

Samson, M., Klis, F. M., Sigon, C. A. M., and Stegwee, D. (1983). *Planta* **159,** 322–328.

Scala, J., Schwab, D. W., and Simmons, E. (1968). *Amer. J. Bot.* **55,** 649–657.

Schnepf, E. (1961). *Flora* **151,** 73–87.

Schnepf, E. (1963). *Flora* **153,** 23–48.

Schnepf, E. (1964) *Protoplasma* **58,** 193–219.

Schnepf, E. (1969a). *Protoplasma* **67,** 185–194.

Schnepf, E. (1969b). *Protoplasma* **67,** 195–203.

Schnepf, E. (1969c). *Protoplasma* **67,** 205–212.

Schnepf, E. (1969d). *Protoplasmatologia* **8,** 8–22.

Schnepf, E. (1974). In *Dynamic Aspects of Plant Ultrastructure*, (A. W. Robards, ed.) pp. 331–357. McGraw-Hill London.

Schnepf, E. (1977). *Apidologie* **8,**295–304.

Schwab, D. W., Simmons, E., and Scala, J. (1969). *Amer. J. Bot.* **56,** 88–100.

Schwaiger, H. and Tanner, W. (1979). *Eur. J. Biochem.* **102,** 375–381.

Selvendran, R. R. and O'Neill, M. A. (1982). *Encyclopedia of Plant Physiology* Vol. 13A, pp. 515–583.

Sharon, N. and Lis. H. (1981). *Chemical & Engineering News* **59,** 21–44.

Siekevitz, P. and Zamecnik, P. C. (1981). *J. Cell Biol.* **91,** 53s–65s.

Smith, M. A. (1981a). *Plant Physiol.* **68,** 956–968.

Smith, M. A. (1981b). *Plant Physiol.* **68,** 964–968.

Snider, M. W. and Robbins, P. W. (1981). *Meth. Cell Biology* **23,** 89–100.

Struck, D. K., Sinta, P. B., Lane, M. D., and Lennarz, W. J. (1978). *J. Biol. Chem.* **253,** 5332–5337.

Tabas, I. and Kornfeld, S. (1978). *J. Biol. Chem.* **253,** 7779–7786.

Tabas, I. and Kornfeld, S. (1979). *J. Biol. Chem.* **254,**11655–11663.

Tanaka, M., Sato, K., and Uchida, T. (1981). *J. Biol. Chem.* **256,** 11397–11400.

Tanaka, M., Shibata, M., and Uchida, T. (1980). *Biochim. Biophys. Acta.* **616,** 188–198.

Tanzer, M. C., Rowland, F. N., Murray, L. W., and Kaplan, J. (1977). *Biochem. Biophys. Acta* **500,** 187–196.

Tartakoff, A. M. and Vassalli, P. (1977). *J. Exp. Med* **146,** 1332–1345.

Tartakoff, A. and Vassalli, P. (1978). *J. Cell Biol.* **79,** 694–707.

Thiéry, J.-P. (1967). *J. Microsc.* **6,** 987–1018.

Timell, T. E. (1964). *Adv. Carbohyd. Chem.* **19,** 247–302.

Timell, T. E. (1965). *Adv. Carbohyd. Chem.* **20,** 409–483.

Unzelmann, J. M. and Healey, P. L. (1974). *Protoplasma* **80,** 285–303.

Van Der Woude, W. J., Morré, D.J., and Bracker, C. E. (1971). *J. Cell Sci.* **8,** 331–351.

van Holst, G.-J., Klis, F. M., De Wildt, P. J. M., Hazenberg, C. A. M., Buijs, J., and Stegwee, D. (1981). *Plant Physiol.* **68,** 910–913.

Varner, J. E. and Mense, R. M. (1972). *Plant Physiol.* **49,** 187–189.

Villemez, C. L (1971). *Biochem. J.* **121,** 151–159.

Villemez, C. L., Lin, T. Y., and Hassid, W. Z. (1965). *P.N.A.S.* (*USA*) **54,** 1626.

Vögel, A. (1960). *Schweiz. Forstver. Beih.* **30,** 113–122.

von Heijne, G. and Blomberg, C. (1979). *Eur. J. Biochem.* **97** 175-–181.

Vuust, J. and Piez, K. A. (1972). *J. Biol. Chem.* **247,** 856–862.

Waksman, A., Hubert, P., Cremel, G., Rendon, A. and Burgun, C. (1980). *Biochim. Biophys. Acta* **604,** 249–296.

Wallach, D., Kirshner, N., and Schramm, M. (1975). *Biochem. Biophys. Acta* **375,** 87–105.

Walther, P. and Blobel, G. (1981a). *J. Cell Biol.* **91,** 551–556.

Walther, P. and Blobel, G. (1981b). *J. Cell Biol.* **91,** 557–561.

Weinstock, M. and Leblond, C. P. (1974). *J. Cell Biol.* **60,** 92–127.

Whaley, G. (1975). *The Golgi Apparatis*, Cell Biology Monographs, Vol. 5 Springer Verlag, Vienna.

Whistler, R. L. and Richards, E. L. (1970). In *The Carbohydrates*, (W. Pigman and D. Horton, eds.), Vol. IIA, pp. 447–469. Academic Press, New York.

Whur, P., Herscovics, A., and Leblond, C. P. (1969). *J. Cell Biol.* **43,** 289–311.

Wienecke, K., Glas, R., and Robinson, D. G. (1982). *Planta* **155,** 58–63.

Wright, K. (1975). *Phytochemistry* **14,** 759–763.

Wright, K. and Northcote, D. H. (1974). *Biochem. J.* **139,** 525–534.

Wright, K. and Northcote, D. H. (1976). *Protoplasma* **88,** 225–239.

Wright, K., Northcote, D. H., and Davey, R. (1976) *Carbohydr. Res.* **47,** 141–150.

Yariv, J., Lis, H., and Katchalski, E. (1967). *Biochem. J.* **105,** 1c–3c.

Yomo, H. (1960). *Hakko Kyokaishi* **18,** 600–602.

Zagury, D., Uhr, J. W., Jamieson, J. D., and Palade, G. E. (1970). *J. Cell Biol.* **46,** 52–63.

11

SYNTHESIS, ACCUMULATION, AND BREAKDOWN OF STORAGE PRODUCTS

There should be no need to explain the importance, even to the layman, of studies concerned with elucidating the metabolism of reserve products in plants. Investigations, particularly with respect to seeds, have received considerable impetus over the last decade as a result of methodological advances in protein and nucleic acid biochemistry. Indeed this is an area in which plant cell biologists and biochemists are beginning to catch up with their colleagues in the animal field in terms of the successful application of highly refined techniques.

The nutritional value of seeds depends upon the relative amounts of starch, protein, or fat stored in them. These substances may be present in the endosperm, which surrounds the developing embryo (albuminous seeds), or in the embryo itself (usually the cotyledons in exalbuminous seeds), which develops at the cost of the endosperm before dessication. Whereas the economically important cereal plants are of the former type, the equally valuable dicotyledenous plants may be of the albuminous (e.g., castor beans) or exalbuminous (e.g., legumes) type. Although the production of starch is a property of plastids, and will therefore not be considered in this book, the formation and catabolism of reserve proteins and fats are intimately associated with components of the endomembrane system.

11.1. RESERVE PROTEINS

Seed proteins are arbitrarily classified according to their solubility. After Osborne (1895) we differentiate the albumins (water soluble) from the globulins (soluble in NaCl), the prolamines (soluble in alcohol), and the glutelins (soluble in acid or alkali). In the monocotyledons (mainly the cereals) the reserve proteins are usually the prolamines and glutelins, although in one important exception (oats) globulins account for 80% of the storage proteins (Konzak, 1977). As a rule the prolamines are the more prevalent proteins in cereals and are named after the latin name of the genus; for example, in maize "Zein" is composed of 19 kd-and 22-kd polypeptides (Lee et al., 1976). In rice, however, the major reserve proteins are the glutelin polypeptides of 23 kd and 39 kd molecular weight (Tanaka et al., 1980).

In the dicotyledons the most intensively studied species are the legumes and here the major reserve proteins belong to the globulin group. The reserve proteins are of two types: vicilin and legumin, heterogeneous complexes of 7S and 12S, respectively. Vicilin, from beans (*Phaseolus vulgaris*), also termed *phaseolin*, consists of 52-, 49-, and 46-kd polypeptides in a ratio of 2:2:1 (Bollini and Chrispeels, 1978). In contrast, the vicilin fraction from peas (*Pisum sativum*) contains at least 13 different polypeptides: the major ones are 75, 50 (twice), 30, and 18 kd and the minor ones 70, 49, 34, 25, 14 (twice), 13, and 12 kd (Thomson et al., 1980). Legumin from the same source has 12 polypeptides (six are 40 kd and six are 20 kd—Derbyshire et al., 1976). Vicilin and legumin are not always present in equal amounts in the cotyledons of legumes; for example, in *Ph. vulgaris* legumin is present in only very small quantities.

Many of the storage protein polypeptides have covalently linked sugar residues, in the main mannose and glucosamine. This is true for some of the vicilin polypeptides (Badenoch-Jones et al., 1981; Racusen and Foote, 1971) and for zein polypeptides (Burr, 1979) but is not the case for legumin.

The various reserve proteins are usually stored in more or less globular deposits (<10 μm in diameter) surrounded by a single membrane. These are termed *protein bodies* (PBs) and have been recognized cytologically for over a century (see Guilliermond, 1941, for the older literature). However, only with the availability of immunocytochemical methods has it been possible to really confirm the localization of reserve proteins in the PBs (Craig, Millerd, and Goodchild, 1980; Graham and Gunning, 1970). There is a great variability in size and types of inclusion in PBs (Pernollet, 1978), and in cereals one must be careful to distinguish between the glutelin-rich PBs of the aleurone layer from the otherwise prolamine-rich PBs of the endosperm.

In addition to the storage proteins there are also a large number of enzymes (see Section 11.4. and Chapter 12) and phytic acid (in salt form) and a variety of more or less toxic substances present in the PBs. Of the latter, the most important for this section are the lectins, which can make up almost 10% of the total protein stored in seeds (Liener, 1976). One of these, phytohemagglutinin (PHA), has been extensively studied in terms of its synthesis and intracellular transport (Section 11.3).

The following two sections on protein body formation will be restricted to answering two questions: which endomembranes are involved in their formation and what role, if any, do posttranslational modifications play in the choice of route between sites of synthesis and deposition? This is somewhat selective and the reader is therefore referred to a number of

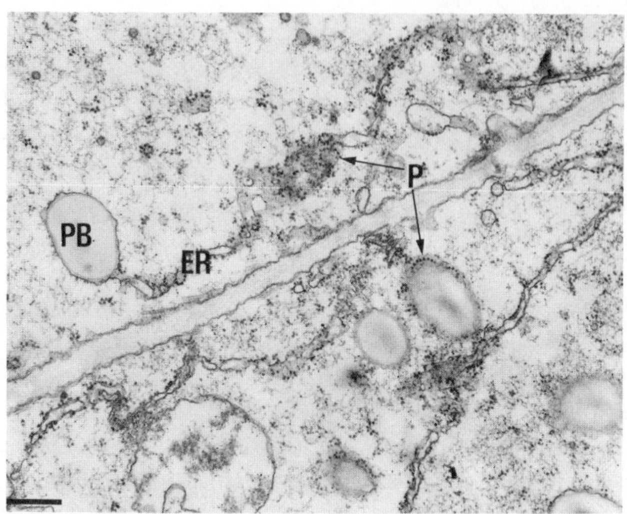

Figure 11.1. Section through adjacent portions of two 19-day-old maize endosperm cells. Continuities between protein body (PB) and ER membranes are visible (arrows). Note also the presence of polysomes (P) at the surface of both these membranes. Bar = 0.5 μm. Published micrograph of Larkins and Hurkman (1978).

excellent review articles (Larkins, 1981; Lott, 1980; Miége, 1982; Müntz, 1982; Weber and Neumann, 1980) for other aspects and further details.

11.2. PROTEIN BODY FORMATION IN CEREALS

Depending on the cereal in question, PBs may be completely membrane bound (e.g., rice, maize) or lie, with some degree of membrane association, more or less free in the cytoplasm. In the former case the membrane is usually derived from the ER. There are several lines of evidence which support this statement. First, numerous EM studies show quite convincingly that the PB membrane is studded with ribosomes (Figure 11.1) (Bechtel and Juliano, 1980; Khoo and Wolf, 1970). Analysis of polysomes from isolated PBs and from rER has proved their identical nature both in terms of their size and in terms of the polypeptides they are capable of synthesizing *in vitro* (Larkins and Hurkman, 1978). Second, it is quite clear that there are direct attachments to the ER during the formation of the PBs (see above references). Third, the ER marker enzyme CCR can also be measured in fractions containing PBs (Cameron-Mills et al., 1978; Larkins and Hurkman, 1978; Miflin et al., 1981).

Occasionally, in electron micrographs one sees PBs in cereal grain endosperm which do not have a complement of ribosomes on their limiting membrane. Here it is difficult to know whether the ribosomes in question have been released because the PB is completed or whether the membrane has a non-ER origin. In rice grains there is one type of PB whose membrane does not appear to come from the ER but rather has a vacuolar origin as is the case for the PBs in legumes (Section 11.3). According to Bechtel and Juliano (1980), electron-dense deposits, putatively representing reserve proteins, are seen at the *trans*-face of dictyosomes and in released GA vesicles. These then presumably fuse with the vacuole membrane which gradually becomes filled with reserve protein. A subsequent subdivision of these vacuoles into many smaller PBs, as in legumes, does not occur. It is important to note that in rice ER-derived PBs exist alongside these other PBs (Figure 11.2) and that two classes of PBs have been isolated: one (PB I) containing principally the prolamines and the other (PB II) being made up of the glutellin polypeptides (Tanaka et al., 1980). In the opinion of the latter authors the membranes of PB II are formed from the GA.

PBs from wheat and barley endosperm are either membrane-free or are only partially associated with ER membranes (Campbell et al., 1981; Jennings et al., 1963; Miflin and Shewry, 1979; Miflin et al., 1980; Parker, 1980). The absence of a complete membrane covering to the PBs in these two cereals is supported by digestion experiments with protease-K (Miflin and Burgess, 1982). In contrast to PBs from maize or legume PBs, isolated barley and wheat PBs are rapidly degraded by the addition of protease-K. Interestingly, the digestion is enhanced when the PBs are isolated under high Mg^{2+} conditions. From work on animal systems it is known that proteases cannot penetrate sealed membrane vesicles (Snider and Robbins, 1981), so that in the opinion of Miflin and Burgess (1982) wheat and barley storage proteins are initially sequestered in the lumen of the ER, but in contrast to other cereals, continued deposition results in the release of the aggregates into the cytoplasm.

For those PBs which result from a swelling, or in the case of wheat and barley a rupture, of the ER, transport from site of synthesis to site of deposition does not exist. Thus posttranslational modifications of synthesized polypeptides take place at the same locus as the translation itself.

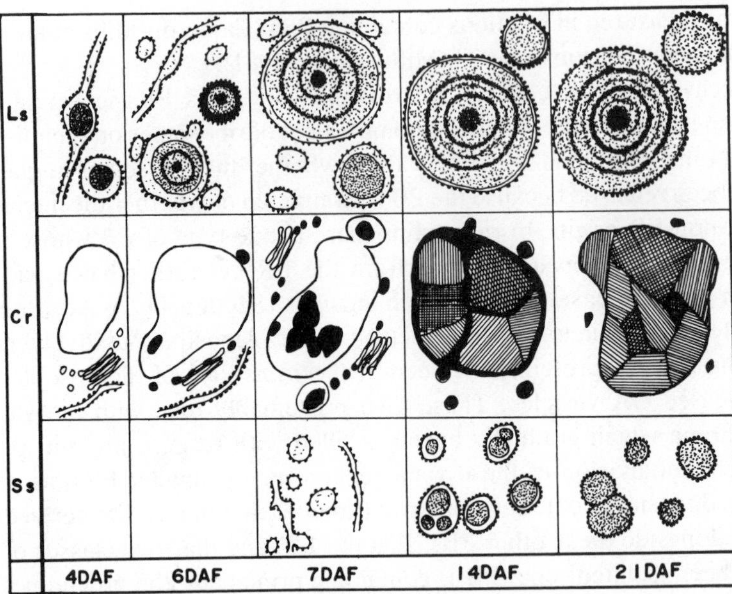

Figure 11.2. Diagram indicating the types and their intracellular origin of protein bodies found in the developing rice caryopsis. (DAF = days after fertilization; Ls = large spherical protein body; Cr = crystalline protein body; Ss = small spherical protein body). Published drawing of Bechtel and Juliano (1980).

The first *in vitro* translation experiments using maize polysomes indicated the synthesis of polypeptides identical in size with the native zein of the PBs (Burr and Burr, 1976; Larkins and Dalby, 1975). Later work (Larkins and Hurkman, 1978) with more refined techniques showed that the initial synthesized polypeptides were 1.1 and 2 kd heavier than the native ones. This is in agreement with Blobel's signal hypothesis (see Section 10.1) whereby the translated protein chain is extended by a leader or signal sequence. Evidence for the expected proteolytic cleavage of these N-terminus peptides has been obtained by Larkins and Hurkman (1978) who carried out the *in vitro* translation with rER vesicles. The synthesis of polypeptides larger than those occurring in the PB has also been demonstrated for barley (Brandt and Ingversen, 1978; Cameron-Mills and Ingversen, 1978; Cameron-Mills et al., 1978). Although the sequestration of these hordein polypeptides within the rER vesicles used in translation experiments was inferred from their inaccessibility to exogenously added proteases; a chain shortening was not seen in these experiments.

The removal of the signal sequence is a cotranslational rather than posttraslational modification of the synthesized polypeptide (Section 10.1). A true posttranslational proteolytic cleavage of a cereal reserve protein has so far only been demonstrated for rice (Yamagata et al., 1982). Here the 23- and 39-kd glutelin polypeptides are synthesized initially as a 57-kd species. Whether this latter polypeptide originally had a terminal signal sequence is unclear. With the exception of one type of PB in the rice grain, all we can say about the posttranslational glycosylation of cereal reserve proteins is that if it occurs, it takes place in the ER.

11.3. PROTEIN BODY FORMATION IN LEGUMES

In legumes the PBs are found in cells of the cotyledons rather than endosperm tissue. After a period of cell division the cotyledon cells enlarge and become highly vacuolate (see Figure 11.3a). Immediately preceeding the onset of reserve protein synthesis, there is a subdivision of these vacuoles into smaller ones (Bain and Mercer, 1966b; Briarty et al., 1969; Harris and Boulter, 1976). The exact time point after fertilization (post anthesis, in days) when the reserve proteins are detectable cytologically and biochemically is variable and depends both on the species involved (see, e.g., Figure 1 in Müntz, 1982) and the environmental conditions used (see, e.g., Chrispeels, Higgins, Craig, and Spencer, 1982).

As a rule the first cytological events are visible roughly a week after fertilization, and take the form of the appearance of small osmiophilic deposits which line the TP (see Figure 11.3c). The deposits grow in size and number, but then, approximately halfway through seed development, lose their globular character and peripheral location and give way to a more diffuse and evenly distributed matrix in the vacuoles (see Figure 11.4a). In the final stages of protein synthesis the cytological picture is one of a cytoplasm filled with membrane-bound PBs but without vacuoles (see Figures 11.3c and 11.4b). This sequence of events has been established on numerous occasions for legumes (Bain and Mercer, 1966a; Boulter, 1981; Briarty et al., 1969; Craig, Goodchild, and Hardham, 1979; Harris and Boulter, 1976; Neumann and Weber, 1978; Öpik, 1968), as well as for other dicotyledonous plants (Bergfeld et al., 1980; Davey and Van Staden, 1978; Rest and Vaughan, 1972), and is clearly different from

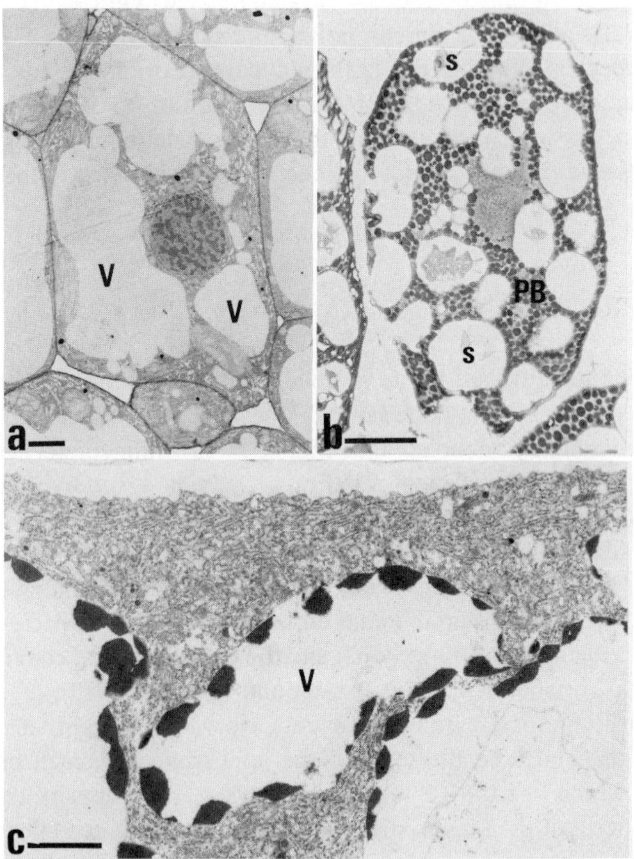

Figure 11.3. (a) Section through a young pea cotyledon cell before reserve protein deposition has begun. Large vacuoles (V) are present. Bar = 2 μm. (b) Pea cotyledon cell after reserve protein synthesis has been completed (22 days after flowering). Vacuoles are no longer visible and the cytoplasm is packed with starch-filled amyloplasts (s) and thousands of protein bodies (PB). Bar = 20 μm. (c) Section through a pea cotyledon cell in the early phase of reserve protein deposition (here 12 days after flowering). Osmiophilic deposits at the periphery of the vacuoles (V) are visible. Bar = 2 μm. Published micrographs of (a) Boulter, 1981; (b) Craig et al., 1979b; (c) Craig et al., 1979a.

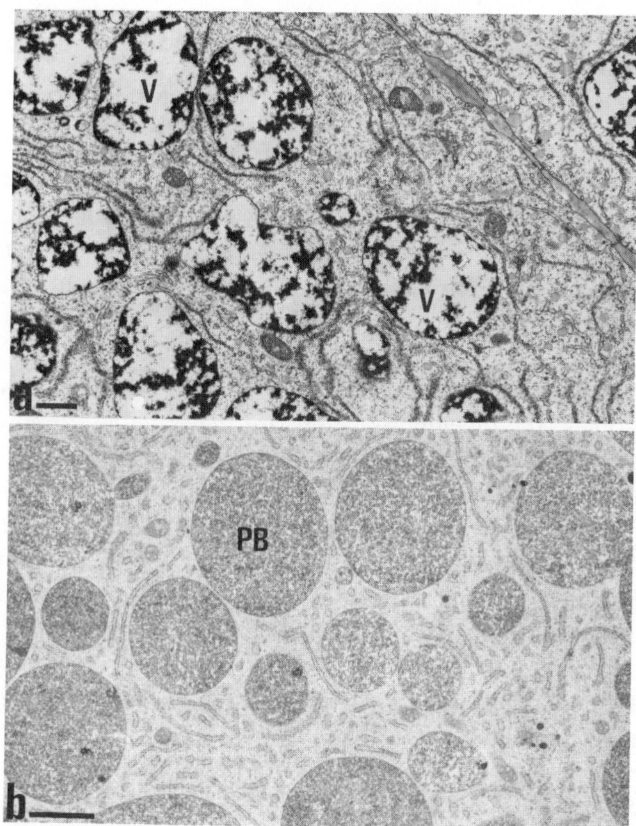

Figure 11.4. (a) Section through a pea cotyledon cell during the middle phase of reserve protein deposition. The vacuoles (V) are smaller, more numerous, and have a diffuse osmiophilic content. (b) Pea cotyledon cell in the closing stages of reserve protein synthesis. The characteristic protein bodies (PB) are now present in great numbers. Bar = 1 μm. Published micrographs of Boulter (1981).

the situation with cereals, with the possible exception of one type of PB in rice grains (see above).

Although it has been doubted (Harris and Boulter, 1976) that the PB membrane is derived from the TP of the original vacuoles, the immunocytochemical demonstration of reserve protein aggregates during the early stages of protein synthesis (Craig et al., 1980) makes this now more or less certain. The transformation of a relatively small number of vacuoles to more than 100,000 PBs per cell can only occur through subdivision of the larger volume vacuoles. In the opinion of Yoo and Chrispeels (1980), this process occurs at the same time as the deposition of protein into the vacuole and is essentially a large-scale pinocytosis. Irrespective of the timing of the vacuole subdivision with respect to the entry of reserve proteins, there is a considerable increase of membrane associated with it. Craig et al. (1979) have estimated that whereas the volume increase is only 200%, the surface–volume ratio increases 55 times. Although the authors were unable to provide information as to the origin of this extra membrane, it is clearly the result of some sort of membrane flow or transfer mechanism.

Since the PBs, like the vacuoles from which they are derived, do not bear ribosomes, the question is where are the proteins synthesized and how do they reach the vacuoles? More specifically, what is the role(s) of the ER and GA in this process? The first evidence for participation of the ER in legume reserve protein synthesis was obtained by autoradiography of cotyledon slices incubated *in vivo* with ^{14}C-leucine (Bailey et al., 1970). After initial incorporation into the ER, radioactivity was found over the PBs after a chase-out. Additional cytological evidence for the presence of reserve proteins in the ER comes from the application of ferritin-conjugated antibodies against phaseolin in frozen sections of *Ph. vulgaris* cotyledon cells (Baumgartner et al., 1980).

In vitro translation experiments using polysomes from a number of legume cotyledons have been successfully carried out (Beachy et al., 1978; Evans et al., 1979; Higgins and Spencer, 1977; Sun et al., 1975, 1978), but until the work of Bollini and Chrispeels (1979) it was unclear whether or not the polysomes in question were originally membrane, that is, ER bound. In the meantime, through the precise application of density gradient and immunological techniques, Chrispeels and colleagues have been able to confirm the participation of the ER in the synthesis of legumin and vicilin in *P. sativum* (Chrispeels, Higgins, Craig, and Spencer, 1982; Chrispeels, Higgins, and Spencer, 1982) of phaseolin in *Ph. vulgaris* (Bollini et al., 1982), and of PHA also in *Ph. vulgaris* (Chrispeels and Bollini, 1982). Figures 11.5 and 11.6 are representative of the results obtained in these studies. Normally the PBs from immature cotyledons tend to burst during the homogenization of cotyledon tissue and release their reserve proteins into the soluble portion of the homogenate which remains therefore in the "sample volume" at the top of the gradient after centrifugation (Figure 11.5a). Separation of these liberated proteins from the cytomem-

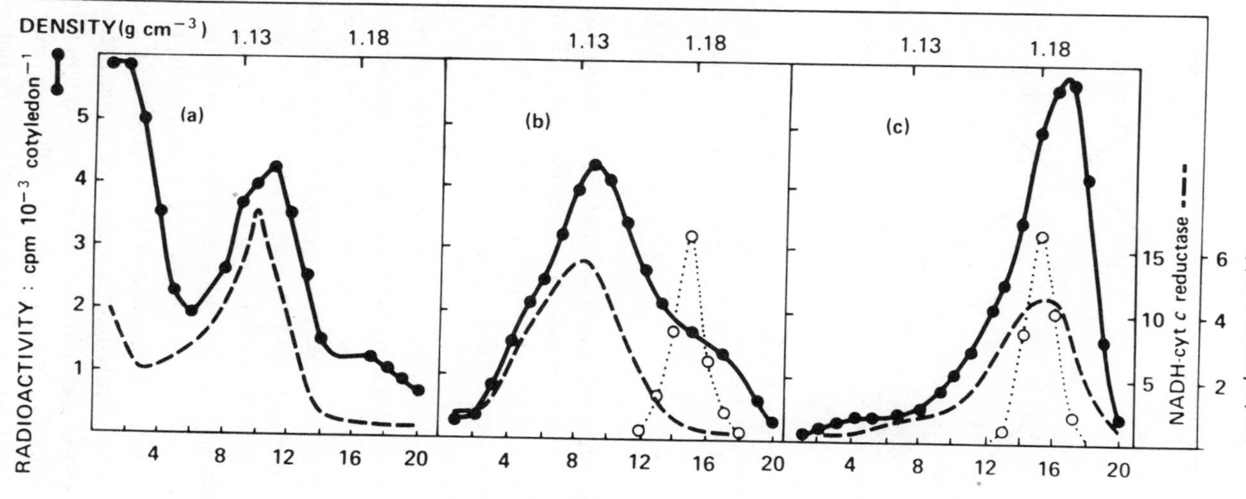

Figure 11.5. Distribution profiles of marker enzymes and radioactivity incorporated *in vivo* into proteins in isopycnic sucrose gradients made from homogenates of developing pea cotyledons. In (a) the total homogenate was applied as a gradient overlay whereas in (b) and (c) the soluble proteins in the homogenate were separated prior to centrifugation by passing through Sepharose 4B. In (a) and (b) gradients were prepared without, in (c) with 3 m*M* $MgCl_2$. Published results of M. J. Chrispeels et al. (1982a).

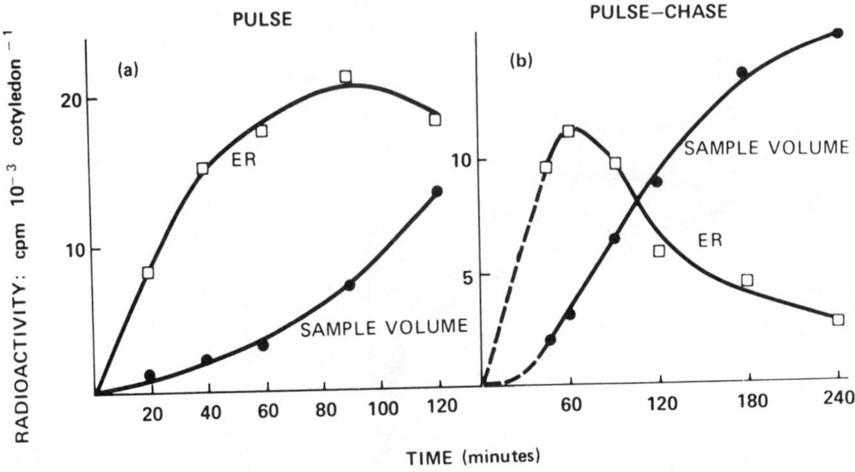

Figure 11.6. Pulse (a) and pulse-chase (b) labeling with ^{14}C-amino acids during the middle phase of reserve protein synthesis in pea cotyledons. Reserve proteins in ER and PB ("sample volume") fractions were isolated by immunoaffinity chromatography. Published results of Chrispeels et al. (1982a).

branes prior to centrifugation results in a unimodal distribution of both membrane associated *in vivo* incorporated radioactivity and ER marker enzyme (CCR) activity. From Mg^{2+}-shifting experiments (compare b and c in Figure 11.5) it is quite apparent that the ER is the site of reserve protein synthesis in legumes.

The ER is labeled not only when incubations are carried out with ^{14}C-amino acids, but also when radioactive glucosamine or mannose are used. Corroborating this latter feature is the fact that the glycosyl transferase activities for these two sugars are principally located in ER fractions from *P. sativum* cotyledons (Nagahashi and Beevers, 1978). That the radioactivity incorporated into the ER fraction *in vivo* is not just loosely associated through the adherence of PB-released reserve proteins but is instead internally localized in the lumen of the ER is to be inferred from three lines of evidence. First, ER-associated radioactivity in these investigations is accessible to proteolytic digestion only in the presence of detergent. Second, very short (<45 min) incubation periods whereby PBs do not become labeled still show the detection of ER-associated radioactivity. Finally, radioactivity associated with the ER chases out into the soluble fraction at the top of the gradient.

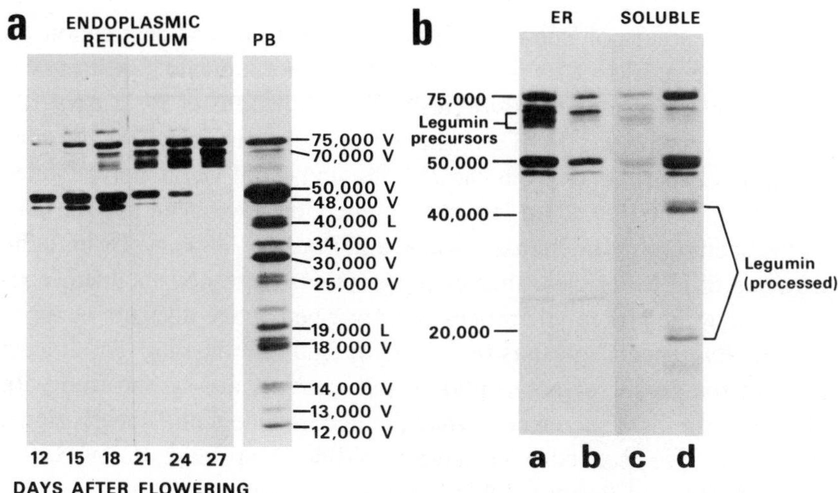

Figure 11.7. (a) SDS–PAGE separation of ER-containing polypeptides synthesized *in vivo* at different stages in the development of pea cotyledons. For comparison is, at the right, a normal (coomassie blue) stained separation of the reserve polypeptides present in PBs isolated from cotyledons 24 days after flowering. Mr of vicilin (V) and legumin (L) polypeptides are given. (b) Fluorograph demonstrating a product–precursor relationship for PB polypeptides arising from the ER. Because of their fragility during seed development, the vacuole–PB polypeptides are released into the "soluble supernatant" during homogenization. Lanes a and c represent a pulse; lanes b and d a chase. Published photograph of M. J. Chrispeels et al. (1982a,b).

PAGE analysis of the polypeptides contained in the ER fraction from *P. sativum* cotyledons which were previously isolated on immunoaffinity gels show their similarity to some of the polypeptides present in the PBs from mature cotyledons (Chrispeels, Higgins, and Spencer, 1982). Present are a group of closely related polypeptides with MWs between 60 and 65 kd as well as bands at 75, 70, 50, and 49 kd (see Figure 11.7b). In later stages of cotyledon development the latter two polypeptides are no longer found in ER fractions (Figure 11.7a). The former represents precursors of the 40- and 20-kd polypeptides of legumin and the latter belong to the vicilin group of polypeptides. According to Chrispeels and colleagues, the posttranslational cleavage of legumin and vicilin (the latter into 34-, 30-, 25-, 18-, 14-, 13-, and 12-kd polypeptides) occurs after they have left the ER (Chrispeels, Higgins, Craig, and Spencer, 1982; Chrispeels, Higgins, and Spencer, 1982). In the former case this processing is complete, but in the case of vicilin it is incomplete with considerable amounts of the original precursor polypeptides present in the PBs at seed dessication. In contrast, the results of Bollini et al. (1982) do not indicate a posttranslational cleavage of phaseolin synthesized in cotyledons of *Ph. vulgaris*.

The *in vivo* results from *P. sativum* are also supported by *in vitro* translation experiments. Thus the 75-, 50-, and 49-kd vicilin polypeptides as well as a putative 60-kd legumin polypeptide have been successfully synthesized (Croy, Gatehouse, Evans, and Boulter, 1980a,b). Despite the presence of rER vesicles, smaller polypeptides were not obtained, confirming that the proteolytic processing described above does not occur in the ER. Evidence also exists from *in vitro* studies indicating that at least some of the presursor polypeptides in *P. sativum* are synthesized with additional terminal sequences, which, in accordance with Blobel's signal hypothesis, are cleaved by a membrane-bound protease (Croy, Gatehouse, Tyler, and Boulter, 1980; Higgins and Spencer, 1981).

Similar elegant experiments have also been performed with one of the lectins in *Ricinus* endosperm PBs ("agglutinin type I"—Roberts and Lord, 1981a). *In vitro* translation experiments with poly A-mRNA from *Ricinus* alone gave rise to two polypeptides: 33.5 kd and 69 kd. When carried out in the presence of rER vesicles the products had MWs of 32 kd and 66–69 kd. These results are to be interpreted as indicating the removal of a leader sequence with (in the case of the 66–69-kd species) cotranslational glycosylation. The 32-kd polypeptides represent the so-called A chains and the 66–69-kd polypeptides a precursor form of the B chains. *In vitro* pulse-chase experiments suggest the posttranslational cleavage of the latter into 37-kd polypeptides, although it is unclear where this event takes place.

As is the case with one of the PB types in rice grains (see above), ultrastructural observations (Briarty, 1980; Dieckert and Dieckert, 1972, 1976; Boulter, 1981) suggest that the GA is along the transport route from ER to PB in legumes. These reports draw attention to the presence of osmiophilic material in mature secretory vesicles at the *trans*-face of dictyosomes in cotyledon cells during reserve protein synthesis (e.g., see Figure 11.8). The equally osmiophilic nature of the deposits of reserve protein in the vacuoles has been the basis for this suggestion. In addition there is evidence (though not as convincing as in the case of the ER) from immunocytochemical studies for the presence of phaseolin in GA cisternae from *Ph. vulgaris* (Baumgartner et al., 1980).

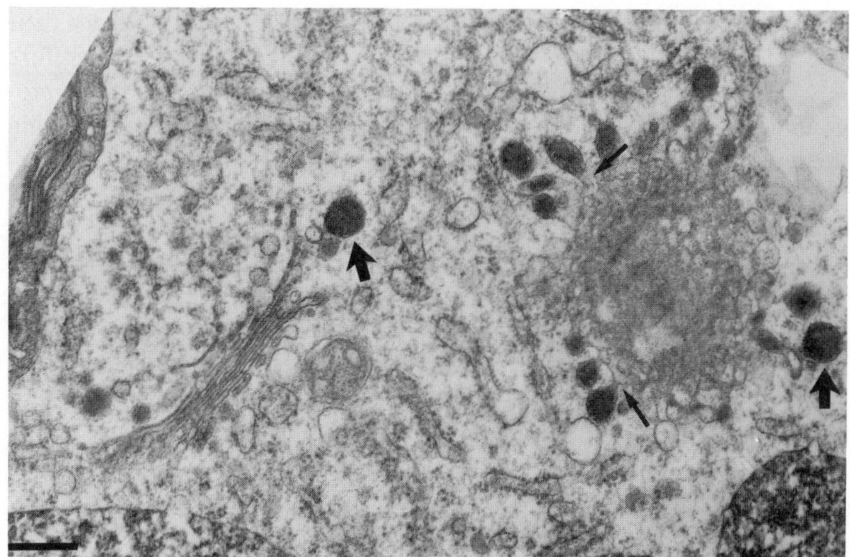

Figure 11.8. Section through a pea cotyledon cell in the later stages of PB formation. Vesicles (arrows) with osmophilic contents are clearly seen at the periphery of the dictyosomes. Direct attachments can also be seen (small arrows). Bar = 2 μm. Unpublished micrograph of N. Harris.

Since the GA is recognized as the site of posttranslational modification for secretory (glyco) proteins (see Section 10.1) it is logical to see this function being performed here for storage proteins too. Certainly both *in vivo* and *in vitro* glucosamine and mannose incorporation in *P. sativum* cotyledons and *Ricinus* endosperm, which are indicative of core glycosylation reactions, are localized in the ER (Chrispeels, Higgins, Craig, and Spencer, 1982; Mellor and Lord, 1979; Mellor et al., 1980; Nagahashi and Beevers, 1978). Moreover, Chrispeels (1983) has now shown that fucose incorporation (a terminal glycosylation event) into PHA is localized in the GA of developing *Ph. aureus* cotyledons.

However, not all of the proteins found in PBs are glycosylated: legumin in *P. sativum* is not, whereas some of the vicilin polypeptides are (Hurkman and Beevers, 1980; Badenoch-Jones et al., 1981); two of the four subunits in *Ricinus* agglutinin I are glycosylated (Cawley and Houston, 1979), and so on. Furthermore, double-label experiments (Bollini et al., 1982) have shown that the phaseolin polypeptides of *Ph. vulgaris* are not subjected to a deglycosylation and reglycosylation en route from their site of synthesis to their site of deposition in the PBs. Is it therefore possible that a dichotomy of ways to the PB exist? If so, those programmed for glycosylation in the GA must already be segregated within the lumen of the ER from those which go directly to the PB.

11.4. MOBILIZATION OF PROTEIN RESERVES DURING GERMINATION

The PBs formed during seed development survive seed dessication and are present at the beginning of germination. More or less in reversal of the events during seed development the PBs give rise to vacuoles: a feature which has been recognized for some 60 years (Dangeard, 1922) and which has been confirmed many times by electron microscopical investigations (see Ashton, 1976, for a review). Depending on the species in question the PBs may at first swell, their dense contents becoming disperse and gradually disappearing, giving rise to numerous vacuoles which then fuse with one another (Bain and Mercer, 1966b; Lott and Volmer, 1973; Srivastava and Paulson, 1968; Yoo, 1970) or alternatively the PBs fuse before their internal degradation (Briarty et al., 1970; Horner and Arnott, 1965; Öpik, 1966). With respect to the breakdown of reserve proteins in the PBs, there are a number of important questions to be answered; only two will be addressed here; namely, what are the enzymes responsible and where do they come from?

As might be expected, the enzymes involved in hydrolysis of the reserve proteins contained in PBs are both endo- and exo- (carboxy- and amino-) peptidases. Usually their activity is measurable 1–2 days after the beginning of germination (Ashton, 1976), and according to Mikala and Kohlemainen (1972), it is the endopeptidases which become active first, producing more or less soluble peptides which are then attacked by the exopeptidases. The detection of these enzymes in isolated PBs is not without difficulty, for the PB membrane becomes increasingly susceptible to disruption as germination proceeds. Thus, the majority of studies involving the demonstration of peptidase activity have been carried out with imbibed but ungerminated seeds. Isolated PBs from such seeds have been shown to contain almost all of the endopeptidase (usually acid-) activity in the tissue (Harris and Chrispeels, 1975; Matile, 1968; Ory and Henningsen, 1969; Schnarrenberger et al., 1972; Yatsu and Jacks, 1972). In addition, carboxypeptidase activity has also been localized in PBs or in the vacuoles they give rise to (Baumgartner and Chrispeels, 1976; Harris and Chrispeels, 1975; Nishimura and Beevers, 1978). Finally, there exists the *in situ* demonstration through immunofluorescence microscopy of the major acid endopeptidase (vicilin peptohydrolase) in *Ph. aureus* PBs (Baumgartner et al., 1978).

With respect to their origin there are two possiblities for the proteolytic enzymes found in the PBs: either they are synthesized *de novo* at the beginning of germination and are inserted into the PBs or they are already present in the PBs at the beginning of germination, having been translated before seed dessication.

There are few reliable studies dealing with the question of *de novo* synthesis of proteolytic enzymes during seed germination. Baumgartner et al. (1978) have shown that antibodies to endopeptidase do not react with extracts prepared from 1-day-imbibed *Ph. aureus* seeds. Enzyme activity was first measurable after 2–3 days and was maximal after 5 days of germination. Density and radioactive labeling experiments (Chrispeels et al., 1976) indicated *de novo* synthesis of the responsible endopeptidase, which was also confirmed by the prevention of its appearance through cycloheximide application. For exopeptidase activity there are conflicting reports. Thus, whereas dipeptidase activity in squash cotyledons appears to be synthesized *de novo* (Jacobsen and Varner, 1967), actinomycin D and cycloheximide treatment of germinating pea seeds failed to prevent the development of exopeptidase activity (Ellemann, 1974).

There is as yet no biochemical evidence as to whether these enzymes are synthesized on cytosolic or ER-bound ribosomes, although there is suggestive ultrastructural evidence for the latter. Chrispeels et al. (1976) and Harris et al. (1975) have presented sections of germinating cotyledons in which ends of ER cisternae are seen to swell up into small vesicles. Similar vesicles are occasionally seen fusing with the PB membrane. Such vesicles would then represent primary lysosomes and the PBs, after fusion with them, secondary lysosomes according to the terminology applicable for animal cells (see Chapter 12). A compartmentalization of this type is clearly advantageous as method of protecting cytoplasmic proteins from random degradation.

11.5. SEED OILS AND FATS

In general oil- or fat-rich seeds tend to be smaller than their counterparts, which store predominantly protein or carbohydrate; furthermore, they always appear to contain more PBs than starch-containing amyloplasts (Bewley and Black, 1978). Usually the lipids which are stored in seeds are triacylglycerols (triglycerides: esters of glycerol and long-chain fatty acids), although there are some cases where glycolipids are present in large amounts. The most common fatty acids in seeds are unsaturated, mainly oleic (18:1), linoleic (18:2) or α-linolenic (18:3) acids, but saturated ones [e.g., palmitic (14:0) or lauric (10:0)] may also figure prominently (Gurr, 1980; Hitchcock and Nichols, 1971). According to the latter authors, the seed oils of plants grown in temperate climates tend to be more unsaturated than those grown in warmer climates. Another factor appears to be light which, in seeds with a relatively translucent pericarp and/or testa (e.g., the Leguminosae or Cruciferae) results in a larger proportion of α-linolenic acid whose synthesis is associated with plastid membranes (Stumpf, 1980).

The various triacylglycerols are stored in the form of oil or lipid bodies (LBs) which may reach 6 μm in diameter. In the literature there is some confusion over terminology, since the terms *spherosomes* and *oleosomes* have also been used for these structures. Based on staining reactions, Sorokin (1967) maintained that spherosomes are present in both lipid storing and non-storing cells, whereas LBs are restricted to the former. Similarly, Gurr (1980) distinguishes between spherosomes, which are normally smaller, from the larger LBs. On the other hand, Yatsu et al. (1971), by introducing the generalized term *oleosome* have questioned such a differentiation. In this section the term LB will be adopted, which is more typical for seed storage tissues.

Based on the observation that in mature seeds there are thousands of individual LBs which are closely packed together but without any fusion profiles, it is natural to assume that there must be some sort of limiting layer to the LB. Nevertheless, there are those who have doubted this (e.g., Sorokin, 1967; Smith, 1974). The work of Yatsu and Jacks (1972) involving the "defatting" of isolated LBs has however shown that the boundary of the LB is a phospholipid monolayer with embedded proteins (i.e., a half-unit membrane). This follows up the suggestion of Schwarzenbach (1971) that the LB develops from an ER membrane (see below) and is also supported by freeze-fracturing data (Buttrose, 1971). Phospholipid and protein analysis of LBs (Allen et al., 1971; Jacks et al., 1967; Yatsu et al., 1971) provide values that are in accordance with a half-membrane covering, although it must be mentioned that there are also cases where the values lie in excess of these (Gurr et al., 1974).

Before starting on the biochemistry and cytology of LB formation, it is worthwhile to bear in mind the situation with respect to fat storage in animals. Essentially a storage in the sense that the fats are later broken down in the same cells in which they are stored does not occur in animal cells (see, however, the so-called brown adipose tissue of hibernating animals). The stored lipids are usually transported away from the cells in which they are stored. This is so in the case of both white adipose tissue and of fat droplets in lactating gland cells. Neither the single-lipid globule characteristic of adipocytes nor the fat droplet of lactating epithelial cells appear membrane delineated in the cells in which they are formed. Thus the production of LBs and their breakdown during germination is a particular feature of the plant cell.

11.6. LIPID BODY FORMATION

The biochemistry of triacylglycerol synthesis may be conveniently dealt with in two sections: the synthesis of long-chain fatty acids and their transfer to glycerol phosphate to form di- and triacylglycerols. The former aspect has been excellently reviewed by Stumpf (1977, 1980) and Beevers (1982a,b) and a comparison of these works indicates the major advances which have occurred in this field over the last few years.

Unlike animal cells, fatty acid synthesis in plants does not take place in the cytosol (see also Chapter 15). Early work indicated the localization of these reactions in a heavy organelle. Thus, working with castor beans (*Ricinus*) endosperm, Yamada et al. (1974) have shown that a 10,000*g* fraction is capable of synthesizing fatty acids from sucrose. Similarly, Weaire and Kekwick (1975) using avocado mesocarp have found that the total capacity for fatty acid biosynthesis from acetate resides in a 2000*g* fraction. The subcellular distribution of the enzymes responsible for the conversion of sucrose to the fatty acids in *Ricinus* endosperm has also been determined. With the exception of the initial phosphorylation step, proplastids have been shown to contain all the enzymes necessary to convert sucrose via pyruvate to the fatty acids (Simcox et al., 1977). Finally, the demonstration of Ohlrogge et al. (1979) that the total cellular complement of acyl carrier protein is to be found in the plastid confirms the role of this compartment in fatty acid biosynthesis in plant cells in general and developing seed cells in particular. Thus, it is now accepted that the two principal steps in the synthesis of fatty acids, namely the "*de novo*" system by which palmitoyl–acyl carrier protein is formed from acetyl–and malonyl–CoA and the "elongation" system which involves the addition of a further C_2 unit to form stearoyl–acyl carrier protein are both localized in the stroma of the plastid.

According to Stumpf (1980), palmitic, stearic, and after oxidation of the latter, oleic acids leave the plastid. Although there is uncertainty as to the exact site of further oxidation of oleic acid to linoleic and α-linoleic acids, there is recent evidence that the ER may be involved. Previously one had assumed that, starting with glycerol-3-phosphate, phosphohydrolase and fatty acid acyl transferases were required to produce triacylglycerols in the glycerol phosphate pathway (Gurr, 1980). Indeed a number of studies have demonstrated the presence of such enzyme activities in microsomal or ER-rich fractions from developing seed tissues (e.g., Barron and Stumpf, 1962; Moore et al., 1973; Stymne and Appelquist, 1978; Vick and Beevers, 1977). According to Slack et al. (1979), however, the induction of further bonds into oleic acid can only occur when the latter is incorporated into phosphatidyl choline. These reactions nevertheless do take place in the ER (Stovart et al., 1982), which is well known to be one of the major sites of phospholipid biosynthesis (see Chapter 15). The ER is also implicated in those reactions required to produce unusual fatty acids in LBs; thus Galliard and Stumpf (1966) have shown that the hydroxylation reaction responsible for the synthesis of ricinoleic acid from oleic acid is localized in the microsomal fraction of *Ricinus* endosperm homogenates.

Ultrastructural observations on developing seeds support the above biochemical studies. According to Frey-Wyssling et al. (1963), small vesicles, which they termed *spherosomes*, are pinched off from the ER. Through the accumulation of triacylglycerols these grow into the LBs. This original scheme has been modified by Schwarzenbach (1971) and more recently by Wanner et al. (1981) (See Figure 11.9). Taking into account the half-membrane covering to the LB, these authors have suggested that the triacylglycerols are not deposited in the lumen of the ER

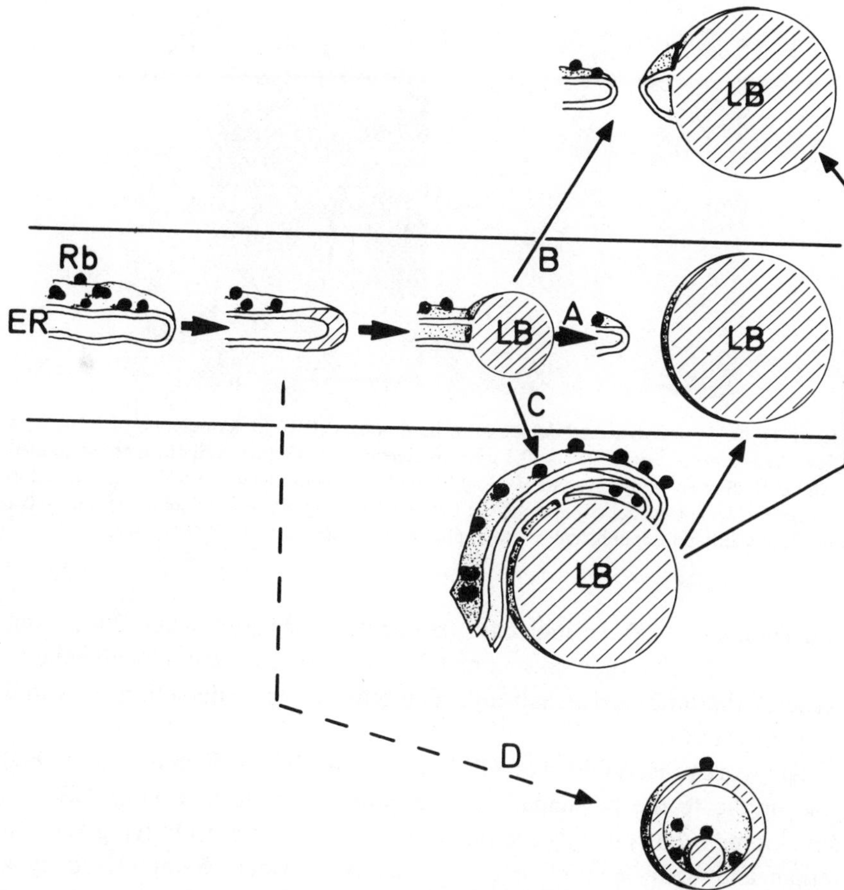

Figure 11.9. Postulated schemes for lipid body (LB) development from the ER according to Wanner et al. (1981). In A the LB is formed without a remaining piece of ER; in B detachment results in a residual piece of attached ER; in C multiple connections with the ER are present during development; in D lipid-rich vesicles are formed rather than a LB.

but rather between the lipid bilayers of the ER. This is supported by direct attachments between ER and LBs (see, e.g., Figure 11.10a). In one particular case, that of the watermelon (*Citrullus vulgaris*), rER attachments are still visible at seed maturity and apparently (see Section 11.7) also survive seed dessication. These attachments are not to be confused with the so-called lipid-rich membranes which are sometimes seen as appendages to LBs, especially in germinating seeds of the Brassicaceae (Wanner

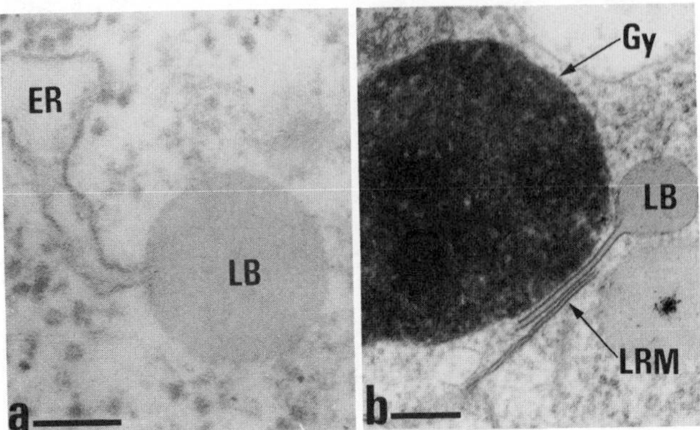

Figure 11.10. Formation and depletion of lipid bodies (LB) (a) Association of a LB with ER in *Neurosposa crassa*. Bar = 0.1 μm. (b) Germinating (3 days) cell of rapeseed stained with DAB showing a glyoxysome (Gy) and "lipid-rich membranes" (LRM) which develop as a result of the unequal collapse of the LBs surrounding half-unit membrane. Bar = 0.1 μm. Published and unpublished micrographs of G. Wanner and R. R. Theimer.

and Theimer, 1978). In such cases the appendages are thicker (20–23 nm, see Figure 11.10b) than normal membranes and represent a localized collapse of the limiting LB half-unit membranes due to depletion of stored triacylglycerols.

The work of Bergfeld et al. (1978) as well as that of Wanner et al. (1981) also implicate the participation of plastids in the formation of LBs. In developing seeds plastids are often associated with closely lying ER cisternae and with developing LBs. They are further characterized by a conspicuous undulation of both inner and outer envelope membranes which Wanner et al. (1981) interpret as indicating the intense export of fatty acids from the plastids.

11.7. LIPID BODY BREAKDOWN: THE ORIGIN OF GLYOXYSOMES

The utilization of triacylglycerols present in the LBs or oil-rich seeds is intimately associated with the biogenesis of glyoxysomes. As already mentioned (Chapter 4), the enzyme complement of these organelles is partly responsible for the reactions involved in converting lipid to sucrose (gluconeogenesis). The rise and fall in their amounts and activities during

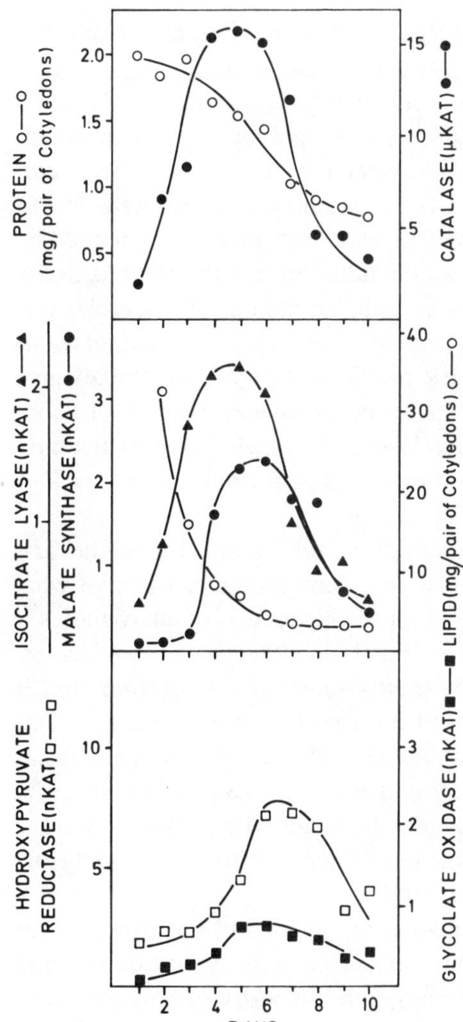

Figure 11.11. Changes in glyoxysomal enzyme activities and lipid content of watermelon seedlings during germination in the dark. Published results of Wanner et al. (1982).

seed germination (see Figure 11.11) belongs to the most intensively studied plant phenomena (Beevers, 1975, 1980, 1982a,b).

Although it is usual to consider the starting point for reactions associated with glyoxysomes as being the presence of free fatty acids, the latter must first be released from the triacylglycerols by the action of lipases (Galliard, 1980). The majority of seed lipases have alkaline pH optima (Huang and Moreau, 1978; Rosnitschek and Theimer, 1980), but the frequently used castor bean system has both alkali and acid lipases whose activity are prominent at different stages of germination (Muto and Beevers, 1974). Work on the localization of lipolytic activity in germinating seeds shows clearly the noninvolvement of the cytosol. Early studies (Ory et al., 1968) indicated the LB half-membrane as being the site of acid lipase activity in castor beans. This was confirmed and extended by Muto and Beevers (1974) who showed in addition that the alkaline lipase (unlike the acid lipase capable only of utilizing monoacyl glycerols) was primarily associated with the glyoxysome membranes, although a significant amount was also present in ER fractions. In contrast, the alkaline lipase of rape seeds does not appear to be associated with the ER but rather with the "lipid-rich membrane" appendages which are particularly characteristic of this species (Theimer and Rosnitschek, 1978; see also Section 11.6).

These latter observations implicating the role of the ER in the first stages of storage lipid breakdown sets the scene for a discussion of what might be termed the "classical model for the biogenesis of microbodies." Originally proposed by De Duve and Baudhuin (1966) for animal tissues, this model envisages the microbody (peroxisome) as arising from the ER by a process of budding-off followed by a period of enlargement. This concept has also been adopted for the case of the glyoxysome and has been promulgated mainly by Beevers and associates (e.g., Beevers, 1978, 1979; Lord, 1980). In its strictest form, however, (i.e., both membrane and enzymic content of the glyoxysome are derived from the ER) it is no longer tenable (Kindl, 1982a; Lord, 1982).

The initial observations of Redman et al. (1972) and Lazarow and De Duve (1973a,b), which showed that labeled catalase is first detected in the cytosol of liver cells before appearing in the peroxisomes, have now found their counterpart in studies on glyoxysomes (Frevert and Kindl, 1978, 1980; Frevert et al., 1980; Kindl, 1982a; Kindl, Schiefer, and Löffler, 1980). Furthermore, the possibility that these cytosolic pools might have originally been present in the ER lumen and were released upon homogenization into the supernatant fraction has been shown to be unlikely as a result of control experiments involving ER content markers (Kindl, 1982b; Lazarow et al., 1982).

In the case of the enzyme malate synthase, participation of the ER has been invoked. Thus, in addition to the glyoxysomal form Gonzalez and

Beevers (1976) and Gonzalez (1982) have shown the presence of a peak of malate synthase in sucrose density gradients of castor bean homogenates which bands at the same position as do ER markers. This has been interpreted in terms of the classical model as being a precursor form which is synthesized cotranslationally at the ER, and glycosylated there before being sequestered into the glyoxysome (Lord and Bowden, 1978; Mellor et al., 1978). However, two lines of evidence suggest caution with this interpretation. First, experiments of Köller and Kindl (1978) and Kindl, Köller, and Frevert (1980) have shown that in cucumber seeds there is a highly aggregated homomeric form of malate synthase which possesses sedimentation characteristics superficially similar to ER vesicles, but which can, under appropriate conditions, be separated from the ER. Second, there is now uncertainty as to whether malate synthase is regularly glycosylated (Bergner and Tanner, 1981).

The results of *in vitro* translation experiments also speak against cotranslational synthesis of glyoxysomal enzymes. Suggesting the absence of a leader peptide sequence are the syntheses of isocitrate lyase (Kindl, 1982a; Weir et al., 1980), malate synthase (Kruse et al., 1981), and catalase (Kindl, 1982a), all of which show MWs identical to the native glyoxysomal ones. Furthermore, in accordance with similar results from animal systems (Goldman and Blobel, 1978), *in vitro* translations of the appropriate plant mRNA in the presence of ER vesicles gave no indication of internalization or cotranslational modification (Roberts and Lord, 1981b). As judged by *in vitro* import experiments, posttranslational modifications other than insertion into the glyoxysome apparently do not take place for the above-mentioned examples (Kruse et al., 1981; Zimmermann and Neupert, 1980). This is not the case with malate dehydrogenase, which has a subunit size, upon *in vitro* translation, 7.5 kg greater than the native glyoxysomal species (Hock and Gietl, 1982; Walk and Hock, 1978). This is much too large to be interpreted as a leader sequence. Import experiments (Hock and Gietl, 1982) with the *in vitro* translated enzyme were successful but glyoxysome-internalized native-size enzyme was only obtained after a subsequent protease treatment, indicating the cleavage *in situ* of the large extra polypeptide sequence during or at the end of the entry into the glyoxysome.

Whereas it can now be regarded as certain that the enzyme complement of the glyoxysome is translated on cytosolic rather than ER-bound ribosomes, the role of the ER as a membrane donor is not so clear. There are occasional micrographs in the literature where continuities between ER and microbody are depicted (Frederick et al., 1968; Goeckermann and Vigil, 1975; Vigil, 1970) and there are even more cases where an "association" of microbody with the ER has been noted (see Gerhardt, 1978, and Chapter 4, for literature). Such micrographs are indeed suggestive of an ER origin for the glyoxysomal membrane and find support in some biochemical measurements [e.g., similarities in phospholipid composition (Donaldson and Beevers, 1977) and antigenic properties (Bowden and Lord, 1976; Hock 1974)]. Difficulties are, however, encountered when glyoxysomal and ER membrane polypeptide patterns are compared. Some authors (Bowden and Lord, 1976; Brown et al., 1976) present evidence for their identical nature while others (Kruse and Kindl, 1982; Kindl, 1982b) refute this. Unfortunately, in some of these papers the inadequate removal of the matrix and other contaminating proteins may have clouded the picture somewhat. This is borne out at least in the case of the rat liver peroxisome whose rigorously purified membrane has a strikingly different polypeptide pattern when compared to that of similarly treated ER membranes (Fujiki et al., 1982; Lazarow et al., 1982). Nevertheless, it is possible that at least one protein is common to both ER and glyoxysomal membranes, namely the NADH–cytochrome *b* oxidoreductase whose activity is measurable in purified preparations of both organelles (Donaldson et al., 1981). On the other hand, since it has recently been shown in animal cells that this enzyme protein is translated on free polysomes (Okada et al., 1982), an ER-independent insertion into the glyoxysomal membrane is also possible.

The inconclusive nature of the results pertaining to an ER origin for peroxisomal membranes in the liver has led Lazarow and co-workers (1980, 1982) to postulate that peroxisomes arise by subdivision and growth of preexisting ones. A similar suggestion has also been made for fungal glyoxysomes (Zimmermann and Neupert, 1980). In the case of oil-rich seeds there is evidence for the presence of glyoxysomes which are present before, during, and after seed dessication (e.g., Hock, 1974; Kindl, 1982b; Miernyk and Trealease, 1981). These can be recognized at the beginning of germination (see Figure 11.12a) but are present in such small numbers that it is hard to visualize all subsequent glyoxysomes being derived from them. In this respect a most revealing cytochemical study on the origin of glyoxysomes in germinating watermelon seeds has just been published (Wanner et al., 1982), which in my opinion does much to clear the con-

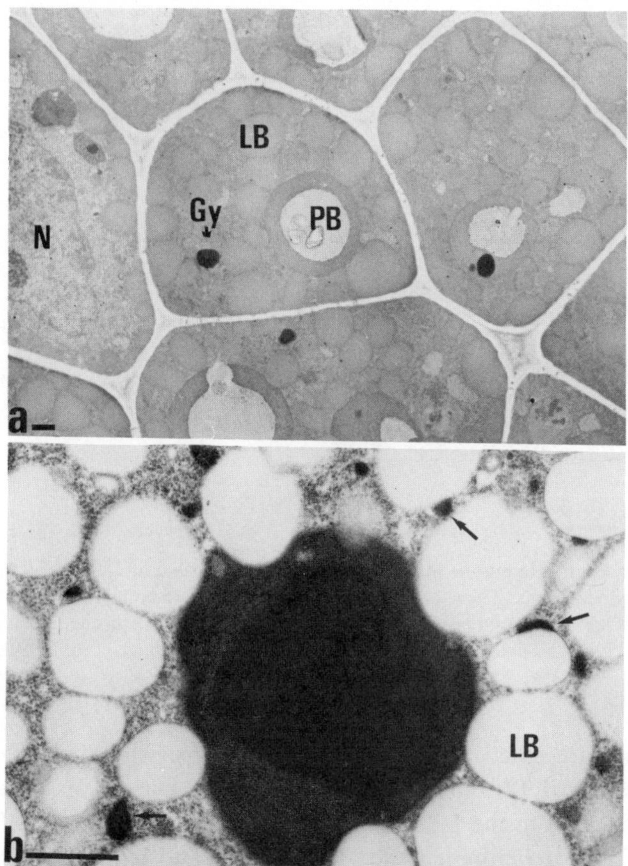

Figure 11.12. (a) Low-power view of watermelon cotyledon cells 1–2 days after imbibition. The cells are packed with lipid bodies (LB) and protein bodies (PB). Nuclei (N) and a few glyoxysomes (Gy) which have survived seed dessication are also visible. (b) Initial stages of LB breakdown are seen in cells of watermelon cotyledons after 3–4 days germination. DAB-positive structure (arrows) are visible between and attached to the LBs. Bar = 1 μm. Published and unpublished micrographs of G. Wanner et al. (1982).

fusion as to the role of the ER. While this paper does not substantiate the "classical" model for microbody biogenesis, it does confirm a participation of the ER in the form of derivatives which survive dessication.

Using the DAB method (see Chapter 4) for the detection of catalase, Wanner et al. (1982) have been able to show that in addition to the few glyoxysomes which survive dessication two other structures react posi-

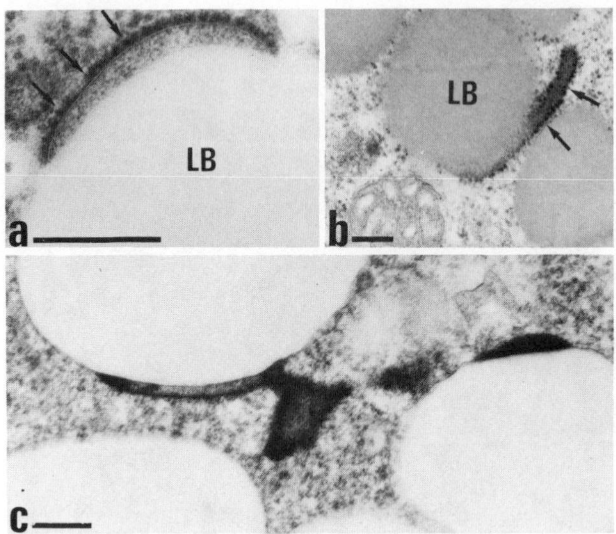

Figure 11.13. (a,b) Membrane attachments to LBs are visible in 3–4-day germinated watermelon cotyledon cells. These are studded with ribosomes (arrows). (c) The lumen of the membrane attachments stain positively with DAB, indicating the presence of catalase. Bars = 0.25 μm. Published and unpublished micrographs of G. Wanner et al. (1982).

tively. Both are small, more or less flattened vesicles which bear ribosomes (Figure 11.12). One type lies free in the cytoplasm, whereas the other is attached to the LBs and represents a remnant of the ER carried over from seed development (see Section 11.6 and Figure 11.13). As germination proceeds, the free ER vesicles also become attached to the surface of the LBs and then both these and the LB appendages lose their ribosomes. Gradually, and at the expense of the LB, these "proglyoxosome" structures enlarge and in so doing engulf the LB trapping portions of the cytoplasm (Figure 11.14). During the enlargement additional contacts between the LB and the proglyoxysome are created. As a result of the irregular enlargement, glyoxysomes with a pleomorphic structure are found produced. These ultrastructural changes are summarized in Figure 11.15. Clearly, due to the loss of ribosomes and their growth around the LBs, it can be expected that the resultant glyoxysomal membrane will be different from its ER originator, at least in its protein make-up. Thus, the controversy as to whether or not the ER is involved in glyoxysome biogenesis is very conveniently solved with this scheme. Of course, not every

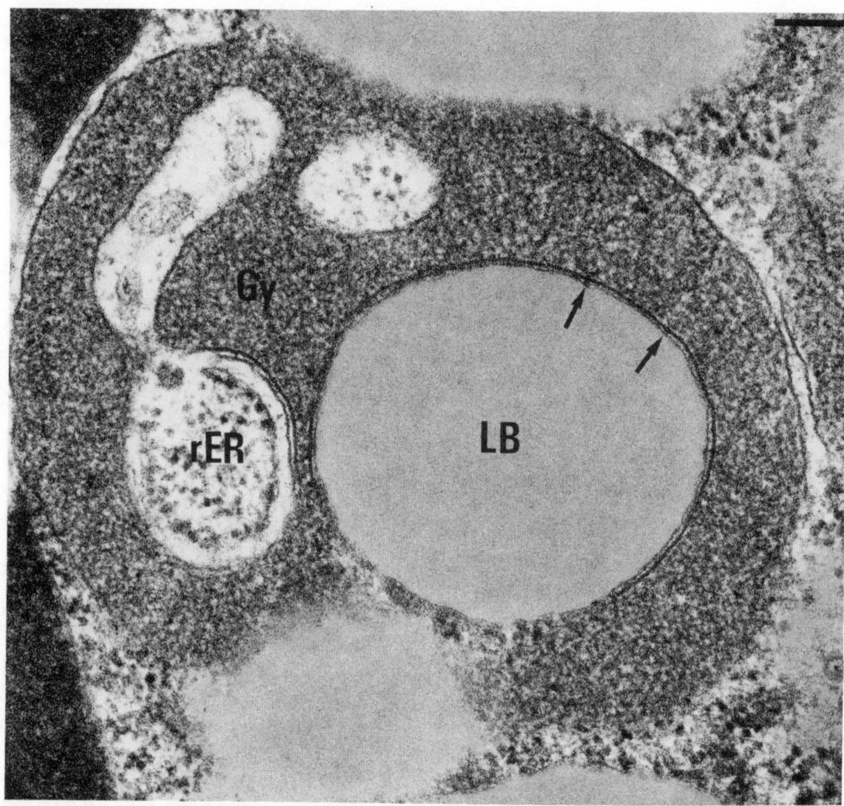

Figure 11.14. An intimate relationship between glyoxysome (Gy) and lipid body (LB) is to be recognized in 5–6-day old watermelon cotyledon cells. Additional contact sites between LB and glyoxysome are seen (arrows) as well as invaginated cytoplasm (in one case with a piece of rER) which arises as a result of the unequal growth of the glyoxysome and the concomitant shrinkage in volume of the LB. Bar = 0.25 μm. Published micrograph of G. Wanner et al. (1982).

oil-rich seed possesses LBs with attached ER remnants (Wanner et al., 1981), but since ER is always present in the cytoplasm in small amounts having survived dessication (see also Chapter 16), it might be justifiable to generalize on these observations.

Although sucrose density gradient analysis (Theimer, 1982) has confirmed the cytochemical data, the origin of the catalase in these "proglyoxysomes" is not yet known. According to current opinion, this should be posttranslationally inserted. But what is the function of the ribosomes

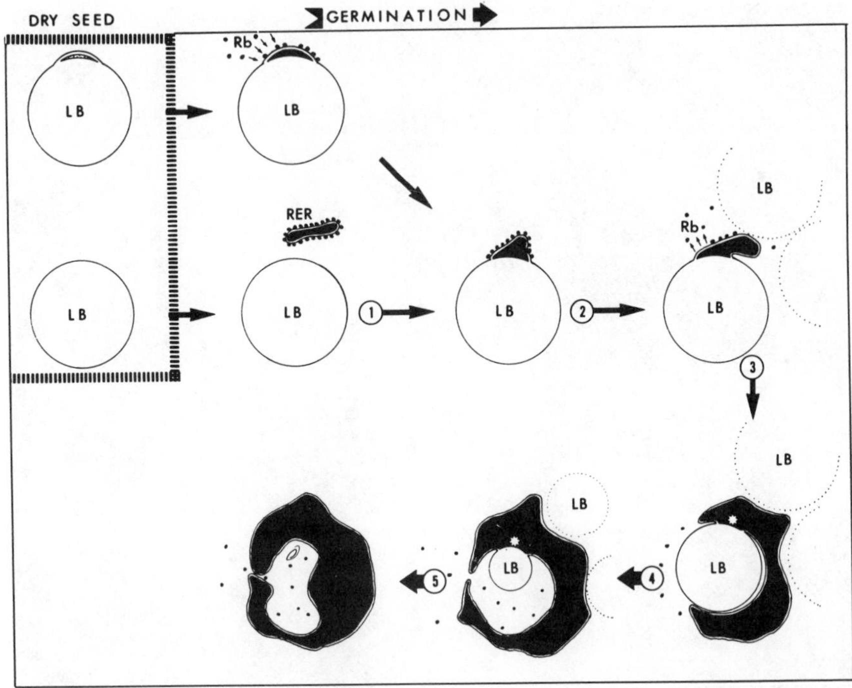

Figure 11.15. A summary of glyoxysome biogenesis is depicted in this scheme of Wanner et al., 1982, who describe the individual stages as follows: (1) Fusion of rER cisternae with lipid bodies leads to a structure homologous to an appendix (ER remnant from lipid body formation) with attached ribosomes. (2) Membrane property changes are signalized by the detachment of ribosomes. (3) Enlargement of glyoxysome around lipid body leads to a pleomorphic structure. The white star indicates the primary site of fusion. (4) The cavity remaining after lipid utilization is filled with cytoplasm and eventually begins to collapse.

initially present on the LB–ER appendages? Do they synthesize specific glyoxysomal proteins, or perhaps receptors for the recognition and transport of cytosolic proteins into the glyoxysomes, or is it possible that glyoxysomal enzyme synthesis starts cotranslationally but changes very quickly as the ribosomes are shed from the developing proglyoxysomes?

REFERENCES

Allen, C. F., Good, P., Mollenhauer, H. H., and Totten, C. (1971). *J. Cell Biol.* **48**, 542–546.

Ashton, F. M. (1976). *Ann. Rev. Plant Physiol.* **27,** 95–117.

Badenoch-Jones, J., Spencer, D., Higgins, T. J. V., and Millerd, A. (1981). *Planta* **153,** 201–209.

Bailey, C. J., Cobb, A., and Boulter, D. (1970). *Planta* **95,** 103–118.

Bain, J. M. and Mercer, F. V. (1966a). *Austr. J. Biol. Sci.* **19,** 49–68.

Bain, J. M. and Mercer, F. V. (1966b). *Austr. J. Biol. Sci.* **19,** 69–84.

Barron, E. J. and Stumpf, P. K. (1962). *Biochim. Biophys. Acta* **60,** 329–337.

Baumgartner, B. and Chrispeels, M. J. (1976). *Plant Physiol.* **58,** 1–6.

Baumgartner, B., Tokuyasu, K. T., and Chrispeels, M. J. (1978). *J. Cell Biol.* **79,** 10–19.

Baumgartner, B., Tokuyasu, K. T., and Chrispeels, M. J. (1980). *Planta* **150,** 419–425.

Beachy, R. N., Thompson, J. F., and Madison, J. T. (1978). *Plant Physiol.* **61,** 139–144.

Bechtel, D. B. and Juliano, B. O. (1980). *Ann. Bot.* **45,** 503–509.

Beevers, H. (1975). In *Recent Advances in the Chemistry and Biochemistry of Plant Lipids* (T. Galliard and E. I. Mercer, eds.), Vol. 9, pp. 287–299.

Beevers, H. (1978). In *Regulation of Developmental Processes in Plants* (H. R. Schütte and D. Gross, eds.), pp. 159–172. VEB Gustav Fischer Verlag, Jena.

Beevers, H. (1979). *Ann. Rev. Plant Physiol.* **30,** 159–193.

Beevers, H. (1980). In *The Biochemistry of Plants* (P. K. Stumpf, ed.), Vol. 4, pp. 117–130. Academic Press, New York.

Beevers, H. (1982a). In *Biochemistry and Metabolism of Plant Lipids* (J. F. G. M. Wintermans and P. J. C. Kuiper, eds.), pp. 223–235. Elsevier Biomedical, Amsterdam.

Beevers, H. (1982b). *Ann. NY Acad. Sci.* **386,** 243–253.

Bergfeld, R., Hong, Y.-N., Kühnl, T., and Schopfer, P. (1978). *Planta* **143,** 297–307.

Bergfeld, R., Kühnl, T., and Schöpfer, P. (1980). *Planta* **148,** 146–156.

Bergner, U. and Tanner, W. (1981). *FEBS Lett.* **131,** 68–72.

Bewley, J. D. and Black, M. (1978). *Physiology and Biochemistry of Seeds.* Springer Verlag, Berlin.

Bollini, R. and Chrispeels, M. J. (1978). *Planta* **142,** 291–298.

Bollini, R. and Chrispeels, M. J. (1979). *Planta* **146,** 487–501.

Bollini, R., van der Wilden, W., and Chrispeels, M. J. (1982). *Physiol. Plant.* **55,** 82–92.

Boulter, D. (1981). *Adv. Bot. Res.* **9,** 1–31.

Bowden, L. and Lord, J. M. (1976). *Biochem. J.* **154,** 491–499.

Brandt, A. and Ingversen, J. (1978). *Carlsberg Res. Comm.* **43,** 451–469.

Briarty, L. G. (1980). *J. Exp. Bot.* **31,** 1387–1398.

Briarty, L. G., Coult, D. A., and Boulter, D. (1969). *J. Exp. Bot.* **20,** 358–372.

Briarty, L. G., Coult, D. A., and Boulter, D. (1970). *J. Exp. Bot.* **21,** 513–524.

Brown, R. H., Bowden, L., and Lord, J. M. (1976). *Planta* **130,** 95–96.

Burr, B. (1979). In *Seed Protein Improvement in Cereals and Grain Legumes*, pp. 159–162. Proc. Int. Symp. FAO/FAEA, Vienna.

Burr, B. and Burr, F. A. (1976). *P.N.A.S.* **73,** 515–519.

Buttrose, M. S. (1971). *Planta* **96,** 13–26.

Cameron-Mills, V. and Ingversen, J. (1978). *Carlsberg Res. Comm.* **43,** 471–489.

Cameron-Mills, V., Ingversen, J., and Brandt, A. (1978). *Carlsberg Res. Comm.* **43,** 91–102.

Campbell, W. P., Lee, J. W., O'Brien, T. P., and Smart, M. G. (1981). *Austr. J. Plant Physiol.* **8,** 5–19.

Cawley, D. B. and Houston, L. L. (1979). *Biochim. Biophys. Acta* **581,** 51–62.

Chrispeels, M. J. (1983). *Planta* **158,** 140–151.

Chrispeels, M. J. and Bollini, R. (1982). *Plant Physiol.* **70,** 1425–1428.

Chrispeels, M. J., Baumgartner, B., and Harris, N. (1976). *P.N.A.S. (USA)* **73,** 3168–3172.

Chrispeels, M. J., Higgins, T. J. V., Craig, S., and Spencer, D. (1982). *J. Cell Biol.* **93,** 5–14.

Chrispeels, M. J., Higgins, T. J. V., and Spencer, D. (1982). *J. Cell Biol.* **93,** 306–313.

Craig, S., Goodchild, D. J., and Hardham, A. R. (1979). *Austr. J. Plant Physiol.* **6,** 81–98.

Craig, S., Goodchild, D. J., and Millerd, A. (1979). *J. Histochem. Cytochem.* **27,** 1312–1316.

Craig, S., Millerd, A., and Goodchild, D. J. (1980). *Austr. J. Plant Physiol.* **7,** 339–351.

Croy, R. R. D., Gatehouse, J. A., Evans, I. M., and Boulter, D. (1980a). *Planta* **148,** 49–56.

Croy, R. R. D., Gatehouse, J. A., Evans, I. M., and Boulter, D. (1980b). *Planta* **148,** 57–63.

Croy, R. R. D., Gatehouse, J. A., Tyler, M., and Boulter, D. (1980). *Biochem. J.* **191,** 509–516.

Dangeard, P. (1922). *Comptes Rendus Acad. Sci.* **174,** 319.

Davey, F. E. and van Staden, J. (1978). *Z. Pflanzenphysiol.* **89,** 259–271.

De Duve, C. and Baudhuin, P. (1966). *Physiol. Rev.* **46,** 323–357.

Derbyshire, E., Wright, D. J., and Boulter, D. (1976). *Phytochem.* **15,** 3–24.

Dieckert, J. W. and Dieckert, M. C. (1972). In *Symposium: Seed Proteins* (G. E. Inglett, ed.), pp. 462–489. AVI, Westport, CT.

Dieckert, J. W. and Dieckert, M. C. (1976). *J. Food. Sci.* **41,** 475–482.

Donaldson, R. P. and Beevers, H. (1977). *Plant Physiol.* **59,** 259–263.

Donaldson, R. P., Tully, R. E., Young, O. A., and Beevers, H. (1981). *Plant Physiol.* **67,** 21–25.

Ellemann, T. C. (1974). *Biochem. J.* **141,** 113–118.

Evans, M., Croy, R. R. D., Hutchinson, P., Boulter, D., Payne, P. I., and Gordon, M. E. (1979). *Planta* **144,** 455–462.

Frederick, S. E., Gruber, P. J., Vigil, E. L., and Wergin, W. P. (1968). *Planta* **81,** 229–252.

Frevert, J. and Kindl, H. (1978). *Eur. J. Biochem.* **92,** 35–43.

Frevert, J. and Kindl, H. (1980). *Eur. J. Biochem.* **107,** 79–86.

Frevert, J., Köller, W., and Kindl, H. (1980). *Hoppe-Seyler's Z. Physiol. Chem.* **361,** 1557–1565.

Frey-Wyssling, A., Grieshaber, E., and Mühlethaler, K. (1963). *J. Ultr. Res.* **8,** 506–516.

Fujiki, Y., Fowler, S., Shio, H., Hubbard, A. L., and Lazarow, P. B. (1982). *J. Cell Biol.* **93,** 103–110.

Galliard, T. (1980). In *The Biochemistry of Plants* (P. K. Stumpf, ed.), Vol. 4, pp. 85–116. Academic Press, New York.

Galliard, T. and Stumpf, P. K. (1966). *J. Biol. Chem.* **241,** 5806–5812.

Gerhardt, B. (1978). *Cell Biology Monographs*, Vol. 5. Springer Verlag, Wien.

Goeckermann, J. A. and Vigil, E. L. (1975). *J. Histochem.* **23,** 957–973.

Goldman, B. M. and Blobel, G. (1978). *P.N.A.S. (USA)* **75,** 6066–6070.

Gonzalez, E. (1982). *Plant Physiol.* **69,** 83–87.

Gonzales, E. and Beevers, H. (1976). *Plant Physiol.* **57,** 406–409.

Graham, T. A. and Gunning, B. E. S. (1970). *Nature* **228,** 81–82.

Guilliermond, A. (1941). *The Cytoplasm of the Plant Cell*, Chronica Botanica Co., Waltham, MA.

Gurr, M. I. (1980). In *The Biochemistry of Plants* (P. K. Stumpf, ed.), Vol. 4, pp. 205–248. Academic Press, New York.

Gurr, M. I., Blades, J., Appelby, R. S., Smith, C. G., Robinson, M. P., and Nichols, B. W. (1974). *Eur. J. Biochem.* **43,** 281–290.

Harris, N. and Boulter, D. (1976). *Ann. Bot.* **40,** 739–744.

Harris, N. and Chrispeels, M. J. (1975). *Plant Physiol.* **56,** 292–299.

Harris, N., Chrispeels, M. J., and Boulter, D. (1975). *J. Exp. Bot.* **26,** 544–554.

Higgins, T. J. V. and Spencer, D. (1977). *Plant Physiol.* **60,** 655–661.

Higgins, T. J. V. and Spencer, D. (1981). *Plant Physiol.* **67,** 205–211.

Hitchcock, C. and Nichols, B. W. (1971). Academic Press, New York.

Hock, B. (1974). *Planta* **115,** 271–280.

Hock, B. and Gietl, C. (1982). *Ann. NY Acad. Sci.* **386,** 350–361.

Horner, H. T. and Arnott, H. J. (1965). *Amer. J. Bot.* **52,** 1027–1038.

Huang, A. H. C. and Moreau, R. A. (1978). *Planta* **141,** 111–116.

Hurkman, W. J. and Beevers, L. (1980). *Planta* **150,** 82–88.

Jacks, T. J., Yatsu, L. Y., and Altschul, A. M., (1967). *Plant Physiol.* **42,** 585–590.

Jacobsen, J. V. and Varner, J. E. (1967). *Plant Physiol.* **42,** 1596–1600.

Jennings, A. C., Morton, R. K., and Palk, B. A. (1963). *Austr. J. Biol. Sci.* **16,** 366–374.

Khoo, V. and Wolf, J. J. (1970). *Amer. J. Bot.* **57,** 1042–1050.

Kindl, H. (1982a). *Int. Rev. Cytol.* **80,** 193–229.

Kindl, H. (1982b). *Ann. NY Acad. Sci.* **386,** 314–328.

Kindl, H., Köller, W., and Frevert, J. (1980). *Hoppe-Seyler's Z. Physiol. Chem.* **361,** 465–467.

Kindl, H., Schiefer, S., and Löffler, H.-G. (1980). *Planta* **148,** 199–207.

Köller, W. and Kindl. H. (1978). *FEBS Lett.* **88,** 83–86.

Konzak, E. F. (1977). *Advances in Genetics* **19,** 407–582.

Kruse, C. and Kindl, H. (1982). *Ann. NY Acad. Sci.* **386,** 499–501.

Kruse, C., Frevert, J., and Kindl, H. (1981). *FEBS Lett.* **129,** 36–38.

Larkins, B. A. (1981). In *The Biochemistry of Plants* (A. Marcus, ed.), Vol. 6, pp. 449–489. Academic Press, New York.

Larkins, B. A. and Dalby, A. (1975). *Biochem. Biophys. Res. Comm.* **66,** 1048–1054.

Larkins, B. A. and Hurkman, W. J. (1978). *Plant Physiol.* **62,** 256–263.

Lazarow, P. B. and De Duve, C. (1973a). *J. Cell Biol.* **59,** 491–506.

Lazarow, P. B. and De Duve, C. (1973b). *J. Cell Biol.* **59,** 507–524.

Lazarow, P. B., Robbi, M., Fujiki, Y., and Wong, L. (1982). *Ann. NY Acad. Sci.* **386,** 285–300.

Lazarow, P. B., Shio, H., and Robbi, M. (1980). In *Biological Chemistry of Organelle Formation* (T. Bücher, W. Sebald, and H. Weiss, eds.), pp. 187–206. Springer Verlag, Berlin.

Lee, K. H., Jones, R. A., Dalby, A., and Tsai, C. Y. (1976). *Biochem. Genet.* **14,** 641–650.

Liener, I. E. (1976). *Ann. Rev. Plant Physiol.* **27,** 291–319.

Lord, J. M. (1980). *Subcell. Biochem.* **7,** 171–211.

Lord, J. M. (1982). *Ann. NY Acad. Sci.* **386,** 362–376.

Lord, J. M. and Bowden, L. (1978). *Plant Physiol.* **61,** 266–270.

Lott, J. N. A. (1980). In *The Biochemistry of Plants* (N. E. Tolbert, ed.), Vol. 1, pp. 589–624. Academic Press, New York.

Lott, J. N. A. and Vollmer, C. M. (1973). *Protoplasma* **78,** 255–271.

Matile, P. H. (1968). *Z. Pflanzenphysiol.* **58,** 365–368.

Mellor, R. B. and Lord, J. M. (1979). *Planta* **146,** 147–153.

Mellor, R. B., Bowden, L., and Lord, J. M. (1978). *FEBS Lett.* **90,** 275–278.

Mellor, R. B., Roberts, L. M., and Lord, J. M. (1980). *J. Exp. Bot.* **31,** 993–1003.

Miége, M.-N. (1982). In *Encyclopedia of Plant Physiology*, Vol. 14A (D. Boulter and B. Parthier, eds.), pp. 291–345.

Miernyk, J. A. and Trealease, R. N. (1981). *Plant Physiol.* **67,** 341–346.

Miflin, B. J. and Burgess, S. R. (1982). *J. Exp. Bot.* **33,** 251–260.

Miflin, B. J. and Shewry, P. R. (1979). In *Cereals* (D. C. Laidman and R. Wyn Jones, eds.), pp. 239–273. Academic Press, London.

Miflin, B. J., Burgess, S. R., and Shewry, P. R. (1981). *J. Exp. Bot.* **32,** 199–219.

Miflin, B. J., Mathews, J. A., Burgess, S. R., Faulks, A. J., and Shewry, P. R. (1980). In *Genome Organization and Expression in Plants* (C. J. Leaver, ed.), pp. 233–244. Plenum Press, New York.

Mikala, J. and Kohlemainen, L. (1972). *Planta* **104,** 167–177.

Moore, T. S., Lord, J. M., Kagawa, T., and Beevers, H. (1973). *Plant Physiol.* **52,** 50–53.

Müntz, K. (1982). In *Encyclopedia of Plant Physiology*, Vol. 14A (D. Boulter and B. Parthier, eds.), pp. 505–558.

Muto, S. and Beevers, H. (1974). *Plant Physiol.* **54,** 23–28.

Nagahashi, J. and Beevers, L. (1978). *Plant Physiol.* **61,** 451–459.

Neumann, D. and Weber, F. (1978). *Biochem. Physiol. Pflanzen* **173,** 167–180.

Nishimura, M. and Beevers, H. (1978). *Plant Physiol.* **62,** 44–48.

Ohlrogge, J. B., Kuhn, D. N., and Stumpf, P. K. (1979). *Proc. Nat. Acad. Sci. (USA)* **76,** 1194–1198.

Okada, Y., Frey, A. B., Guenthner, T. M., Oesch, F., Sabatini, D. D., and Kreibisch, G. (1982). *Eur. J. Biochem.* **122,** 393–402.

Öpik, H. (1966). *J. Exp. Bot.* **17,** 427–439.

Öpik, H. (1968). *J. Exp. Bot.* **19,** 64–76.

Ory, R. L. and Henningsen, K. W. (1969). *Plant Physiol.* **44,** 1488–1498.

Ory, R., Yatsu, L. Y., and Kirchner, H. W. (1968). *Arch. Biochem. Biophys.* **123,** 255–264.

Osborne, T. B. (1895). *J. Amer. Chem. Soc.* **17,** 539–567.

Parker, M. L. (1980). *Ann. Bot.* **46,** 29–36.

Pernollet, J. C. (1978). *Phytochem.* **17,** 1473–1480.

Racusen, D. and Foote, M. (1971). *Can. J. Bot.* **49,** 2107–2111.

Redman, C. B., Grab, D. J., and Irukulla, R. (1972). *Arch. Biochem. Biophys.* **152,** 496–501.

Rest, J. A. and Vaughan, J. G. (1972). *Planta* **105,** 245–262.

Roberts, L. M. and Lord, J. M. (1981a). *Eur. J. Biochem.* **119,** 31–41.

Roberts, L. M. and Lord, J. M. (1981b). *Eur. J. Biochem.* **119,** 43–49.

Rosnitschek, I. and Theimer, R. R. (1980). *Planta* **148,** 193–198.

Schnarrenberger, C., Oeser, A., and Tolbert, N. E. (1972). *Planta* **104,** 185–194.

Schwarzenbach, A. M. (1971). *Cytobiologie* **4,** 145–147.

Simcox, P. D., Reid, E. E., Canvin, D. T., and Dennis, D. T. (1977). *Plant Physiol.* **59,** 1128–1132.

Slack, C. R., Roughan, P. G., and Browse, J. (1979). *Biochem. J.* **179,** 649–656.

Smith, C. G. (1974). *Planta* **119,** 125–142.

Snider, M. D. and Robbins, P. W. (1981). *Meth. Cell Biology* **23,** 89–100.

Sorokin, H. P. (1967). *Amer. J. Bot.* **54,** 1008–1016.

Srivastava, L. M. and Paulson, R. E. (1968). *Can. J. Bot.* **46,** 1447–1453.

Stumpf, P. K. (1977). In *Lipids and Lipid Polymers in Higher Plants* (M. Tevini and K. Lichtenthaler, eds.), pp. 75–84. Springer Verlag, Berlin.

Stumpf, P. K. (1980). In *The Biochemistry of Plants* (P. K. Stumpf, ed.), Vol. 4, pp. 177–204. Academic Press, New York.

Stovart, K., Stymne, S., and Glad, G. (1982). In *Biochemistry and Metabolism of Plant Lipids* (J. F. G. M. Wintermans and P. J. C. Kuiper, eds.). Elsevier Biomedical, Amsterdam.

Stymne, S. and Appelquist, L.-A. (1978). *Eur. J. Biochem.* **90,** 223–229.

Sun, S. M., Buchbinder, B. U., and Hall, T. C. (1975). *Plant Physiol.* **56,** 780–785.

Sun, S. M., Mutschler, M. A., Bliss, F. A., and Hall, T. C. (1978). *Plant Physiol.* **61,** 918–923.

Tanaka, K., Sugimoto, T., Ogawa, M., and Kasai, Z. (1980). *Agric. Biol. Chem.* **44,** 1633–1639.

Theimer, R. R. (1982). Abstr. N2.462. Jahrestagung der Dtsch Bot. Ges (Freiburg).

Theimer, R. R. and Rosnitschek, I. (1978). *Planta* **139,** 249–256.

Thomson, J. A., Schroeder, H. E., and Tassie, A. M. (1980). *Austr. J. Plant. Physiol.* **7,** 271–282.

Vick, B. and Beevers, H. (1977). *Plant Physiol.* **59,** 459–463.

Vigil, E. L. (1970). *J. Cell Biol.* **46,** 435–454.

Walk, R. A. and Hock, B. (1978). *Biochem. Biophys. Res. Comm.* **81,** 636–643.

Wanner, G. and Theimer, R. R. (1978). *Planta* **140,** 163–169.

Wanner, G., Formanek, H., and Theimer, R. R. (1981). *Planta* **151,** 109–123.

Wanner, G., Vigil, E. L., and Theimer, R. R. (1982). *Planta* **156,** 314–325.

Weaire, P. J. and Kekwick, R. G. O. (1975). *Biochem. J.* **146,** 425–437.

Weber, E. and Neumann, D. (1980). *Biochem. Physiol. Pflanzen* **175,** 279–306.

Weir, E. M., Riezman, H., Grienenberger, J.-M., Becker, W. M., and Leaver, C. J. (1980). *Eur. J. Biochem.* **112,** 409–477.

Yamada, M., Usami, Q., and Nakajima, K. (1974). *Plant & Cell Physiol.* **15,** 49–58.

Yamagata, H., Sugimoto, T., Tanaka, K., and Kasai, Z. (1982). *Plant Physiol.* **70,** 1094–1100.

Yatsu, L. Y. and Jacks, T. J. (1972). *Plant Physiol.* **49,** 937–947.

Yatsu, L. Y., Jacks, T. J., and Hensarling, T. P. (1971). *Plant Physiol.* **48,** 675–682.

Yoo, B. Y. (1970). *J. Cell Biol.* **45,** 158–171.

Yoo, B. Y. and Chrispeels, M. J. (1980). *Protoplasma* **103,** 201–204.

Zimmermann, R. and Neupert, W. (1980). *Eur. J. Biochem.* **112,** 225–233.

12

THE VACUOLE AND COMPARTMENTATION

12.1. THE VACUOLE AS A MULTIFUNCTIONAL COMPARTMENT

Functionally speaking, the vacuole is one of the most versatile of organelles. In certain respects it is also probably one of the best examples for demonstrating and discussing compartmentalization. The substances found in vacuoles of higher plant cells are multifarious and encompass a range of both organic (see Table 12.1) and inorganic types. Of the latter both cations and anions may be present and the literature is full of examples correlating their movement to and from the vacuole with some physiological event (e.g., stomatal and other movements) (see Haupt and Feinleib, 1979, for a collection of articles on these events). This type of compartmentalization might be termed *facultative* in the sense that it is reversible. The vacuole is also involved in the obligatory compartmentalization of a range of substances which are either toxic to metabolic reactions or deleterious to structures present within the cytosol. Plants do not possess a central excretory organ and they must dispose of their waste substances internally. Matile (1982) is therefore not far short of the mark when he says that plants live dangerously.

Some substances which are sequestered in the vacuole are not in themselves toxic but they become so through contact with other substances localized elsewhere in the cell. This is the case in clover (*Melilotus alba*) leaves where the β-glucosidase responsible for the conversion of vacuolar coumarinyl glucoside to coumarin is localized in the cell wall (Oba et al., 1981). Another example is the separation of substrate (glucosinolates) and activator (ascorbic acid) from the enzyme (myrosinase) capable of forming sinigrin (mustard oil) from them. According to Grob and Matile (1980) and Matile (1980), the enzyme is localized in the wall and ER lumen and the substrate and activator in the vacuole. There are also examples for a "tissue compartmentalization" involving vacuolar materials. Thus, Kojima et al. (1979) have demonstrated the vacuolar localization of the cyanogenic glucoside dhurrin in epidermis cells of *Sorghum bicolor* and have shown that the enzymes (β-glucosidase and hydroxynitrile lyase) which cause a release of cyanide when they come into contact with it are instead present in the mesophyll cells. In all of these cases the enzymatic conversion of vacuole-deposited substances occurs through homogenization (chewing), a piquant feature of radishes and clover but a deadly property of some other plants.

TABLE 12.1. ORGANIC CONTENTS AND PROPERTIES OF PLANT CELL VACUOLES

Substance	Category	Duration in the vacuole	Function
a. Amino acids[a] b. Carboxylic acids[b]	Intermediary products of metabolism	Temporary	Metabolic Pools (Homeostasis)
c. Alkaloids[c] d. Glucosides (Cardiac,[d] Coumaric,[e] Cyanogenic,[f] etc.) e. Glucosinolates[g] f. Saponins[h] g. Tannins[i]	Secondary products of metabolism	Permanent	Detoxification of the cytosol (Excretion)
h. Acid phosphatase,[j] -protease,[k] etc.	Hydrolytic enzymes	Permanent	Digestion (lysis)
i. Vicilin, Legumin, Zein, Hordein, etc.	Proteins	Transient	Seed maturation and germination (storage)
j. Sucrose,[l] Fructose,[m] etc.	Carbohydrate	Permanent (?)	Storage
k. Betanin[n] Anthocyanin,[o] etc.	Pigments	Permanent (?)	?

[a] Wagner, 1979.
[b] Buser and Matile, 1977; Grob and Matile, 1980; Wagner, 1981.
[c] Saunders, 1979.
[d] Löffelhardt et al., 1979.
[e] Alibert et al., 1982.
[f] Saunders and Conn, 1978.
[g] Matile, 1980.
[h] Matile, 1982.
[i] Kartusch, 1975.
[j] Boller and Kende, 1979.
[k] Heck et al., 1981.
[l] Doll et al., 1979.
[m] Moskowitz and Hrazdina, 1981.
[n] Leigh and Branton, 1976.
[o] Sasse et al., 1979.

The above examples are cases where the substrates are vacuolized, but hydrolytic enzymes (Chapter 6) are also vacuolized and can create considerable damage upon homogenization (Scherer and Morré, 1978; Chapter 18). In contrast to substances which are more or less permanently localized in the vacuole, there are others which are only temporarily stored there. One of the best, if not the best, examples of this is the rhythmic fluctuation in organic acid content in vacuoles of succulents. This is a well-known phenomenon (crassulacean acid metabolism, CAM—Osmond, 1978) and involves a dark (night) fixation of CO_2 into oxaloacetate and subsequent conversion of this into malate which is then sequestered in the vacuoles. Not only must this organic acid be stored to act as a CO_2 source for "normal" photosynthesis in the light (day), it has to be removed from its site of synthesis (the cytosol) because it inhibits the enzyme responsible for fixation (feedback inhibition of phosphoenol pyruvate carboxylase).

The amount of malate which can be stored overnight is really quite large (in the order of 0.1 *M*) and it would appear that when it gets into the vacuole, it is maintained there against a high concentration gradient (Buser and Matile, 1977). Working with isolated vacuoles from *Bryophyllum daigremontianum,* Buser-Sutter et al. (1982) have provided evidence for the existence of a permease which facilitates the exchange diffusion of malate across the TP. Alone this is incapable of causing a net accumulation of malate but could when coupled to an electrogenic proton pump (Lüttge and Ball, 1979). Although such pumps have now been established for the TP (see Chapter 6), they have not yet been demonstrated for TP vesicles or vacuoles isolated from CAM plants. Moreover, in contrast to the TP–ATPase described above, ATP does not stimulate malate uptake nor does the protonophore CCCP prevent malate accumulation (Buser-Sutter et al., 1982).

12.2. SYNTHESIS AND SEGREGATION OF LYSOSOMAL ENZYMES

Lysosomes are a well-characterized organelle of animal cells (De Duve and Wattiaux, 1966; Holtzman) and per definition are rich in acid hydrolases. Cytochemical investigations (e.g., Bainton and Farquhar, 1970; Cohn et al., 1966) indicated some time ago that elements of the ER and

GA are involved in the production of such enzymes. These studies have found their culmination in the proposal that lysosomes are formed from a region of interaction of ER and GA membranes at the *trans*-face of the Golgi complex, termed GERL (Novikoff, 1976, and Section 5.4. this volume).

In agreement with results obtained on secretory proteins (see Section 10.1), the synthesis of lysosomal enzymes also occurs on ER-bound polysomes. *In vitro* translation experiments with mRNA coding for cathepsin D and β-glucuronidase have shown that, in the absence of microsomal vesicles, precursors are synthesized with molecular weights about 2 kd larger than in their presence (Erickson and Blobel, 1979; Rosenfeld et al., 1982). Translation experiments carried out in the presence of tunicamycin have also shown that glycosylation of these enzymes occurs before the polypeptide chains are released into the ER lumen (Rosenfeld et al., 1982). Pulse-chase studies have indicated that a posttranslational processing occurs for some lysosomal enzymes e.g., β-hexosaminidase, α-glucosidase, and cathepsin D (Hasilik and Neufeld, 1980; Rosenfeld et al., 1982), leading to an active, mature form considerably smaller than the initially synthesized polypeptide. The proteolytic cleavages involved take place over a period of several hours and presumably take place after their arrival in the lysosome (Strawser and Touster, 1980; Gieselmann et al., 1983).

Since both secretory proteins and lysosomal enzymes are synthesized cotranslationally and, at least initially, follow the same transport route, one would imagine that there must be some kind of sorting mechanism guaranteeing for the export of the one and the retention of the other. It is logical to assume that this occurs within the cell, but there are observations which have led to the so-called secretion-recapture hypothesis (Hickman and Neufeld, 1972) after which the sorting-out occurs at the PM. Two lines of evidence have been the basis for this postulate. The first is that there are a number of cell lines in which a considerable proportion of the synthesized lysosomal enzyme complement is released into the culture medium (Neufeld et al., 1975, 1977). The second is that many cells are capable of internalizing lysosomal enzymes through pinocytosis (Neufeld et al., 1975; Nicol et al., 1974; von Figura and Kresse, 1974). This latter feature was originally recognized in cells derived from humans having a genetic disorder resulting in the deficiency of lysosomal enzymes. Fibroblasts from patients suffering from I-cell disease were, however, capable of taking up acid hydrolases from normal fibroblasts, while being unable to do so for their own secreted enzymes. The fact that normal fibroblasts could not take up I-cell hydrolases suggested that normal acid hydrolases were "marked" in such a way that they could be recognized by specific receptors present at the cell surface of both normal and diseased cells (Sahagian et al., 1980). Diseased cells, on the other hand, produced enzymes without this marker, thereby preventing their recapture.

It is now more or less certain that this "marker" consists of phosphomannosyl groups present on the oligosaccharide side chains of the enzyme (Hasilik et al., 1980; Natowicz et al., 1979; von Figura and Klein, 1979), and it has been shown that I-cell fibroblasts lack the enzyme responsible for the phosphorylation (Hasilik et al., 1981). The phosphate group is added in the form of PGlcNAc from UDPGlcNAc (Reitman and Kornfeld, 1981) and the subsequent removal of GlcNAc by a phosphodiesterase has been localized in GA or GERL membranes (Waheed et al., 1981). The identification, by immunological methods, of receptors at the PM for the lysosomal enzymes (Fischer et al., 1980a; Hasilik et al., 1981; Rome et al., 1979) supports the idea that lysosomal enzymes are exported in much the same way as secretory proteins and are then recaptured pinocytotically (Kaplan et al., 1977; Sando and Neufeld, 1977).

A number of observations are, however, at hand which necessitate a revision of the secretion–recapture hypothesis. Mannose-6-phosphate is known to block the receptor-mediated uptake of lysosomal enzymes (Distler et al., 1979). However, when cells are grown in its presence, intracellular enzyme levels are not appreciably reduced, which should have been the case if the enzymes must obligatorily leave the cell (Vladuhu and Ratiazzi, 1979; von Figura and Weber, 1978). Second, experiments with amines (e.g., chloroquine) (Gonzalez-Noriega et al., 1980) have led to the conclusion that intracellular receptor sites must also exist. These substances enhance the secretion of normal hydrolases, deplete the number of available PM receptor sites, and inhibit pinocytotic uptake of exogenous enzyme. They are believed to achieve these effects by raising the pH in the lumen of the lysosome to a value that does not favor the dissociation of the enzymes from their lysosomal membrane receptors (Shoh and Pooli, 1978). Phosphomannosyl receptors are also present on endo- as well as plasma membranes (Fischer et al., 1980a,b), so that when receptors for lysosomal enzymes begin to accumulate in the lysosome as a result of high pH conditions, they become depleted in number at other sites. As a result their absence in the GA or GERL leads to their not being sorted out from secretory proteins and their increased discharge at the cell surface, with less chance of being recovered.

Current views on the segregation and transport of lysosomal enzymes take into account the ability to recapture released enzymes at the cell surface but do not regard it as being the only or predominant way for these enzymes to reach the lysosomes (Steinman et al., 1983). Instead the majority of such enzymes are separated from secretory proteins at receptor regions in *trans*-face GA or GERL membranes and directed via vesicles to the lysosome. After dissociation in the lysosome the receptors are recycled back either directly or via the PM. If sufficient receptors are not available at these stations, or if the enzymes are synthesized without or with imperfect oligosaccharide side chains, they are released from the cell as if they were secretory proteins (Rosenfeld et al., 1982).

12.3. THE ORIGIN OF VACUOLAR ENZYMES IN PLANT CELLS

We have already seen that vacuoles in plant cells contain many different types of hydrolytic enzymes (Chapter 6). The vacuole as a lytic compartment, as Matile (1975) has suggested, is therefore analogous to the lysosome of animal cells. In the previous section we dwelt upon the molecular biological aspects of hydrolytic enzyme synthesis and transport in animal cells. The purpose of this section is to see whether similar mechanisms exist in the case of plant cells.

At the very start one must say that extremely little information is available. Cytologically speaking, the organelles (ER, GA) are present and there is cytochemical evidence that they are involved in vacuole formation (see Chapters 5 and 6). Presumably, the enzymes are synthesized by ER-borne ribosomes, but to date there is no direct biochemical evidence for either this or for their further routing to the higher plant vacuole.

With respect to the breakdown of PBs in seeds during germination (see Section 11.4), I have already mentioned the fact that the responsible proteolytic enzymes are present in the PB. Similar enzymes are, however, already present during the deposition phase in seed maturation. The relatively slow processing of storage proteins and the localization of this event in the PB (Chrispeels et al., 1982) is reminiscent of the situation with lysosomal enzymes (Hasilik and Neufeld, 1980; Rosenfeld et al., 1982) and with the yeast vacuolar enzyme carboxypeptidase Y (Hasilik and Tanner, 1978). Therefore, as Chrispeels et al. (1982) have stated, PB-localized acid hydrolases "might be responsible for endoproteolytic cleavage of the reserve protein polypeptides after their arrival in the protein bodies. This nicking or processing may be the consequence of existing in a lysosomal compartment and could represent the first stages in the remobilization of this protein which mainly occurs during seed germination."

If we consider storage proteins as being equivalent to lysosomal enzymes, then the lysosomal analogy for the PB vacuole is quite convincing. However, just as is the case with the animal lysosome, one must ask whether the enzymes responsible for processing are segregated spatially (implicating different routing) or temporally (sent ahead of or after the other proteins) from the major proteins which become compartmentalized in the vacuole/lysosome or whether it is an activation problem (the processing enzymes traveling the same route but only becoming active within the vacuole/lysosome).

Vacuoles, if the analogy to lysosomes is accepted, could be regarded as primary lysosomes. However, they are also capable of phagocytosing organelles. This feature is implicit in Marty's proposal for vacuole biogenesis (see Chapter 6) and is also visible in PBs during seed germination (Herman et al., 1981). Thus, perhaps it is better to consider vacuoles as being more equivalent to secondary lysosomes and the (ER?) vesicles, which transport hydrolytic enzymes to them, as being the primary lysosomes.

As we have already seen (Section 12.2), animal cells have the ability to recapture lysosomal enzymes lost from the cell. Whether plant cells can also do this is a matter of conjecture. Certainly, the structural prerequisite for a receptor-mediated endocytosis, namely coated pits in the PM (see Chapter 8), are at hand. Furthermore, the cell wall is well known to contain lytic enzymes (see Matile, 1975, and Labavitch, 1981, for literature), some of which are involved in extension growth and differentiation. Parrish (1975a,b) has presented evidence for both a vacuolar as well as a cell wall location for peroxidase and acid hydrolase activities. Is this just a superficial similarity to the situation in animal cells or are plant cells also able to recover lytic enzymes from the extracellular milieu?

REFERENCES

Alibert, G., Boudet, A. M., and Rataboul, P. (1982). In *Plasmalemma and Tonoplast: Their Functions in the Plant Cell* (D. Marme and R. Hertel, eds.), pp. 193–200. Elsevier, Amsterdam.

Bainton, D. F. and Farquhar, M. G. (1970). *J. Cell Biol.* **45,** 54–73.

Boller, T. and Kende, H. (1979). *Plant Physiol.* **63,** 1123–1132.

Buser, C. and Matile, P. (1977). *Z. Pflanzenphysiol.* **82,** 462–466.

Buser-Sutter, C., Wiemken, A., and Matile, P. (1982). *Plant Physiol.* **69,** 456–459.

Cohn, Z. A., Hirsch, J. G., and Fedorko, M. E. (1966). *J. Exp. Med.* **123,** 757–766.

Chrispeels, M. J., Higgins, T. J. V., and Spencer, D. (1982). *J. Cell Biol.* **93,** 306–313.

Distler, J., Hieber, V., Sahagan, G., Schmickei, R., and Jourdian, G. W. (1979). *Proc. Natl. Acad. Sci. USA* **76,** 4235–4239.

Doll, S., Rodier, F., and Willenbrink, J. (1979). *Planta* **144,** 407–411.

De Duve, C. and Wattiaux, R. (1966). *Ann. Rev. Physiol.* **28,** 435–492.

Erickson, A. H. and Blobel, G. (1979). *J. Biol. Chem.* **154,** 11771–11774.

Fischer, H. D., Gonzalez-Noriega, A., and Sly, W. S. (1980a). *Fed. Proc.* **39,** Abstr. Vol. 3.

Fischer, H. D., Gonzalez-Noriega, A., and Sly, W. S. (1980b). *J. Biol. Chem.* **255,** 5069–5074.

Fischer, H. D., Gonzalez-Noriega, A., Sly, W. S., and Morré, D. J. (1980b). *J. Biol. Chem.* **255,** 9608–9615.

Gieselmann, V., Pohlmann, R., Hasilik, A., and von Figura, K. (1983). *J. Cell Biol.* **97,** 1–5.

Gonzalez-Noriega, A., Grubb, J. H., Talkad, V., and Sly, W. S. (1980). *J. Cell Biol.* **85,** 839–852.

Grob, K. and Matile, P. (1980). *Z. Pflanzenphysiol.* **98,** 235–243.

Hasilik, A. and Neufeld, E. F. (1980). *J. Biol. Chem.* **255,** 4937–4945.

Hasilik, A. and Tanner, W. (1978). *Eur. J. Biochem.* **85,** 599–608.

Hasilik, A., Voss, B., and von Figura, K. (1981). *Exp. Cell Res.* **133,** 23–30.

Hasilik, A., Washeed, A., and von Figura, K. (1981). *Biochem. Biophys. Res. Commun.* **98,** 761–767.

Holtzman, E. (1976). *Lysosomes: A Survey,* Springer-Verlag, New York.

Hasilik, A., Klein, U., Waheed, A., Strecker, G., and von Figura, K. (1980). *Proc. Natl. Acad. Sci. USA* **77,** 7074–7078.

Haupt, W. and Feinleib, M. E. (eds.). (1979). Encyclopedia of Plant Physiology, Vol. 7, 1–731.

Heck, U., Martinoia, E., and Matile, P. (1981). *Planta* **151,** 198–200.

Herman, E. M., Baumgartner, B., and Chrispeels, M. J. (1981). *Eur. J. Cell Biol.* **24,** 226–235.

Hickman, S. and Neufeld, E. F. (1972). *Biochem. Biophys. Res. Commun.* **49,** 992–999.

Kaplan, A., Achord, D. T., and Sly, W. S. (1977). *Proc. Nat. Acad. Sci. USA* **74,** 2026–2030.

Kojima, M., Poulton, J. E., Thayer, S. S., and Conn, E. E. (1979). *Plant Physiol.* **63,** 1022–1028.

Labavitch, J. (1981). *Ann. Rev. Plant Physiol.* **32,** 385–406.

Leigh, R. A. and Branton, D. (1976). *Plant Physiol.* **58,** 656–662.

Löffelhardt, W., Kopp, B., and Kubellea, W. (1979). *Phytochem.* **18,** 1289–1291.

Lüttge, U. and Ball, E. (1979). *J. Membrane Biology* **47,** 401–422.

Matile, P. (1975). *The Lytic Compartment of Plant Cells,* Cell Biol. Monographs vol. 1, Springer-Verlag, Vienna.

Matile, P. (1980). *Biochem. Physiol. Pflanzen* **175,** 722–731.

Matile, P. (1982). *Physiol. Veg.* **20,** 303–310.

Moskowitz, A. H. and Hrazdina, G. (1981). *Plant Physiol.* **68,** 686–692.

Natowicz, M. R., Chi, M. M. Y., Lowry, O. H., and Sly, W. S. (1979). *Proc. Natl. Acad. Sci. USA* **76,** 4322–4326.

Neufeld, E. F., Lim, T. W., and Shapiro, L. J. (1975). *Ann. Rev. Biochem.* **44,** 357–376.

Neufeld, E. F., Sando, G. N., Garvin, A. J., and Rome, L. H. (1977). *Supramol. Struct.* **6,** 95–101.

Nicol, D. M., Lagunoff, D., and Pritzl, P. (1974). *Biochem. Biophys. Res. Commun.* **59,** 941–946.

Novikoff, A. B. (1976). *Proc. Natl. Acad. Sci. USA* **73,** 2781–2787.

Oba, K., Conn, E. E., Canut, H., and Boudet, A. M. (1981). *Plant Physiol.* **68,** 1359–1363.

Osmond, C. B. (1978). *Ann. Rev. Plant Physiol.* **29,** 379–414.

Parrish, R. W. (1975a). *Planta* **123,** 1–13.

Parrish, R. W. (1975b). *Planta* **123,** 15–31.

Reitman, M. D. and Kornfeld, S. (1981). *J. Biol. Chem.* **256,** 4275–4281.

Rome, L. H., Weissman, B., and Neufeld, E. H. (1979). *Proc. Natl. Acad. Sci. USA* **76,** 2331–2334.

Rosenfeld, M. G., Kreibich, G., Popov, D., Kato, K., and Sabatini, D. D. (1982). *J. Cell Biol.* **93,** 135–143.

Sando, G. N. and Neufeld, E. F. (1977). *Cell* **12,** 619–627.

Sasse, F., Backs-Hüsemann, D., and Barz, W. (1979). *Z. Naturforsch.* **34c,** 848–853.

Saunders, J. A. (1979). *Plant Physiol.* **64,** 74–78.

Saunders, J. A. and Conn, E. E. (1978). *Plant Physiol.* **61,** 154–157.

Scherer, G. F. E. and Morré, D. J. (1978). *Plant Physiol.* **62,** 933–937.

Shoh, O. and Pooli, B. (1978). *Proc. Natl. Acad. Sci. USA* **75,** 3327–3331.

Steinman, R. M., Mellman, I. S., Mueller, W. A., and Cohn, Z. A. (1983). *J. Cell Biol.* **96,** 1–27.

Strawser, L. D. and Touster, O. (1980). *Rev. Physiol. Biochem. Pharmacol.* **87,** 169–210.

Vladuhu, G. D. and Ratiazzi, M. (1979). *J. Clin. Invest.* **63,** 595–601.

von Figura, K. and Kresse, H. (1974). *J. Clin. Invest.* **53,** 85–90.

von Figura, K. and Weber, E. (1978). *Biochem. J.* **176,** 943–950.

von Figura, K. and Klein, U. (1979). *Eur. J. Biochem.* **94,** 347–354.

Wagner, G. J. (1979). *Plant Physiol.* **64,** 88–93.

Wagner, G. J. (1981). *Plant Physiol.* **67,** 591–593.

Waheed, A., Pohlman, R., Hasilik, A., and von Figura, K. (1981). *J. Biol. Chem.* **256,** 4150–4152.

13

RECOGNITION PHENOMENA

Plants occupy the same habitats (water, land) as animals and are therefore exposed to the same environmental factors. Changes in the environment are registered (*perception*) and translated into some kind of signal (*transduction*) which is then passed on to a site either within the cell or in another cell (*transmission*), where a set of events (*responses*) are invoked upon its arrival. Although the links in this chain exist for both animals and plants, there are fundamental differences in the nature of these links. Not all environmental factors are equally perceived; for example, animals are not as sensitive to changes in the light regime as plants are. Signal types are often chemically unrelated in plant and animal cells, and because of their differing anatomical construction, both transmission and response assume different forms and extremes. Thus, whereas hormones or nerve impulses are relayed through blood or nervous systems, respectively, in animal cells, the comparable long-distance transport channels in multicellular plants are the xylem and phloem. Moreover, due to their rigid cell walls and the relatively small amounts of contractile protein (the actomyosin system), plants are restricted in their ability to move in response to environmental changes. Instead the response in plants is expressed much more in terms of growth and morphogenesis.

As might be expected, membranes are involved in this chain of events, particularly as organelles of perception and as targets (either primary or secondary) for transmitted signals. The purpose of this chapter is then to examine, using a few selected examples, how and which membranes in plant cells are involved in these regulatory events.

13.1. GRAVIPERCEPTION

Plants respond to changes in their orientation with respect to gravity by differential growth along the vegetative axis. This is termed *gravitropism* and has been a phenomenon of great interest to botanists and plant physiologists for over a century (see, e.g., Nemeč, 1901; Volkmann and Sievers, 1979). Roots are regarded as positively geotropic because they grow in the direction of the gravitational field; shoots, in contrast and for obvious reasons, are termed *negatively gravitropic*. Only minimal changes in the parameters of the gravitational field are necessary to elicit a growth response, with roots being somewhat more sensitive than shoots. Thus in a 1g field, 0.5°–10° changes in orientation for less than 30 sec are sufficient (Volkmann and Sievers, 1979).

13.1.1. Sites and Organelles Involved in Perception

Roots whose calyptra (cap) have been carefully dissected away are no longer capable of perceiving changes in gravity (Juniper et al., 1966). Since decapped roots still continue growing, the sites of perception and response cannot be identical in this case. On the other hand, graviperception in shoots is not a feature exclusive of the apex, since it is still possible to elicit a response after excision of several millimeters of the terminal tissue (Johnsson et al., 1971). Nevertheless, it has been shown by stripping or peeling experiments (Firn and Digby, 1977; Firn et al., 1978) that the perceptive region in negatively geotropic tissues is confined to the outermost cell layers.

Graviperception requires that a particle in the cells of the sensitive region must change its position as a result of a change in the orientation or intensity of the gravitational field. It can be calculated so that the minimum radius of such a particle in a 1g field is 0.1 μm (Pollard, 1971). This rules out the participation of a number of nonmembranous organelles (e.g., ribosomes, microtubules) in this process. Furthermore, the particle must be sufficiently dense to begin sedimenting or show a perceptible change in sedimentation behavior within a presentation time of less than 30 sec (Audus, 1962). As a result of these two prerequisites, the field of candidates is essentially narrowed down to starch-filled plastids of the amyloplast type (density ~1.5 $g \cdot cm^{-3}$).

The concept that amyloplasts behave as statoliths is not new (see, e.g., Haberlandt, 1902), but convincing proof of this has only been available for 10–15 years. Amyloplast size and starch content can be genetically and hormonally regulated with corresponding effects on the ability to graviperceive (Filner et al., 1970; Iversen, 1969). Moreover, graviperception recovery in decapped roots is associated with the development of amyloplasts from proplastids in the root meristem (Barlow and Grundweg, 1974).

Gravity-perceiving cells are usually termed *statocytes* because of the presence of amyloplasts (20–30). In shoots such cells are highly vacuolate and the amyloplasts tend to be gathered together in a cushion of cytoplasm at the distal cell pole (Osborne and Wright, 1977). Unfortunately, such

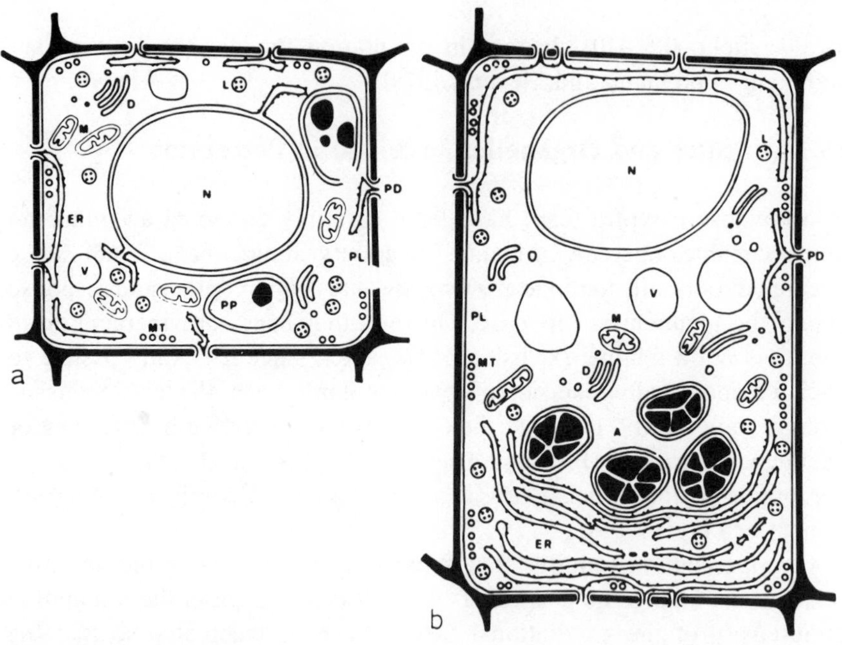

Figure 13.1. Schematic representations of meristematic (a) and statocyte (b) cells in the root cap of cress. ↓ g denotes the direction of the gravitational field. The nucleus (N) is at the proximal, the stacked ER at the distal pole. A = amyloplast; D = dictyosome; ER = endoplasmic reticulum; L = lipid droplet; M = mitochondria; MT = microtubule; PD = plasmodesmata; PP = proplastid; V = vacuole. Published drawing of Volkmann and Sievers (1979).

cells have not been adequately investigated ultrastructurally with respect to graviperception, but it would appear from the lack of organelles in the cytoplasmic cushion that the amyloplasts interact with the PM in generating the signal transduction. In contrast, root cap statocytes have been the subject of intensive investigations (see Volkmann and Sievers, 1979). In the well-studied cress (*Lepidium sativum*) root cap there are four parabolically shaped layers of statocytes situated between the meristem and the peripheral secretory tissue. The statocytes possess only small vacuoles and, when examined from positively geotropic roots, exhibit a characteristic polar distribution (see Figures 13.1b and 13.2a). Typically, the PM at the distal cell pole is covered with several layers of rER cisternae upon which the amyloplasts are seen to lie. Not all of the ER are stacked in this manner; some individual cisternae are recognizable at the proximal

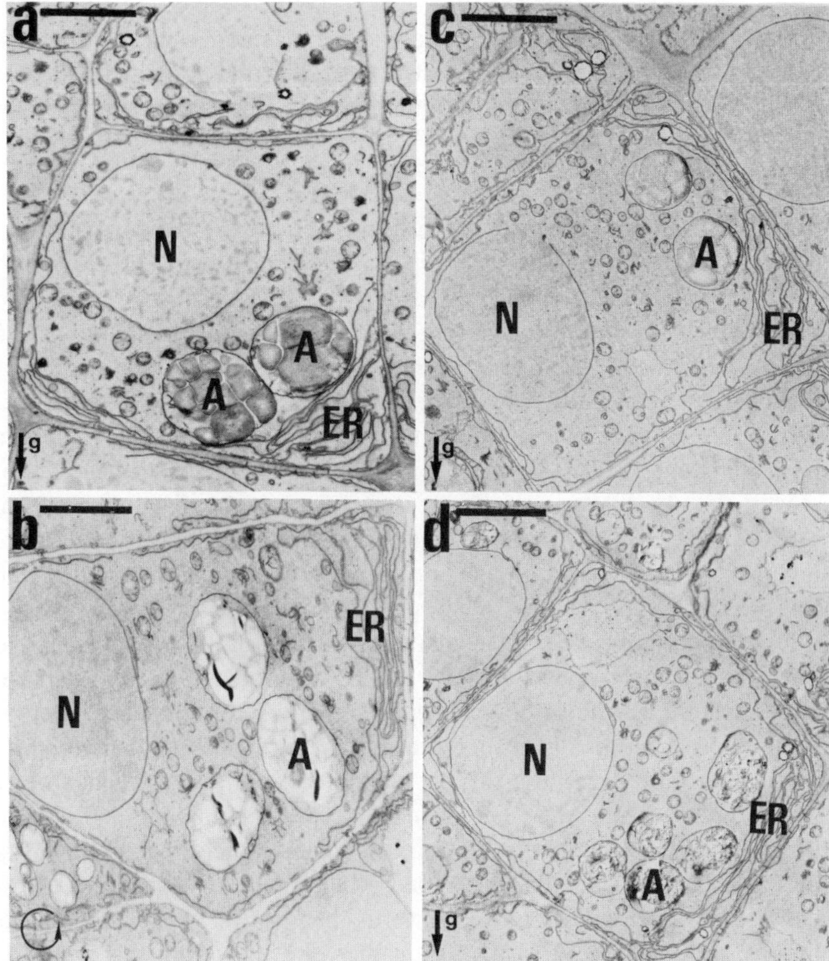

Figure 13.2. Ultrastructural aspects of graviperception in statocytes of cress roots. (a) Statocyte from the central region of the statenchyma of a positively geotropic root, fixed after 44 hr of growth. (b) Statocyte from the peripheral region of the statenchyma of a root germinated for 44 hr on a horizontal rotating clinostat. (c,d) Statocytes from the upper (c) and lower (d) flank regions of the statenchyma of a positively geotropic root which has been briefly tilted 90° to the direction of the gravitational field before fixation. In a the amyloplasts (A) lie on a stack of ER at the distal cell pole (bottom). In b the ER at the distal cell pole (right) has been formed but amyloplasts are centrally located. Whereas in c (distal cell pole to the right of the pictures) the amyloplasts no longer press onto the ER stack, they still do so in d. $KMnO_4$ fixations; N = nucleus; ↓g denotes the direction of the gravitational field with respect to the root axis; signifies that the root has been exposed to clinostatic rotation for the whole of its development. Bar = 5 μm. Unpublished micrographs of W. Hensel and A. Sievers.

pole and these are usually in contact via the plasmodesmata with ER in neighboring cells. The nucleus is also located at the proximal cell pole and the other organelles (dictyosomes, mitochondria, vacuoles) are located more or less centrally (Sievers and Volkmann, 1972, 1977).

13.1.2. Changes in Statocyte Ultrastructure Resulting from Gravistimulation

In the vertical position (i.e., during positive gravitropism) statocytes on either side of the root axis are essentially mirror images of one another. If cress roots are tilted 90° with respect to the direction of the gravitational field, the ER stack remains at the distal cell pole (Volkmann, 1974; Sievers and Volkmann, 1977). However, whereas the amyloplasts in the statocytes of the physically upper flank of the root appear to fall away from the ER stack, contact is maintained (albeit at the cell corners) between amyloplasts and ER in statocytes of the lower flank (compare c and d in Figure 13.2).

Other roots have been investigated ultrastructurally with respect to their gravitropic response, but the results are not as clear-cut as with cress. Thus the ER in statocytes from positively geotropic maize roots is distributed symmetrically around the cell. Nevertheless, the amyloplasts do not abut directly on the distal PM: there are only a few (much less than in cress) intervening ER cisternae present (Juniper and French, 1970). Tilting through 90° also causes a sedimentation of the amyloplasts, and these latter authors have claimed in addition that a stacking of ER occurs which then slowly moves towards the position of the amyloplasts (Juniper and French, 1973). Unfortunately, the authors of this paper do not distinguish between upper and lower flank statocytes.

13.1.3. The Mechanism of Graviperception

The ultrastructural observations just described suggest that amyloplast–ER interactions are central to graviperception. That this is indeed so can be demonstrated with the help of a clinostat. The continual rotation produced by this apparatus results in an omnilateral gravistimulation; that is, the orientation of the gravitational field with respect to the plant axis is continually changing and is therefore the same from all directions. Roots and shoots under such conditions grow parallel to the axis of clinostatic rotation.

When cress seeds are allowed to germinate on a clinostat rotating at an angle of 90° to a $1g$ field, the roots which emerge and grow possess statocytes having a "normal" organelle stratification (i.e., with a stack of ER cisternae at the distal pole and the nucleus at the proximal pole). The amyloplasts, however, are located centrally (see Figure 13.2 and Sievers et al., 1976). If, on the other hand, seedlings which have germinated under normal positively geotropic conditions are exposed to 2–3 hr of horizontal clinostatic rotation, the polar distribution of organelles in the statocytes is completely lost and they are randomly distributed throughout the cell (Hensel and Sievers, 1979). Prolonged rotation (>20 hr) leads to considerable damage in statocyte cell structure through autolysis (Hensel and Sievers, 1980). Depending upon the extent of the damage such roots have a reduced ability to recover and respond to unilateral gravistimulation. From these two experiments one can infer that (a) organelle polarity in statocytes is genetically determined and (b) the first unilateral gravity-induced contact between amyloplast and ER "primes" a root for further gravitational stimulation.

The first of these inferences is supported by observations on the plagiogravitropic lateral roots of cress whose growth is maintained at an angle of 65° with respect to the gravitational field. Whereas the statocytes in the upper flank show an organelle distribution corresponding to that of a lower flank statocyte in a positively geotropic main root which has been tilted through 90° (e.g., Figure 13.2d)—that is, with amyloplasts lying on a stack of ER in the corner between distal periclinal and anticlinal cell walls—those in the lower flank have their amyloplasts lying below rather than above the stacked ER (Sievers and Schmitz, 1973). The second inference finds support in the clinostat experiments of Hensel and Sievers (1981). These authors have examined the effect of interrupting the growth of clinostat-germinated cress seeds by short-term (10-min) exposures to unilateral gravity. If the roots are placed vertically (i.e., in the positively geotropic position) for this period, contact between amyloplasts and the distal stack of ER is made possible. Clinostatic rotation for a further 20 hr results in partial hydrolysis of the statocytes. This contrasts with that which ensues after the roots have been held in an inverted position. In this case ER and amyloplasts are held apart and subsequent prolonged clinostatic rotation does not result in ultrastructural changes.

Although it is abundantly clear that ER–amyloplast contact is a prerequisite for graviperception in roots and that the perception itself involves a modification of this status, the actual nature of ER–amyloplast interaction remains a mystery. Certainly the bodily movement or translocation of the amyloplasts is, as such, not responsible, which differentiates this type of graviperception from that which occurs in the *Chara* rhizoid (Sievers and Volkmann, 1979). Sievers and Volkmann (1972) have suggested that the perception mechanism *sensu strictu* is a change in pressure distribution, but there is no direct proof for this.

13.2. RED LIGHT PERCEPTION

Exposure to red light (~660 nm) results in a large number of effects which are visible both as a morphological (i.e., growth) response and are detectable biochemically (Smith, 1975). Since the effects are so numerous and often unrelated to one another, there are difficulties in providing a universal mechanism for this light control. Indeed in terms of the red light regulation of enzyme activity, there is evidence that control sites at the levels of transcription, translation, and posttranslation may exist (Schopfer, 1977). In this section, however, an attempt will be made to answer the question as to which role membranes play in the perception of red light rather than to give a general discourse on the physiology of red light effects in plants.

As we now know, red light is not only absorbed by the chlorophylls and used for photosynthesis but also by a specific pigment termed *phytochrome*. This is a conjugated protein with the chromophore portion being a linear tetrapyrrole similar to that of phycocyanin, the accessory photosynthetic pigment in some algal chloroplasts (Briggs and Rice, 1972; Rüdiger, 1980). There appears to be one chromophore per protein monomer. Until recently the protein portion was considered to consist of two 120-kd monomers (Pratt, 1982), but it has now been shown that the correct size is 124 kd (Viestra and Quail, 1982). The 6-kd difference represents a peptide segment cleaned by endopeptidase activity, and its loss leads to an alteration in the spectral properties of the pigment (Viestra and Quail, 1983). Although the amino acid composition of phytochrome is similar from a number of sources, there are significant antigenic differences between monocotyledenous and dicotyledenous types (Cordonnier and Pratt, 1982). It has been claimed that phytochrome is glycosylated (4% w/w—Roux et al., 1975) but this has been challenged (Boeshore and Pratt, 1981).

When exposed to red light of 664 nm phytochrome is photoconverted to a form with a maximum absorbance at 724 nm. Since sequential irradiation with red and far-red light cancels out the effects of red light, phytochrome in the red form (Pr) is termed *physiologically active* and in the far-red form (Pfr) *inactive* (Briggs and Rice, 1972).

Because it is more difficult to detect in green plants, phytochrome is usually extracted from etiolated tissue. When this is done under green light conditions phytochrome behaves as a soluble protein (Pr^{sol}). The photoconversion of phytochrome by irradiation with red light leads to its pelletability by as much as 80% (Marmé et al., 1973). This conversion may be carried out *in vivo* prior to homogenization or *in vitro* using the homogenate. The former is a property of all plants so far studied, whereas the *in vitro* pelletability is more restricted in its occurrence (see Marmé, 1974). Since the photoinduced *in vivo* pelletability occurs well within the shortest presentation time for red light responses (i.e., less than 10 sec—Pratt and Marmé, 1976), it is appropriate to ask whether this pelletability involves membrane binding, and if so, which membrane?

13.2.1. Characteristics of Phytochrome Pelletability

In vivo and *in vitro* photoinduced pelletability require or are enhanced by the presence of 10–20 m*M* concentrations of divalent cations, respectively (see Marmé, for review). If this is present in the extraction media, the phytochrome remains soluble (Pfr^{sol}). When irradiated with red light in the presence of 20 m*M* $CaCl_2$, purified Pr^{sol} becomes aggregated (measured as an increase in turbidity). Subsequently, far-red light reverses this effect (Yamamoto et al., 1980). These aggregates are capable of associating (binding) with membranes as well as with 31S ribonucleoprotein particles (Quail, 1975). If photoinduction *in vivo* with red light is followed by a dark period before homogenization, pelletability declines with first-order kinetics. This dissociation is temperature dependent and does not take place at temperatures under 13 °C (Pratt and Marmé, 1976).

Experiments with purified ^{35}S-phytochrome (Pratt, 1980) have shown that (a) exogenously added phytochrome does not exchange with phytochrome which has already been photoinduced to bind *in vivo* and (b) binding of radioactive phytochrome to resuspended microsomal pellets *in vitro* is independent of its form (Pfr, Pr) and of the type of irradiation given prior to homogenization. Taken together with the observation of Quail and Briggs (1978) that the levels of bound phytochrome are the same for microsomal pellets prepared from mixed homogenates of separately homogenized nonirradiated and irradiated tissues as for pellets prepared from homogenates of mixed tissues, it would seem that "phytochrome induced to pellet by irradiation *in vivo* is already bound prior to tissue homogenization" (Pratt, 1980). The binding itself is energy dependent (Quail and Briggs, 1978) and appears to result in a modification of the protein portion of the phytochrome complex, as judged by differences in gel filtration and electrophoretic properties of Pr^{sol}, and of pelleted Pfr dissociated from the membranes by removal of the divalent cations (Boeshore and Pratt, 1980).

13.2.2. Phytochrome Localization *in Vitro*

The first attempt at localizing pelletable or bound phytochrome using cell fractionation placed it in a mitochondrial fraction (Manabe and Furuya, 1974). Later work has not confirmed this. Using isopycnic sucrose gradients, Marmé (1974) has shown for zucchini hypocotyls that phytochrome bands at a position between ER (smooth) and mitochondria. The peak is coincident with that of the PM marker, NPA binding. A similar distribution has been given for cucumber hypocotyl tissue by Quail et al. (1976) who have shown in addition that it is not due to binding to etioplast membranes. In another paper on this system, Quail and Hughes (1977) have demonstrated that the phytochrome-rich fraction in fractions from discontinuous gradients stain preferentially with the PTC stain.

The above gradient results have been obtained with low Mg^{2+} concentrations which reduce considerably the amounts of bound phytochrome. Gradients run in the presence of 10–20 m*M* Mg^{2+}, however, led to unspecific particle aggregation. To overcome this problem Yu (1975a,b) has used glutaraldehyde in the homogenizing medium and this stabilizes the binding in the absence of divalent cations. Sucrose gradients prepared from such homogenates again show that phytochrome appears to be associated with PM fractions, as indicated by the coincidence with K^+-ATPase activity.

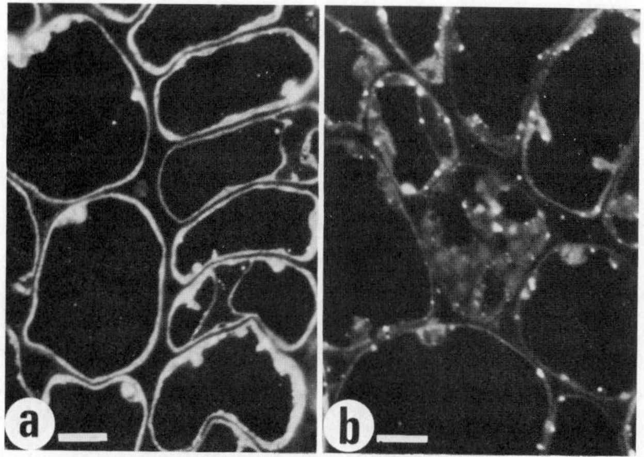

Figure 13.3. Immunofluorescence localization of phytochrome at the light microscopic level. 1-μm thick cryosections of oat coleoptile treated with rhodamine-conjugated antibodies. (a) Etiolated plant. Fluorescence is diffuse throughout the cytoplasm. (b) Etiolated plant given a 5-min exposure to red light before sectioning. Fluorescence is now restricted to small point sources in the cytoplasm. Bar = 10 μm. Published micrographs of Verbelen et al. (1982).

13.2.3. Phytochrome Localization *in Situ*

Pratt and colleagues have applied immunocytological methods in attempting to localize phytochrome (Coleman and Pratt, 1974a,b; Mackenzie et al., 1975). By using conventional embedding media and applying antibody conjugates to the section, it has been established that the great majority of phytochrome in its Pr form is present in the tip region of dark-grown coleoptiles. Within the cells the stain appeared to be fairly evenly distributed throughout the cytoplasm (see Figures 13.3 and 13.4a). In contrast, after irradiation with red light the Pfr form of phytochrome appeared localized as discrete loci in the cytoplasm. Improvements in technique (cryoultramicrotomy; ferritin antibody conjugates) have allowed the loci of Pfr accumulation to be recognized as small vacuoles (see Figures 13.3b, 13.4b; Verbelen et al., 1982). Although this redistribution of staining is very rapid (a matter of minutes), its localization is basically different from that which has been derived from the fractionation studies described above. At the moment this conflict has not been resolved.

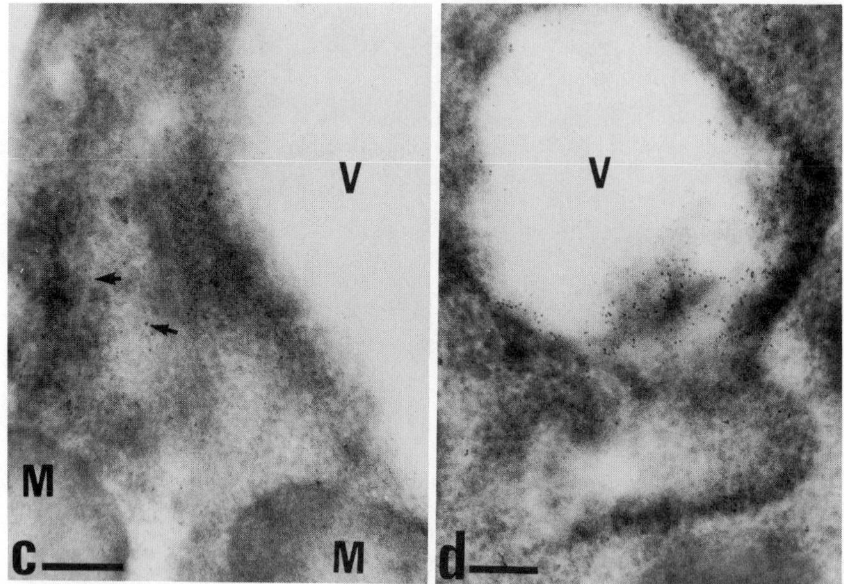

Figure 13.4. Immunofluorescence localization of phytochrome at the electron microscopic level. Thin sections of oat coleoptile tissue treated with ferritin-conjugated antibodies. (a) Portion of cytoplasm from an etiolated plant with two mitochondria (M) and a vacuole (V). Stain (arrows) is generally distributed in the cytoplasm. (b) Etiolated plant given a 5-min exposure to red light before sectioning. The stain is now concentrated in small vacuoles. Bar = 2 μm. Published micrographs of Verbelen et al. (1982).

13.3. BLUE LIGHT PERCEPTION

There are a large number of physiological responses including phototropism which are induced by UV and blue light (Presti and Delbrück, 1978). The action spectra for these photoresponses show a broad band around 370 nm and a multiple peak between 430 and 480 nm, indicating that the pigment responsible could be either of the carotenoid or flavin type (see Briggs, 1980, for references). Due to the similarities between the absorption spectra for flavins and these action spectra and also the fact that redox reactions involving carotenoids are highly unlikely, Munoz and Butler (1975) postulated that a flavin is the primary photoreceptor in all blue light responses and this is coupled to a cytochrome of the *b* type which is "capable of mediating redox changes." In support of this contention these authors have pointed to the similarity in action spectra for cytochrome reduction with that of blue light responses.

Proof that the photoreceptor is indeed a flavin has come from several observations. First, there are some mutants of *Neurospora* which show reduced sensitivity toward blue light and this is correlated with reduced flavin levels (Paietta and Sargent, 1981). There have also been experiments with substances which specifically inhibit (NaN_3—Schmidt et al., 1977; phenylacetic acid—Viestra and Poff, 1981) or enhance (Acifluorfen—Leong and Briggs, 1982) phototropism. In both cases flavins are implicated as the photoreceptor. Furthermore, blue light inducible cytochrome reduction is optimalized by adding free flavins in the form of riboflavin or FAD to the reaction mixture (Goldsmith et al., 1980).

Dichroism studies on fungi have pointed to the PM as the most likely site for the photoreceptor (Jesaitis, 1974) and fractionation studies have confirmed this. Schmidt and Butler (1976) were the first to show that the flavin-mediated photoreactions were confined to the pelletable rather than soluble portions of *Neurospora* homogenates. Brian et al. (1977) carried this observation further and showed that the blue light absorbance change was greatest in a 20,000*g* fraction, a fraction characterized by the putative fungal PM marker Na^+-ATPase. Leong and Briggs (1981) have also localized the photoreceptor in a fraction from maize coleoptiles which is presumably the PM because of its cosedimentation with GS II activity and noncoincidence with markers for ER, GA, and mitochondria. This fraction appears to be identical with that described by Jesaitis et al. (1977) as containing a *b*-type cytochrome. Partial confirmation of the PM as the site for the photoreceptor comes from *in vitro* binding studies with ^{14}C-riboflavin (Hertel et al., 1980). In contrast to the flavin–cytochrome complex of fungi, that of maize coleoptiles can be solubilized with low concentrations (0.1%) of Triton X-100 and still retain full activity (Leong and Briggs, 1981).

13.4. TRANSMISSION AND RESPONSE WITH AUXIN(S) AS THE SIGNAL

The response to a light or gravitational impulse usually results in a differential growth response of the plant in question; that is, one side grows quicker than the other. The major signal responsible for initiating this is the auxin group of plant hormones (Thimann, 1969), which are transported from the region of perception to those cells capable of elongation growth (Goldsmith, 1969). Since this transport is dependent upon physiologically competent tissue, it precludes an apoplastic transport route to the "target" cells. Membranes are involved both in active transport and in signal recognition.

13.4.1. Auxin Transport

The major auxin, indole acetic acid (IAA), can accumulate in the cytoplasm leading to a concentration of 10–100 times that in the intercellular space. Rubery and Sheldrake (1974) and Raven (1975) proposed, on the basis of Mitchell's chemiosmotic theory, that IAA uptake is dependent on a pH gradient at the PM (pH cyt > pH wall). This has now been confirmed experimentally both *in vivo* (Edwards and Goldsmith, 1980; Sussman and Goldsmith, 1981; Rubery, 1978) and *in vitro* with PM vesicles from zucchini hypocotyls (Hertel et al., 1983). According to the original proposal, undissociated IAA or IAA^- was thought to enter nonelectrogenically, but Hertel (1983) has recently put forward evidence for an electrogenic component and has postulated that the entry of each IAA is accompanied by two, instead of one, protons.

Efflux of IAA is believed to occur via a specific anion carrier and is probably electrogenic. This carrier can be blocked by application of 2,3,5-triiodo benzoic acid (TIBA), a powerful inhibitor of auxin transport *in vivo*, and by a number of auxin analogs (so-called phytotropins) (Depta et al., 1983; Goldsmith, 1982; Hertel et al., 1983). Because (a) Ca^{2+} moves against the direction of auxin transport (Arslan-Cerim, 1966; Bangerth, 1979) and (b) TIBA also blocks Ca^{2+} transport in tissue segments (Bangerth, 1979), it is considered likely that the anion carrier is associated with a Ca^{2+} antiporter. Whereas IAA uptake is thought to occur at all points on the PM, the anion carrier is apparently located only at the distal end of each cell (Goldsmith and Ray, 1973; Jacobs and Gilbert, 1983).

13.4.2. Auxin "Receptors"

When auxin(s) reach their "target" cells, they are recognized by binding to some sort of protein. Attempts at localizing such sites by *in vitro* binding tests are fraught with difficulties both of a technical and interpretative nature (see Hertel, 1981; Murphy, 1979; Rubery, 1981; Stoddart and Venis, 1980). Both soluble and membrane-bound proteins have at times been implicated. Of the former much is to be attributed to ribulose bisphosphate carboxylase (Wardrop and Polya, 1980) which is probably released from plastids upon tissue homogenization. Whether binding to this protein *in situ* actually occurs seems doubtful.

Since the original *in vitro* binding studies of Hertel et al. (1972) numerous publications have appeared in which either radioactive IAA, one of its analogs (e.g., 1- or 2-naphthyl acetic acid, NAA), or a phytotropin (e.g., TIBA or naphthylphthalamic acid, NPA) have been added to fractions of density gradients. The majority of the work has been done with maize coleoptiles, but other tissues (e.g., zucchini hypocotyls and protoplasts) have also been used (Vreugdenhil et al., 1979, 1980).

NAA binding to resuspended maize microsomal membranes has a pH optimum of 5.5, is cold-stable but thermolabile, and is subject to inhibition by some factor(s) (probably benzoxazoline—Venis and Watson, 1978) present in the soluble supernatant (Ray et al., 1977). Ray (1977) has shown that NAA binding is principally associated with ER fractions, as judged by cosedimentation with CCR activity and Mg^{2+} shifting experiments. However, due to the broad shoulder of the NAA binding profile, additional sites could not be ruled out. NPA binding, in contrast, was associated with fractions rich in GS II activity, suggesting a PM localization. Subsequent work by Dohrmann et al. (1978) has allowed three binding sites in maize microsomal membranes to be recognized: site I, which has high affinities for both 1- and 2-NAA and is contained in fractions bearing CCR activity; site II, which has a low affinity for 2-NAA and corresponds to fractions bearing GS I activity; and site III, the principal site of IAA and 2,4-dichlorophenoxyacetic acid (2,4-D) binding and is present in fractions containing GS II activity. Site I has been interpreted as being ER, site III as the PM. On the basis of cosedimentation with acid phosphatase activity, these authors have preferred to allocate site II to the TP rather than the GA. As we now know, TP vesicles from maize coleoptiles do not equilibrate so deep in sucrose gradients, and acid phosphatase tends to bind unspecifically to other membranes (see Chapter 6), so that a reallocation of site II is now warranted. Sites I and III of Dohrmann et al. (1978) probably correspond to sites 1 and 2 designated as GA–ER and PM, respectively, as described by Batt and Venis (1976) on the basis of binding tests done on fractions from discontinuous gradients of maize coleoptiles homogenates. In other systems Jacobs and Hertel (1978) have been able to detect site II–NAA but not site I–NAA binding in zucchini hypocotyl tissue. However, IAA binding was, as in maize, a property of the PM fraction. Similarly, Williamson et al. (1977) have claimed on the basis of PTAC staining that IAA binds to the PM of soybean hypocotyl cells.

Site I–NAA binding proteins from maize have been successfully solubilized using weak detergent. Cross and Briggs (1978, 1979) and Venis (1977) have given Mrs of 90 kd and 45 kd, respectively, for the responsible protein. This discrepancy has not yet been resolved.

Various suggestions have been made (see Hertel, 1981, 1983; Rubery, 1981, for summaries) as to the biological roles of the receptor sites described above. PM binding sites are topographically more suited than intracellular sites for the acidification of the cell wall which results from auxin treatment. Claims for an auxin stimulated PM–ATPase activity which may work electrogenically and pump protons have been made (Scherer and Morré, 1978; Scherer, 1981) and are attractive as is the possibility that Ca^{2+} and calmodulin are also involved (Schleicher et al., 1982). Alternatively, the PM binding site might well represent auxin transport rather than a true binding to a receptor protein, a possibility which now finds support in the observations of Hertel et al. (1983). These authors have been able to demonstrate that ^{14}C-IAA is transported into sealed PM vesicles of zucchini and that the inhibitors of *in vivo* auxin transport TIBA and NPA cause an accumulation of IAA by blocking its efflux from the vesicle. Furthermore, Depta et al. 1983 have now shown that ^{14}C-TIBA, in addition to NPA, binds to the PM.

If auxin receptors are primarily situated on endo- rather than plasma membranes, how do they achieve an acidification of the cell wall? Ray (1977) has suggested that binding at the ER might stimulate the pumping of protons into the lumen of this organelle which are then transported via vesicles to the PM. Certainly, the lag time (~10 min) between auxin application to tissue segments and the acidification of the cell wall is in good agreement with known transport times for products from the ER to the PM (see Chapter 10).

Another suggestion highlights again the important regulatory role of Ca^{2+} (Hertel, 1983; Raven and Rubery, 1982). Here it is proposed that Ca^{2+} is released from a membrane compartment into the cytoplasm and this is supposed to elicit two effects: one is to indirectly cause a higher level of proton extrusion from the protoplast, while the other would be to stimulate the exocytosis of secretory vesicles in a similar way to the triggering of synaptic vesicles through Ca^{2+}. Claims have been made for an auxin-stimulated Ca^{2+} release from microsomal vesicles from soybean (Buckhout et al., 1981), but it is not clear to which compartment they belong. Hertel (1983) assumes that the TP is the most likely candidate because of his earlier assertion that site II binding is located there, an assumption that is no longer valid (see above).

REFERENCES

Arslan-Cerim, N. (1966). *J. Exp. Bot.* **17,** 236–240.

Audus, L. J. (1962). In *Soc. Exp. Biol. Symp. XVI* (J. W. L. Beament, ed.), pp. 197–226. Cambridge University Press, Cambridge.

Bangerth, F. (1979). *Ann. Rev. Phytopathol.* **17,** 97–122.

Barlow, P. M. and Grundweg, M. (1974). *Z. Pflanzenphysiol.* **73,** 56–64.

Batt, S. and Venis, M. A. (1976). *Planta* **130,** 15–21.

Boeshore, M. L. and Pratt, L. H. (1980). *Plant Physiol.* **66,** 500–504.

Boeshore, M. L. and Pratt, L. H. (1981). *Plant Physiol.* **68,** 789–797.

Brian, R. D., Freeberg, J. A., Weiss, C. V., and Briggs, W. R. (1977). *Plant Physiol.* **59,** 948–952.

Briggs, W. R. (1980). In *Photoreceptors and Plant Development* (J. A. DeGreet, ed.), pp. 17–28. Antwerpen University Press, Antwerp.

Briggs, W. R. and Rice, H. V. (1972). *Ann. Rev. Plant Physiol.* **23,** 293–334.

Buckhout, T. J., Young, K. A., Low, P. S. and Morré, D. J. (1981). *Plant Physiol.* **68,** 512–515.

Colemann, R. A. and Pratt, L. H. (1974a). *Planta* **121,** 119–131.

Coleman, R. A. and Pratt, L. H. (1974b). *J. Histochem. Cytochem.* **22,** 1039–1047.

Cordonnier, M. M. and Pratt, L. H. (1982). *Plant Physiol.* **70,** 912–916.

Cross, J. W. and Briggs, W. R. (1978). *Plant Physiol.* **62,** 152–157.

Cross, J. W. and Briggs, W. R. (1979). *Planta* **146,** 263–270.

Depta, H., Eisele, K. H., and Hertel, R. (1983). *Plant Sci. Lett.* **31,** 181–192.

Dohrmann, U. C., Hertel, R., and Kowallik, H. (1978). *Planta* **140,** 97–106.

Edwards, K. L. and Goldsmith, M. H. M. (1980). *Planta* **147,** 457–466.

Filner, B., Hertel, R., Steele, C., and Fan, V. (1970). *Planta* **94,** 333–354.

Firn, R. D. and Digby, J. (1977). *Austr. J. Plant Physiol.* **4,** 337–347.

Firn, R. D., Digby, J., and Riley, H. (1978). *Ann. Bot.* **42,** 465–468.

Goldsmith, M. H. M. (1969). In *Physiology of Plant Growth and Development* (M. B. Wilkins, ed.), pp. 127–162. McGraw-Hill, London.

Goldsmith, M. H. M. and Ray, P. M. (1973). *Planta* **111,** 297–314.

Goldsmith, M. H. M. (1982). *Planta* **155,** 68–75.

Goldsmith, M. H. M., Caubergs, R. J., and Briggs, W. R. (1980). *Plant Physiol.* **66,** 1067–1073.

Haberlandt, G. (1902). *Ber. Dtsch. Bot. Ges.* **20,** 189–195.

Hensel, S. and Sievers, A. (1979). *Eur. J. Cell Biol.* **20,** 121.

Hensel, W. and Sievers, A. (1980). *Planta* **150,** 338–346.

Hensel, W. and Sievers, A. (1981). *Planta* **153,** 303–307.

Hertel, R. (1981). *Physiol. Pflanzen* **176,** 495–506.

Hertel, R. (1983). *Z. Pflanzenphysiol.* **112,** 153–67.

Hertel, R., Lomax, T. L., and Briggs, W. R. (1983). *Planta* **157,** 193–201.

Hertel, R., Jesaitis, A. J., Dohrmann, U., and Briggs, W. R. (1980). *Planta* **147,** 312–319.

Hertel, R., Thomson, K.-S., and Russo, V. E. A. (1972). *Planta* **107,** 325–340.

Iversen, T.-H. (1969). *Physiol. Plant.* **22,** 1215–1262.

Jacobs, M. and Gilbert, S. F. (1983). *Science* **220,** 1297–1300.

Jacobs, M. and Hertel, R. (1978). *Planta* **142,** 1–10.

Jesaitis, A. J. (1974). *J. Gen. Physiol.* **63,** 1–21.

Jesaitis, A. J., Heners, P. R., Hertel, R., and Briggs, W. R. (1977). *Plant Physiol.* **59,** 941–947.

Johnsson, A., Rengman, K., and Grahm, L. (1971). *Physiol. Plant.* **25,** 43–47.

Juniper, B. E. and French, A. (1970). *Planta* **95,** 314–329.

Juniper, B. E. and French, A. (1973). *Planta* **109,** 211–224.

Juniper, B. E., Groves, S., Landau-Schachar, B., and Audus, L. J. (1966). *Nature* **209,** 93–94.

Leong, T.-Y. and Briggs, W. R. (1981). *Plant Physiol.* **67,** 1042–1046.

Leong, T.-Y. and Briggs, W. R. (1982). *Plant Physiol.* **70,** 875–881.

Mackenzie, J. M., Coleman, R. A., Briggs, W. R., and Pratt, L. H. (1975). *Proc. Natl. Acad. Sci. USA* **73,** 799–803.

Manabe, K. and Furuya, M. (1974). *Plant Physiol.* **53,** 343–347.

Marmé, D. (1974). *J. Supramol. Struct.* **2,** 751–768.

Marmé, D. (1977). *Ann. Rev. Plant Physiol.* **28,** 173–198.

Marmé D., Boisard, J., and Briggs, W. R. (1973). *Proc. Natl. Acad. Sci. USA* **70,** 3861–3865.

Munoz, V. and Butler, W. L. (1975). *Plant Physiol.* **55,** 421–426.

Murphy, G. J. P. (1979). *Plant Sci. Lett.* **15,** 183–191.

Nemeč, B. (1901). *Jahrb. Wiss. Bot.* **36,** 80–178.

Osborne, D. J. and Wright, J. M. (1977). *Proc. Roy. Soc. Lond. B* **199,** 551–564.

Paietta, T. and Sargent, M. L. (1981). *Proc. Natl. Acad. Sci. USA* **78,** 5573–5577.

Pollard, E. C. (1971). In *Gravity and the Organism* (S. A. Gordon and M. J. Cohen, eds.), pp. 25–34. University of Chicago Press, Chicago.

Pratt, L. H. (1980). *Plant Physiol.* **66,** 903–907.

Pratt, L. H. (1982). *Ann. Rev. Plant Physiol.* **33,** 557–582.

Pratt, L. H. and Marmé, D. (1976). *Plant Physiol.* **58,** 686–692.

Presti, D. M. and Delbrück, M. (1978). *Cell & Environment* **1,** 81–100.

Quail, P. H. (1975). *Planta* **123,** 223–234.

Quail, P. H. and Briggs, W. R. (1978). *Plant Physiol.* **62,** 773–775.

Quail, P. H. and Hughes, J. E. (1977). *Planta* **133,** 169–177.

Quail, P. H., Gallagher, A., and Wellburn, A. R. (1976). *Photochem. Photobiol.* **24,** 495–498.

Raven, J. A. (1975). *New Phytol.* **74,** 163–172.

Raven, J. A. and Rubery, P. H. (1982). In *The Molecular Biology of Plant Development* (H. Smith and D. Grierson, eds.), pp. 28–48. Blackwell Scientific Publications, Oxford.

Ray, P. M. (1977). *Plant Physiol.* **59,** 594–599.

Ray, P. M., Dohrmann, U. C., and Hertel, R. (1977). *Plant Physiol.* **59,** 357–364.

Roux, S. J., Glisansky, S., and Stoker, B. M. (1975). *Plant Physiol.* **35,** 85–90.

Rubery, P. H. (1978). *Planta* **142,** 203–206.

Rubery, P. H. (1981). *Ann. Rev. Plant Physiol.* **32,** 569–596.

Rubery, P. H. and Sheldrake, A. R. (1974). *Planta* **118,** 101–121.

Rüdiger, W. (1980). *Structure Binding* **40,** 101–140.

Scherer, G. F. E. (1981). *Planta* **151,** 434–438.

Scherer, G. F. E. and Morré D. J. (1978). *Biochem. Biophys. Res. Comm.* **84,** 238–247.

Schleicher, M., Iverson, D. B., van Eldik, L. J., and Waterson, D. M. (1982). In *The Cytoskeleton in Plant Growth and Development* (C. Lloyd, ed.), pp. 85–106. Academic Press, London.

Schmidt, W. and Butler, W. L. (1976). *Photochem. Photobiol.* **24,** 77–80.

Schmidt, W., Hart, T., Filner, P., and Poff, K. L. (1977). *Plant Physiol.* **60,** 736–738.

Schopfer, P. (1977). *Ann. Rev. Plant Physiol.* **28,** 223–252.

Sievers, S. and Schmitz, U. (1973). *Planta* **114,** 373–378.

Sievers, A. and Volkmann, D. (1972). *Planta* **102,** 160–172.

Sievers, A. and Volkmann, D. (1977). *Proc. Roy. Soc. London B* **199,** 525–536.

Sievers, A. and Volkmann, D. (1979). *Encyclopedia of Plant Physiology*, Vol. 7, 567–572. Springer Verlag, Berlin.

Sievers, A., Volkmann, D., Hensel, W., Sobick, V., and Briegleb, W. (1976). *Naturwissenschaften* **63,** 343.

Smith, H. (1975). *Phytochrome and photomorphogenesis.* McGraw-Hill, London.

Stoddart, J. L. and Venis, M. A. (1980). In *Encyclopedia of Plant Physiology* Vol. 9 (J. Mamillan, ed.), pp. 445–510. Springer Verlag, Berlin.

Sussman, M. R. and Goldsmith, M. H. M. (1981). *Planta* **151,** 15–25.

Thimann, K. V. (1969). In *Physiology of Plant Growth and Development* (M. B. Wilkins, ed.), pp. 3–45. McGraw-Hill, London.

Venis, M. A. (1977). *Nature* **266,** 268–269.

Venis, M. A. and Watson, P. J. (1978). *Planta* **142,** 103–107.

Verbelen, J. P., Pratt, L. H., Butler, W. L., and Tokuyasu, K. (1982). *Plant Physiol.* **70,** 867–871.

Viestra, R. D. and Poff, K. L. (1981). *Plant Physiol.* **67,** 1011–1015.

Viestra, R. D. and Quail, P. H. (1982). *Proc. Natl. Acad. Sci. USA* **79,** 5272–5276.

Viestra, R. D. and Quail, P. H. (1983). *Plant Physiol.* **72,** 264–267.

Volkmann, D. (1974). *Protoplasma* **79,** 759–783.

Volkmann, D. and Sievers, A. (1979). *Encyclopedia of Plant Physiology*, Vol. 7, 573–600.

Vreugdenhil, D., Burgers, A., and Libbenga, U. R. (1979). *Plant Sci. Lett.* **16,** 115–121.

Vrengdenhil, D., Harkes, P. A. A., and Libbenga, K. R. (1980). *Planta* **150,** 9–12.

Wardrop, A. J. and Polya, G. M. (1980). *Plant Physiol.* **60,** 112–118.

Williamson, I. A., Morré, D. J., and Hess, K. (1977). *Cytobiol.* **16,** 63–71.

Yamamoto, K. T., Smith, W. O., and Furuya, M. (1980). *Photochem. Photobiol.* **32,** 233–239.

Yu, R. (1975a). *Austr. J. Plant Physiol.* **2,** 273–279.

Yu, R. (1975b). *J. Exp. Bot.* **26,** 808–822.

14

MEMBRANES AND CELL DIVISION

This book has been principally concerned with the structure and function of organelles in interphase plant cells. It is now time to consider what roles the endomembranes play in cell division. In so doing one will notice that with the exception of chemical or physical attempts at perturbing cell division, the work involved is purely cytological and for the most part phenomenological.

14.1. MITOSIS

Depending on the state of the nuclear envelope (NE) during the condensation and separation of chromosomes, one may speak of closed (intact NE) or open (dispersed NE) mitoses. Closed mitosis is to be regarded as the more primitive type, as indicated by its presence in lower organisms (Heath, 1980, 1981; Kubai, 1975; Pickett-Heaps and Marchant, 1972). In these types the NE may be as in the interphase condition, completely intact, or possess openings ("fenestrae") at the division poles, through which microtubules from a centriole or centriole-equivalent pass. According to the textbooks, vertebrates and higher plants in contrast show a breakdown of their NEs during mitosis, together with a reassembly of the same around each set of daughter chromosomes during telophase. The impression gained from reading such accounts is that the mitotic spindle is completely void of membranes both internally and externally. As we now know, this is not the case. Indeed, this view appears to have originated out of the light microscopic era since a close inspection of the early paper of Porter and Machado (1960) reveals clearly the extensive penetration of tubular ER into the interior of the spindle of dividing onion root tip cells. Of course, this paper was written in the days before microtubules had been discovered; nevertheless their presence could not have been detected due to the $KMnO_4$ fixation used.

With the introduction of the osmium–potassium ferricyanide fixation technique, it is now possible to achieve a staining of membranes, comparable in intensity to that of $KMnO_4$, while being able to retain microtubules. The method was developed as a better way of portraying sarcoplasmic reticulum in muscle tissue (Forbes et al., 1977), but its specificity is not restricted to ER-type membranes (Schnepf et al., 1982). Nevertheless, Hepler (1980) has successfully applied this technique to dividing barley leaf cells. In these cells (see Figure 14.1) only NE remnants

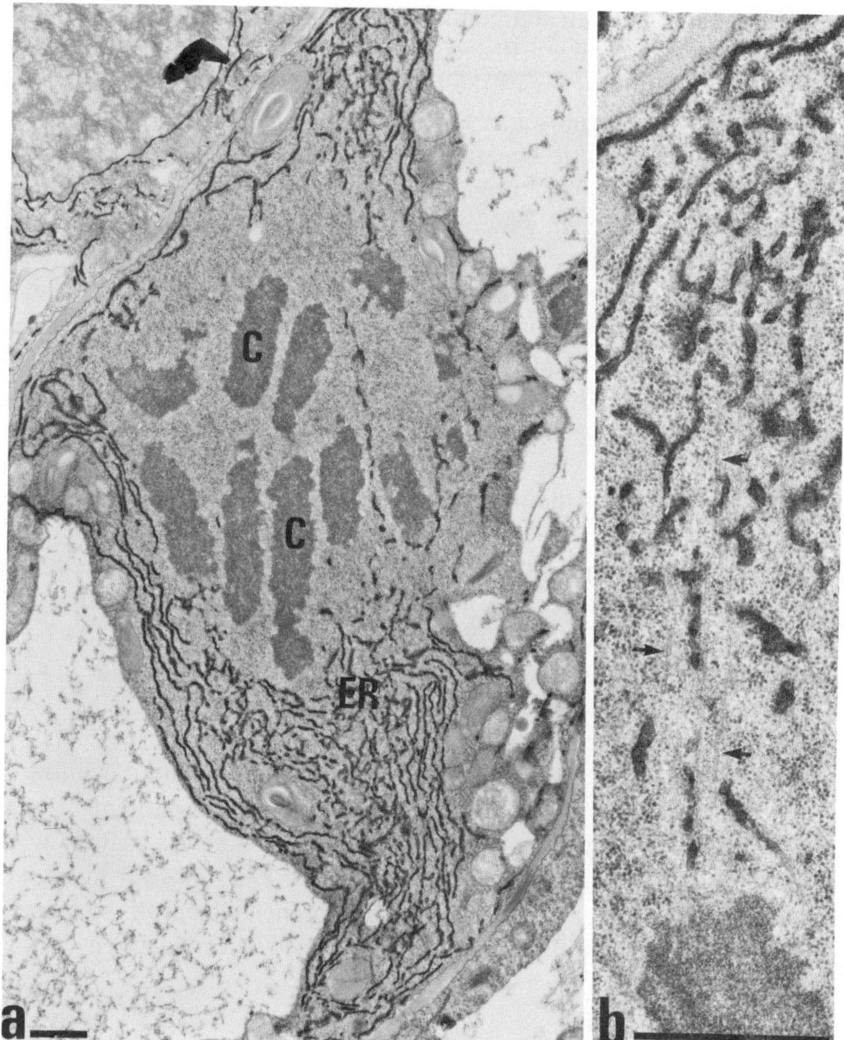

Figure 14.1. (a) Section through a barley leaf cell fixed according to the OsFeCN technique. The chromosomes (C) are in early anaphase. The polar regions of the spindle (asterisks) are occupied by masses of densely staining pieces of ER and NE fragments. (b) Tubular ER intensely stained after OsFeCN fixation penetrate deeply into the spindle, running between microtubules (arrows) almost up to the kinetochore on the chromosome. Bar = 1 μm. Published micrographs of Hepler (1980).

and ER achieve a significant contrast. With this method it is possible to trace elements of tubular ER deep into the mitotic apparatus. They intermingle conspicuously with the microtubules of the spindle and even proceed up to the chromosomes themselves. A similar penetration of sER elements into the mitotic apparatus of animal cells has also been noted (Harris, 1975). Another feature made most apparent by this technique is the accumulation of membrane at the division poles, particularly during anaphase, and also the enclosure of the spindle by NE–ER membrane fragments. Those membranes at the pole have pores or fenestrations of variable (40–100 nm) width. Confirmation of these observations has been given by Hawes et al. (1981), using the ZIO fixation technique. As already mentioned (Chapter 4), the ER is preferentially stained by this method, and the results are analogous to those obtained by Hepler (1980). They are extended, however, by high-voltage electron microscopy of thick sections which underlines the penetration of the spindle by tubular elements of the ER.

The presence of Ca^{2+} in the glutaraldehyde primary fixative and wash stages appears obligatory for a successful staining with the OsFeCN method (Forbes et al., 1977; White et al., 1979). Since Ca^{2+} has been detected in both polar as well as spindle ER by means of *in situ* precipitation with potassium antimonate (Wick and Hepler, 1980), it has been suggested that these membranes act as sequestering sites for such divalent cations. The deleterious role of Ca^{2+} ions in microtubule dynamics in general (see Dustin, 1978) and mitotic spindle function in particular (Petzelt, 1979), together with the demonstration of a spindle-associated Ca^{2+}-ATPase (Mazia et al., 1972; Petzelt and Auel, 1977) make it highly attractive to think that at least the ER membranes which are so closely disposed to the spindle apparatus microtubules do indeed function in Ca^{2+} regulation.

14.2. CYTOKINESIS TYPES IN PLANTS

Cytokinesis, or the separation of daughter protoplasts after nuclear division, in higher plants is basically different from that which occurs in animals. Whereas in animals the existing PM is drawn together between the daughter nuclei through the agency of a peripherally located belt of actin–myosin filaments (“the contractile ring”—Arnold, 1969, 1975;

Schroeder, 1972, 1973), cytokinesis in plant cells is in contrast a centrifugal event achieved by the fusion of vesicles to form new PM. Inevitably, since plant cells are surrounded by a wall, this event is coupled to cell wall deposition and is termd *cell plate production*. This is typical for all higher plants, and a number of algae (e.g., *Chara*) that may be regarded as advanced, also show it. There are instead a number of algae, mainly flagellates, which divide through a progressive, centripetally directed furrowing. Whether microfilaments are responsible for this invagination is not known.

Intimately associated with both forms of cytokinesis are microtubules, some of which are remnants of the mitotic spindle. Typical for higher plants (and animal cells for that matter) is the phragmoplast which signifies an arrangement of microtubules running perpendicular to the plane of cell division. This set of microtubules builds essentially a packed cylinder at the beginning of cytokinesis and gradually moves radially outwards marking the periphery of the growing cell plate. Another form exists, the phycoplast, and consists of microtubules lying parallel to the division plane. As the name suggests, this latter form is restricted to the algae. Whereas higher plants are characterized by a cell plate and a phragmoplast, there are a number of variations in the algae involving combinations of furrowing or cell plate with either phragmoplast or phycoplast microtubule arrangements (see Table 14.1 and Pickett-Heaps, 1975, for details of the various algal examples).

14.3. PHRAGMOPLAST AND CELL PLATE

14.3.1. Morphological Aspects

Light microscopy of living endosperm cells (Bajer, 1965; Bajer and Allen, 1966) established that the phragmoplast begins forming in the region between separating chromosomes during anaphase. Even at this level of resolution its fibrous nature is recognizable. Small nodules develop centrally, swell, and together with small particles that appear at this stage, fuse with one another to produce the cell plate. Electron microscopy has confirmed this sequence of events both in endosperm (Bajer, 1968; Hepler and Jackson, 1968) as well as in other cell types (e.g., Cronshaw and Esau, 1968; Hepler and Newcomb, 1967). It is clear from these investi-

TABLE 14.1. VARIABILITY IN MITOTIC AND CYTOKINETIC FORM IN THE ALGAE[a]

Nuclear Envelope	Centrioles	Cytokinesis Form	Microtubule System	Examples
Intact	Present	Furrowing	Phycoplast	*Chlamydomonas*
Semiintact	Present	Furrowing + cell plate	Phycoplast	*Microspora*
Semiintact	Absent	Furrowing + cell plate	Phragmoplast	*Spirogyra*
Intact	Present	Cell plate	Phycoplast	*Fritschiella*
Dispersed	Present	Furrowing	None	*Klebsormidium*
Dispersed	Present	Furrowing	Phragmoplast	*Coleochaete*
Dispersed	Absent	Cell plate	Phragmoplast	*Chara*

[a] Information taken from Pickett-Heaps (1975).

gations that the microtubules constitute the phragmoplast fibers and that the nodules represent essentially an overlap region of oppositely running microtubules surrounded by some amorphous staining material.

The particles which appear to collect between the phragmoplast are, as might be expected, membranous. Whaley and Mollenhauer (1963) were the first to suggest that dictyosome-derived vesicles fused to form the cell plate. Based on the similar staining of such vesicles and the newly formed middle lamella in $KMnO_4$ fixations, this supposition has found general support under a number of authors, although as O'Brien (1972) noted, this is "a plausible interpretation, not a proven fact." Hepler and Jackson (1968) have claimed that in *Haemanthus* endosperm there are few dictyosomes and they "appear to produce only a few vesicles." However, with the same material, Bajer (1968) has claimed just the opposite. If the vesicles do not come from the GA, they can only come from the ER. Indeed, it is possible, since serial-sectioning has not been done here, that vesicles could instead be transversely sectioned tubular ER. There is no doubt that ER is present in the developing cell plate (see Figure 14.2 and O'Brien, 1972, for references), but until the recent work of Hepler (1982) it could not be adequately distinguished from vesicles or vesicle-fusion profiles. Using the OsFeCN fixation technique, ER elements are well

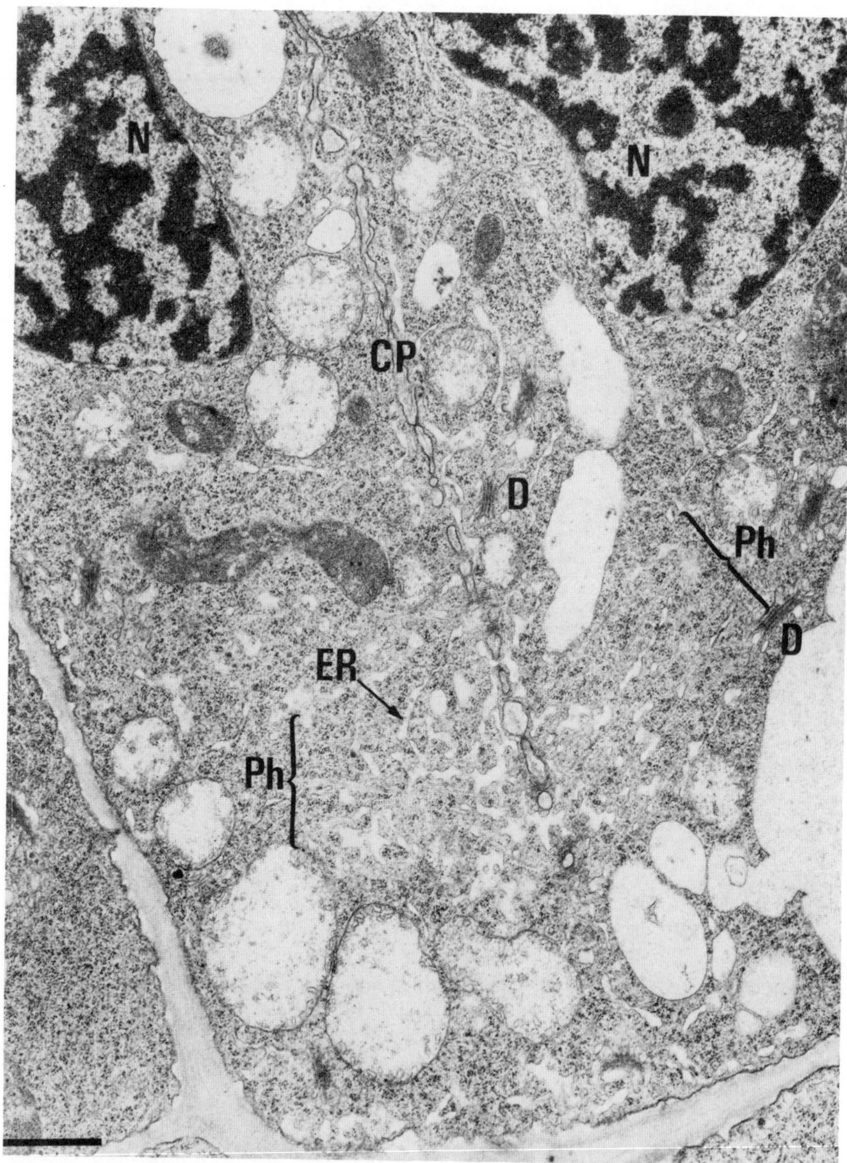

Figure 14.2. Portion of an onion root tip cell at a later stage in cytokinesis. The daughter nuclei (N) are already in the interphase condition and the phragmoplast (brackets, Pb) has almost reached the parent PM. Dictyosomes (D) are visible in this region as well as large amounts of smooth ER, which bisect the developing cell plate (CP) at various points. Bar = 11 μm. Unpublished micrograph of the author.

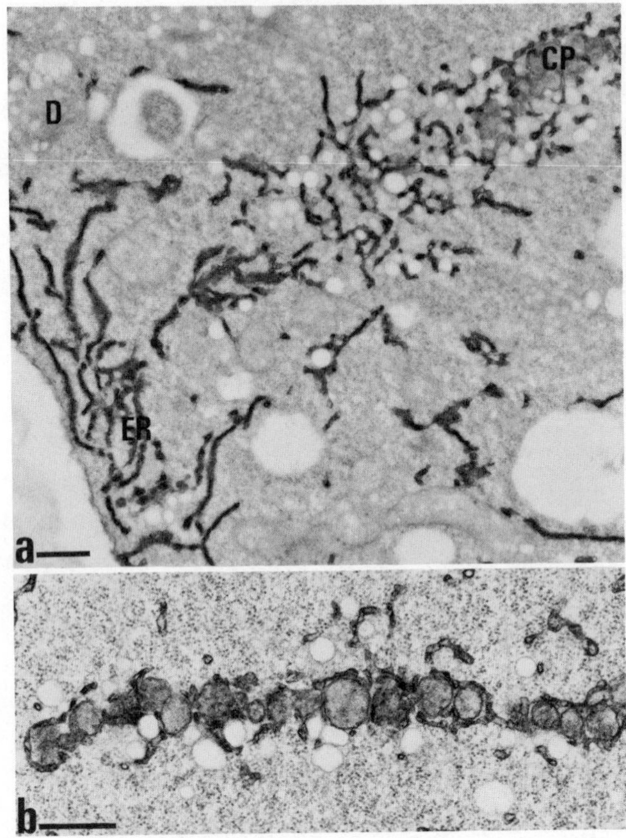

Figure 14.3. (a) Late cytokinesis in a lettuce root tip cell fixed according to the OsFeCN technique. Dictyosome (D) vesicles with electron-opaque contents collect at the margin of the developing cell plate (CP). Also at the margin is a network of tubular ER. (b) High-magnification view of the margin of a developing cell plate. Note the clear difference in staining of the contents of vesicles incorporated into the cell plate from those which lie in the vicinity and have not yet fused. The ER builds a complicated meshwork which appears to hold the vesicles together. Bar = 0.5 μm. Published micrographs of Hepler (1982).

differentiated from other endomembranes, making it abundantly clear that the cell plate is formed from non-ER, presumably dictyosome-derived vesicles (see Figure 14.3a). Nevertheless, the developing cell plate is intimately associated with tubular ER which ramifies between and encases the newly arrived vesicles at the growing periphery of the plate. As the vesicles fuse, portions of the tubular ER become trapped (Figure 14.3b) and these constitute the desmotubules of what will later become plasmodesmata (see Section 7.1).

14.3.2. Inhibition of Cytokinesis

The successful completion of cytokinesis in higher plants (i.e., the fusion of the cell plate with the parent PM) is dependent upon the presence of a wall surrounding the parent cell. As studies on regenerating protoplasts have shown, this wall does not have to be complete to allow a normal cytokinesis to occur (Fowke and Gamborg, 1980; Fowke et al., 1974), but the inhibition of cell wall regeneration through coumarin or 2,6-dichlorobenzonitrile (Meyer and Herth, 1978) prevents cytokinesis. Phragmoplasts and cell plates are initiated but survive only for a short period of time (Fowke and Gamborg, 1980).

Cytokinesis, as opposed to nuclear division, can be selectively inhibited by the addition of a number of substituted methyl xanthines (Pareyre et al., 1979), the most well-known example of which is caffeine. It has been claimed that caffeine inhibits cytokinesis by preventing the fusion of cell-plate-destined vesicles (Lopez-Saez et al., 1966) but there is now LM evidence (given in Gunning, 1982) that caffeine actually induces dissociation of preexisting cell plates. Although its effect seems to be mediated via Ca^{2+} (Becerra and Lopez-Saez, 1978), critical EM studies are now needed to investigate exactly the nature of this inhibition.

14.4. THE PLANE OF CELL DIVISION AND ITS CONTROL

Cell divisions resulting in unequally sized or shaped daughter cells [e.g., in the stomatal complex (Palevitz, 1982)] are not random events but are under cellular control. Although cytokinesis follows upon mitosis, there are numerous lines of evidence which point to a premitotic programming of this event [see Gunning (1982) and Lloyd and Barlow (1982) for recent excellent reviews on this subject].

The future plane of division in premitotic cells of higher plant tissues is characterized by two features: the preprophase band and the phragmosome. The former is a densely packed girdle of microtubules situated in the cell cortex at the future site of the fusion of the cell plate with the parent PM. Originally discovered in thin sections of wheat root cells by Pickett-Heaps and Northcote (1966), its presence has been confirmed on numerous occasions (see Table I in Gunning, 1982) and has also been visualized by immunofluorescence microscopy (Wick et al., 1981). Preprophase bands have not been found in regenerating protoplasts or in suspension-cultured plant cells (Fowke and Gamborg, 1980) nor have they been recorded in algae or fungi, but are present in a widely differing range of angiosperms, pteridophytes, and bryophtes. With the onset of NE breakdown the preprophase band disperses, presumably contributing in part to the mitotic spindle.

The other structure which predicts the future plane of cytokinesis is the phragmosome. This term was coined by Sinnott and Bloch (1941) to describe the collection of cytoplasm that splays out from the premitotic nucleus in vacuolated cells. Recently Lloyd et al. (1982) have been able to show that the nucleus in suspension-cultured carrot cells is held in place by a system of 5–7 nm diameter fibrils (apparently containing neither tubulin nor actin), but whether a similar scaffolding of cytoskeletal elements constitutes the phragmosome remains to be seen.

Gunning (1982) has pointed out that, due to numerous exceptions, "it is invalid to make any global claim that preprophase bands are essential for spatial organization of cytokinesis." He has also drawn attention to the fact that there is next to no information on the relationships, if any, between phragmosome and preprophase band. As a consequence, he prefers to use the term "division site," signifying a site in the cell cortex where the fusion of the cell plate with the PM will eventually take place. In addition to the more well-known preprophase band and phragmosome, Gunning (1982) has marshalled evidence to show that there are other "markers" which denote the division site. For example, plasmolyzed premitotic cells tend to remain attached at the division site and unequal or asymmetric cell divisions are usually predicted by a nuclear migration prior to mitosis.

Once the division site has been established, displacement of the nucleus through centrifugation generally does not lead to a change in the division plane, providing there is sufficient time for recovery prior to the onset of prophase (Busby and Gunning, 1980). Thus in contrast to the nucleus, preprophase bands of microtubules do not appear to be dislodged by such a treatment (Pickett-Heaps, 1969). If cells are centrifuged late in mitosis, cell plate development is such that a curvature ensues in the direction of the division site (Ota, 1961).

Even if the cell plate has already reached the vicinity of the parent PM, fusion only occurs after the cell plate has encountered the division site. This reorientation or "cellular docking" is energy dependent and the actomyosin system appears to be involved (Palevitz, 1980), but both its mechanism and the nature of the dock itself remain unsolved problems.

REFERENCES

Arnold, J. M. (1969). *J. Cell Biol.* **41,** 894–904.

Arnold, J. M. (1975). *Cytobiologie* **11,** 1–19.

Bajer, A. (1965). *Exp. Cell Res.* **37,** 376–398.

Bajer, A. (1968). *Chromosoma* **24,** 383–417.

Bajer, A. and Allen, R. D. (1966). *J. Cell Sci.* **1,** 455–462.

Becerra, J. and Lopez-Saez, J. F. (1978). *Exp. Cell Res.* **111,** 301–308.

Busby, C. H. and Gunning, B. E. S. (1980). *Eur. J. Cell Biol.* **21,** 214–223.

Cronshaw, J. and Esau, K. (1968). *Protoplasma* **65,** 1–24.

Dustin, P. (1978). *Microtubules.* Springer Verlag, Heidelberg.

Forbes, M. S., Plantholt, B. A., and Sperelakis, N. (1977). *J. Ultr. Res.* **60,** 306–327.

Fowke, L. C. and Gamborg, O. L. (1980). *Int. Rev. Cytol.* **68,** 9–51.

Fowke, L. C., Bech-Hansen, C. W., Constabel, F., and Gamborg, O. L. (1974). *Protoplasma* **81,** 189–203.

Gunning, B. E. S. (1982). In *The Cytoskeleton in Plant Growth and Development* (C. Lloyd, ed.), pp. 229–292, Academic Press, London.

Harris, P. (1975). *Exp. Cell Res.* **94,** 409–425.

Hawes, C. R., Juniper, B. E., and Horne, J. C. (1981). *Planta* **152,** 397–407.

Heath, I. B. (1980). *Int. Rev. Cytol.* **64,** 1–80.

Heath, I. B. (1981). *Int. Rev. Cytol.* **69,** 191–221.

Hepler, P. K. (1980). *J. Cell Biol.* **86,** 490–499.

Hepler, P. K. (1982). *Protoplasma* **111,** 121–133.

Hepler, P. K. and Jackson, W. T. (1968). *J. Cell Biol.* **38,** 437–446.

Hepler, P. K. and Newcomb, E. H. (1967). *J. Ultr. Res.* **19,** 498–513.

Kubai, D. F. (1975). *Int. Rev. Cytol.* **43,** 167–227.

Lloyd, C. and Barlow, P. B. (1982). In *The Cytoskeleton in Plant Growth and Development* (D. Lloyd, ed.), pp. 203–228. Academic Press, London.

Lloyd, C. W., Slabas, A. R., Powell, A. J., and Peace, G. W. (1982). *Cell Biol. Int. Rep.* **6,** 171–175.

Lopez-Saez, J. F., Risuena, M. C., and Gimenez-Martin, G. (1966). *J. Ultr. Res.* **14,** 85–94.

Mazia, D., Petzelt, C., Williams, R. O., and Meza, I. (1972). *Exp. Cell Res.* **70,** 325–332.

Meyer, Y. and Herth, W. (1978). *Planta* **142,** 253–262.

O'Brien, T. P. (1972). *Bot. Rev.* **38,** 87–118.

Ota, T. (1961). *Cytologia* **26,** 428–447.

Palevitz, B. A. (1980). *Can. J. Botany* **58,** 773–785.

Palevitz, B. A. (1982). In *The Cytoskeleton in Plant Growth and Development* (C. Lloyd, ed.), pp. 345–376. Academic Press, London.

Pareyre, C., Lasselain, M.-J., and Deysson, G. (1979). *Cytobios.* **26,** 153–166.

Petzelt, C. (1979). *Int. Rev. Cytol.* **60,** 53–92.

Petzelt, C. and Auel, D. (1977). *P.N.A.S. (USA)* **74,** 1610–1613.

Pickett-Heaps, J. D. (1969). *J. Ultr. Res.* **27,** 24–44.

Pickett-Heaps, J. D. (1975). *Green Algae: Structure, Reproduction in Selected Genera,* Sinauer Associates, Sunderland, MA.

Pickett-Heaps, J. D. and Marchant, H. J. (1972). *Cytobios.* **6,** 255–264.

Pickett-Heaps, J. D. and Northcote, D. H. (1966). *J. Cell Sci.* **1,** 109–120.

Porter, K. R. and Machado, R. D. (1960). *J. Biophys. Biochem. Cytol.* **7,** 167–180.

Schnepf, E., Hausmann, K., and Herth, W. (1982). *Histochem.* **76,** 261–271.

Schroeder, T. E. (1972). *J. Cell Biol.* **53,** 419–434.

Schroeder, T. E. (1973). *P.N.A.S. (USA)* **70,** 1688–1692.

Sinnot, E. W. and Bloch, R. (1941). *Amer. J. Bot.* **28,** 225–232.

Whaley, W. G. and Mollenhauer, H. H. (1963). *J. Cell Biol.* **17,** 216–221.

White, D. L., Mazurkiewicz, J. E., and Barnett, R. J. (1979). *J. Histochem. Cytochem.* **27,** 1084–1091.

Wick, S. M. and Hepler, P. K. (1980). *J. Cell Biol.* **86,** 500–513.

Wick, S. M., Seagull, R. M., Osborn, M., Weber, K., and Gunning, B. E. S. (1981). *J. Cell Biol.* **89,** 685–690.

Part 3

Biogenesis and Turnover of Plant Cell Membranes

15

GENERAL PRINCIPLES

15.1 MEMBRANE SYNTHESIS

15.1.1. Proteins

It has been established on numerous occasions and for several different types of animal cell that membrane proteins are synthesized in much smaller amounts than secretory proteins (Meldolesi, 1974; Terris and Steiner, 1975; Winkler, 1977). Indeed, as Franke et al. (1971) have stated, "only a relatively small portion of the membrane protein of rER, GA and NE seems to parallel the kinetics of the vectorial flow of the secretory product." Similar to secretory proteins the great majority of integral membrane proteins, irrespective of their final location, appear to be synthesized by ER-bound polysomes (Palade, 1959). With some few exceptions (e.g., rhodopsin—Schechter et al., 1979), signal sequences are present which are cleaved before the nascent polypeptide has left the polysome (see Lodish et al., 1981, for a review). Cytochrome b_5 is however an example of one integral membrane protein which does not follow this rule. It is synthesized exclusively on free polysomes (Rachubinski et al., 1980) and, as might be expected, is synthesized without a signal sequence. Its presence in membranes other than the ER is therefore not necessarily a result of membrane flow (see below).

Another unusual case has been documented by Elder and Morré (1976). They have shown that ^{14}C-leucine could be incorporated *in vitro* into integral membrane proteins of a rat liver GA fraction to an extent "greater than could be explained by contamination with rough ER." These proteins are present, however, *in vivo* in relatively small amounts in the GA but are comparable to several found in the PM. Interestingly, GA-associated polysomes (not attached, but lying very close) have been recorded for plant cells (Franke et al., 1972; Mollenhauer and Morré, 1974), but the necessary biochemical proof for such a posttranslational insertion has not yet been given for plants.

Depending on their final location, extrinsic proteins (i.e., those associated with the surface of the membrane) are synthesized either on free or ER-bound polysomes. If the protein in question is extracytoplasmically situated, it is synthesized as a secretory protein, for example β-microglobulin in lymphocytes (Dobberstein et al., 1979). In this latter case binding already occurs within the ER lumen and the molecule is then transferred via the GA to the cell surface. In contrast, those located on the cytoplasmic face of a membrane are synthesized on free polysomes. Only few examples of this type have been adequately studied, with the matrix M protein of virus (particularly vesicular stomatitis) infected cells being probably the best known (Knipe et al., 1977).

15.1.2. Lipids

The synthesis of fatty acids (FA) destined for membrane lipids does not occur in the membranes themselves. Whereas FA synthesis in animal cells is carried out by a FA synthetase complex located in the cytosol (Vagelos, 1974), in plants FA are synthesized in plastids and are exported with the help of acyl carrier proteins (Ohlrogge et al., 1979; Roughan et al., 1980). After their esterification with CoA, FA are converted to phosphatidic acid by acyl transferases localized in the microsomal fraction of the cell (Hendry and Possmayer, 1974. Vick and Beevers, 1978). Just which membranes are involved has not yet been demonstrated, but in analogy to the localization of lecithin biosynthetic enzymes (see below), the GA as well as the ER are probably implicated.

In order to prevent extreme surface electronegativity, the majority of the phosphatidic acid is substituted, forming the isoelectric phospholipids phosphatidylcholine (lecithin, PC) and phosphatidylethanolamine (PE). In the synthesis of PC there are three steps, the first of which, catalyzed by choline kinase, is a soluble reaction. The other two reactions, the coupling of phosphorylcholine to cytidine triphosphate and its subsequent transfer to the diglyceride, occur in the membrane (Morré, 1975). The enzymes responsible for catalyzing these latter two steps are localized principally in the ER [see Wilgram and Kennedy (1963) for animal cells and Hoch and Hartmann (1981), Moore et al. (1973), and Morré et al. (1970) for plant cells], but there is clear evidence for their presence in the membranes of the GA in both animal (Jelesma and Morré, 1978) and plant cells (Montague and Ray, 1977; Morré et al., 1970).

The first steps leading to the synthesis of the sterols are also located in the cytosol (Staby et al., 1973). The conversion of the soluble squalene precursors to cholesterol is a many-step reaction which also occurs in microsomal membranes (Hartmann et al., 1973; Hartmann-Bouillon et al., 1979, for plants; Scallen et al., 1974, for animals).

15.2. MEMBRANE TURNOVER

15.2.1. Proteins

Although turnover encompasses both synthesis and degradation and therefore denotes the overall process of renewal, the methods normally employed to study it (isotopic dilution, sequential double-isotope method—Taylor et al., 1973) give more a measure of degradation than of net synthesis. Even then it is sometimes unclear as to whether the results obtained after such methods really do indicate a breakdown at the molecular level (i.e., a proteolysis) or whether a transfer, for example via membrane flow (see below), has taken place. Particularly in studies involving only one individual membrane type, without corresponding data on the total cellular content of the protein in question, is the latter possibility not to be ruled out. Nevertheless, a few generalizations are possible to make. As may be seen from the data presented in Table I of Hubbard (1978) and Table VI of Morré et al. (1979), the half-life times of proteins within a particular membrane can be extremely variable ranging between 3 to more than 100 hr. In addition, the half-life of ER proteins is greater than those of PM proteins. It is also known that cells can regulate their membrane content according to their growth status by altering either synthesis or degradation. Thus, Warren and Glick (1968) have shown that the PM from nongrowing mouse fibroblasts has a degradation rate of sevenfold of that of growing cells.

A problem with the values for protein turnover in the PM is the relative participation of endocytotic internalization. Although recent results suggest the recycling of many PM proteins, undegraded, back to the cell surface from the lysosome or endosome, the hydrolysis of PM proteins in lysosomes certainly can be an important form of turnover in animal cells (see Chapters 8 and 16). The release of ^{135}I-tyrosine into the medium following the lactoperoxidase (LPO) mediated iodination of externally accessible polypeptides perhaps provides a more reliable indication of true molecular degradation of PM proteins. Thus, whereas in mouse cells 50% of the PM is internalized via endocytosis per hour (Steinman et al., 1976), values of only 2% per hour (corresponding to a half-life of about 30 hr) were obtained with the LPO method (Hubbard and Cohn, 1975). As far as plant cells are concerned we can, at the moment, only speculate on the possibility of endocytosis and the fate of internalized membrane proteins.

15.2.2. Lipids

The synthesis and degradation of membrane lipids is also very variable (Gallaher and Blough, 1975; Omura et al., 1967) and is complicated by the existence of transfer reactions mediated by so-called exchange proteins (Kader, 1977). The transfer is energy dependent, does not involve a *de novo* synthesis (Kamath and Rubin, 1973) and is demonstrable *in vitro*. There are proteins responsible for the exchange of phospholipids (Wirtz and Zilversmit, 1969), as well as for sterols (Bloj and Zilversmit, 1977). They have been isolated from both animal (Wirtz et al., 1972) and plant (Tanaka and Yamada, 1979) tissues and appear to have MWs of about 20–30,000 with isoelectric points at low pH. Some are specific for a particular phospholipid (Helmkamp et al., 1974), others apparently not so.

15.3. VESICULAR MEMBRANE MOVEMENTS

15.3.1. Exocytosis and Endocytosis

The synthesis of extracellular macromolecules invariably occurs in an organelle whose lumen is topographically equivalent to the extracellular milieu. This is the case for the NE, the ER, and the GA. The polymers synthesized in these organelles reach the cell surface by the fusion of some sort of vesicle with the PM. This process constitutes exocytosis. This addition of membrane to the PM cannot go uncorrected; clearly some sort of retrieval mechanism or endocytosis must exist and the extent to which it occurs inversely reflects the growing nature of the cell. The term *recycling* has often been used synonymously with endocytosis (see, e.g., Morré et al., 1979) and, for those whose major field of interest lie in exocytotic events, will probably remain so. Others (e.g., Steinman et al., 1983), however, use recycling to describe just the opposite, namely the retrieval of internalized PM. It is important to be aware of this flexibility of expression.

As just used the terms *exocytosis* and *endocytosis* imply the movements of membrane to and from the PM respectively, but in plant cells I think it is justifiable to extend their usage to include the TP as well. Teleologically speaking, the vacuole is an extracellular compartment and macromolecules are also deposited there. When higher plant cells grow, the increase in surface area of the PM is usually kept pace with by the TP. Thus, in plants exocytosis and endocytosis are somewhat more than they are in animal cells, a fact which is often neglected.

There are several questions worth noting that arise out of these definitions of exocytosis and endocytosis because they can be equally well adapted to the various endomembranes: "what is the origin of membrane added to the PM," and "What is the fate of the membrane removed?" "Is the addition or removal of membrane obligatorally coupled with the export or import of macromolecules?" "Does exocytotically inserted membrane remain in the PM or is it rapidly removed via endocytosis without intermixing?" "Can endocytosed membrane be utilized for exocytosis without being degraded?" "Are vesicles the only mode for addition and removal of membrane?"

15.3.2. Membrane Flow and Differentiation versus Vesicle Shuttles

Although the term *membrane flow* was originally introduced by Bennett (1956) to describe the events following phagocytosis and pinocytosis (i.e., endocytosis) it is now more commonly associated with the movement of membrane in the direction of the PM. As stated by Franke et al. (1971), membrane flow is "accomplished by the physical transfer of membrane material from one cell component to another in the course of their formation or normal functioning." In its most extreme form membrane flow could be considered as a bulk transfer of membrane; that is, whole portions of one organelle are passed onto another organelle. Expressed in terms of secretion-associated membrane flow, this would mean that secretory (glyco)proteins are packaged into an ER (transition) vesicle the membrane of which fuses with the GA, passess through the GA, leaves the GA once again as a vesicle, and is finally incorporated into the PM whereby the contents are discharged into the cell exterior. In other words, the PM and GA would be direct derivatives of the ER.

The intracellular transport of secretory or pinocytosed polymers is also possible without involving a flow of membrane from one compartment into another. This can be achieved when the membrane accompanying the transport of such polymers is always returned unchanged to the organelle from which it came. A series of "shuttles" is then created which would prevent the intermixing of membrane from different organelles (Meldolesi, 1974; Palade, 1975; Rothman, 1981).

We thus have two radically different models to explain the intracellular transport of macromolecules and as might be expected neither is valid in the form in which I have portrayed them. The membranes of ER, GA, and PM are significantly different to be characterized as individual membrane types, but there are significant similarities with respect to certain components which allude to a biogenetic relationship between them.

This fact can be made apparent by considering the presence of so-called marker enzymes or proteins in fractions other than those for which they are specific. Thus, a number of enzyme activities which are regarded as being typical for the ER in animal cells [e.g., NADPH-dependent CCR, glucose-6-phosphatase, whose presence in GA fractions has previously been considered as representing contaminating ER fragments (Fleischer, 1974)] are now recognized as *bona fide* GA membrane components (Howell et al., 1978; Ito and Palade, 1978), albeit in smaller proportions. Indeed, apart from the ribosomal binding proteins, the ribophorins (see Chapter 10), it is hard to find a constituent of the ER which is not present in the GA (see Morré et al., 1979, for a review). At the other end of the scale, the PM marker in animal cells, 5′-nucleotidase, has been demonstrated in the GA (Bergeron et al., 1973; Farquhar et al., 1974). Similarly, in plant cells, $K^+(Mg^{2+})$ ATPase, which is regarded by many as a PM marker (see Chapter 7), has now been demonstrated in the GA (Chanson and Taiz, 1983). In the middle, so to speak, the GA is typified in animal cells by thiamine pyrophosphatase (Morré, 1971) which is also detectable in the ER (Cheetham et al., 1971) and to a lesser extent in the PM (Morré and Ovtracht, 1977).

Whereas results similar to those described seem to indicate a selective rather than a bulk transfer of membrane, they do not entirely rule out a shuttling of membrane between organelles. As early as 1959 Palade suggested that small vesicles might be responsible for bringing back to the ER nonsecretory proteins which have inadvertently been carried to the GA. This idea has now been elaborated on by Rothman (1981). He distinguishes between two types of sorting phenomena associated with the GA: one which takes place in the *trans* compartment and is responsible for differentiating PM destined proteins from secretory and lysosomal ones and a second, located in the *cis* compartment, which channels ER proteins (both membrane bound and soluble) via vesicles back to the ER.

Membrane flow and the shuttling of vesicles between organelles are therefore not mutually incompatible theories. The movement of membrane to and from the PM probably involves two selective events. After Morré et al. (1979) they are as follows:

1. *Selective transfer.* Through which some membrane components in an organelle are allowed to leave while others are somehow exluded from this process.
2. *Selective removal.* This may be accomplished by vesicles bringing back that which was lost, but it may also occur chemically through proteolysis or phospholipid transfer.

The next result of these events is that membrane which is passed from organelle A to B to C changes underway. Membrane flow embodies to a certain extent, therefore, the principles of maturation and differentiation.

15.4. NONVESICULAR MEMBRANE MOVEMENTS

The possible transfer of membrane from one organelle to another via units not recognizable at the level of the EM has been suggested many times over the last quarter of a century (e.g., Hokin, 1968; Fawcett, 1962; Morré and Mollenhauer, 1974; Robinson and Kristen, 1982; Schnepf, 1974). Whereas one may continue to speculate as to the form (e.g., micelles, phospholipid exchange), there are cases where it is difficult not to invoke the participation of such mechanisms—for example, in the growth of the primary nucleus in *Acetabularia* (Franke et al., 1975), the postmitotic formation of the nuclear envelope in *Amoeba* (Maruta and Goldstein, 1975), and the inhibitor-induced production of giant dictyosomes in plant cells (see Chapter 17).

REFERENCES

Bennett, H. S. (1956). *J. Biophys. Biochem. Cytol.* **2,** 99–103.

Bergeron, J. J. M., Ehrenreich, J. H., Siekevitz, P., and Palade, G. E. (1973). *J. Cell Biol.* **59,** 73–88.

Bloj, B. and Zilversmit, D. B. (1977). *J. Biol. Chem.* **252,** 1613–1619.

Chanson, A. and Taiz, L. (1983). *Plant Physiol. Suppl.* **72,** 48.

Cheetham, R. D., Morré, D. J., Pawek, C., and Friend, D. S. (1971). *J. Cell Biol.* **49,** 899–905.

Dobberstein, B., Garroff, H., and Warren, G. (1979). *Cell* **17,** 759–769.

Elder, J. H. and Morré, D. J. (1976). *J. Biol. Chem.* **251,** 5054–5068.

Farquhar, M. B., Bergeron, J. J. M., and Palade, G. E. (1974)., *J. Cell Biol.* **60,** 8–25.

Fawcett, D. W. (1962). *Circulation* **26,** 1105–1125.

Fleischer, B. (1974). In *Methods in Enzymology* (S. P. Colowick and N. O. Kaplan, eds.), Vol. 31A, pp. 180–191. Academic Press, New York.

Franke, W. W., Kartenbeck, J., Krien, S., Van Der Woude, W. J., Scheer, U., and Morré, D. J. (1972). *Z. Zellforsch.* **132,** 365–380.

Franke, W. W., Morré, D. J., Deumling, B., Cheetham, R. D., Kartenbeck, J., Jarasch, E. D., and Zentgraf, H. W. (1971). *Z. Naturforsch.* **26b,** 1031–1039.

Franke, W. W., Spring, H., Scheer, U., and Zerban, H. (1975). *J. Cell Biol.* **66,** 681–689.

Gallaher, W. R. and Blough, H. A. (1975). *Arch. Biochem. Biophys.* **268,** 104–114.

Hartmann, M. A., Ferne, M., Gigot, C., Brandt, R., and Benveniste, P. (1973). *Physiol. Veg.* **11,** 209–230.

Hartmann-Bouillon, M. A., Benveniste, P., and Roland, J.-C. (1979). *Biol. Cellulaire* **35,** 183–194.

Helmkamp, G. M., Harvey, M. S., Wirtz, K. W. A., and van Deenen, L. L. M. (1974). *J. Biol. Chem.* **249,** 6382–6389.

Hendry, A. T. and Possmayer, F. (1974). *Biochem. Biophys. Acta* **369,** 156–172.

Hock, K. and Hartmann, E. (1981). *Plant Sci. Lett.* **21,** 389–396.

Hokin, L. E. (1968). *Int. Rev. Cytol.* **23,** 187–208.

Howell, K. E., Ito, A., and Palade, G. E. (1978). *J. Cell Biol.* **79,** 581–589.

Hubbard, A. L. (1978). In *Transport of Macromolecules in Cellular Systems* (S. C. Silverstein, ed.), 363–390. Dahlem Konferenzen, Abakon Verlagsgesellsch, Berlin.

Hubbard, A. L. and Cohn, Z. A. (1975). *J. Cell Biol.* **64,** 461–479.

Ito, A. and Palade, G. E. (1978). *J. Cell Biol.* **79,** 590–597.

Jelesma, C. L. and Morré, D. J. (1978). *J. Biol. Chem.* **253,** 7960–7971.

Kader, J. C. (1977). In *Dynamic Aspects of Cell Surface Organisation* (G. Poste and G. L. Nicholsen, eds.), pp. 127–204. Elsevier, Amsterdam.

Kamath, S. A. and Rubin, E. (1973). *Arch. Biochem. Biophys.* **158,** 312–322.

Knipe, D. M., Baltimore, D., and Lodish, H. F. (1977). *J. Virol.* **21,** 1128–1139.

Lodish, H. F., Braell, W. A., Schwartz, A. L., Strous, G. J. A. M., and Zilberstein, A. (1981). *Int. Rev. Cytol. Suppl.* **12,** 247–307.

Maruta, H. and Goldstein, L. (1975). *J. Cell Biol.* **63,** 180–196.

Meldolesi, J. (1974). *Adv. Cytopharmacol.* **2,** 71–84.

Mollenhauer, H. H. and Morré, D. J. (1974). *Protoplasma* **79,** 333–336.

Montague, M. J. and Ray, P. M. (1977). *Plant Physiol.* **59,** 225–230.

Moore, T. S., Lord, J. M., Kagawa, T., and Beevers, H. (1973). *Plant Physiol.* **52,** 50–53.

Morré, D. J. (1971). *Meth. Enzymol.* **22,** 130–148.

Morré, D. J. (1975). *Ann. Rev. Plant Physiol.* **26,** 441–481.

Morré, D. J. and Mollenhauer, H. H. (1974). In *Dynamic Aspects of Plant Ultrastructure* (A. Robards, ed.), pp. 84–137, McGraw-Hill, London.

Morré, D. J. and Ovtracht, L. (1977). *Int. Rev. Cytol. Suppl.* **5,** 61–188.

Morré, D. J., Kartenbeck, J., and Franke, W. W. (1979). *Biochim. Biophys. Acta* **559,** 72–152.

Morré, D. J., Nyquist, S., and Rivera, E. (1970). *Plant Physiol.* **45,** 800–804.

Ohlrogge, J. B., Kuhn, D. N., and Stumpf, P. K. (1979). *Proc. Natl. Acad. Sci. USA* **76,** 1194–1198.

Omura, T., Siekevitz, P., and Palade, G. E. (1967). *J. Biol. Chem.* **242,** 2389–2396.

Palade, G. E. (1959). In *Subcellular Particles* (T. Hayashi, ed.), pp. 64–83. Ronald Press, New York.

Palade, G. E. (1975). *Science* **189,** 347–358.

Rachubinski, R. A., Verma, D. S. P., and Bergeron, J. J. M. (1980). *J. Cell Biol.* **84,** 705–716.

Robinson, D. G. and Kristen, U. (1982). *Int. Rev. Cytol.* **77:** 89–127.

Rothman, J. E. (1981). *Science* **213,** 1212–1219.

Roughan, P. G., Holland, R., and Slack, C. R. (1980). *Biochem. J.* **188,** 17–24.

Scallen, T. J., Srikantaiah, M. V., Seetharam, B., Hansbury, E., and Garvey, K. L. (1974). *Fed. Proc.* **33,** 1733–1746.

Schechter, I., Burstein, Y., Zerwell, R., Ziv, E., Kantor, F., and Papermaster, D. S. (1979). *Proc. Natl. Acad. Sci.* **76,** 2654–2658.

Schnepf, E. (1974). In *Dynamic Aspects of Plant Ultrastructure* (A. E. Robards, ed.), pp. 331–357. McGraw-Hill, New York.

Staby, G. L., Hackett, W. P., and De Hertogh, A. A. (1973). *Plant Physiol.* **52,** 416–421.

Steinman, R. M., Brodie, S. E., and Cohn, Z. A. (1976). *J. Cell Biol.* **68,** 665–687.

Tanaka, T. and Yamada, M. (1979). *Plant & Cell Physiol.* **20,** 533–542.

Taylor, J. M., Dehlinger, P. J., Dice, J. F., and Schimke, R. T. (1973). *Drug. Met. D.* **1,** 84–91.

Terris, S. and Steiner, D. F. (1975). *J. Biol. Chem.* **250,** 8389–8398.

Vagelos, P. R. (1974). In *MTP International Review of Science, Biochemistry of Lipids I,* T. W. Goodwin, ed.), pp. 100–140. Butterworth, London.

Vick, B. and Beevers, H. (1978). *Plant Physiol.* **62,** 173–178.

Warren, L. and Glick, M. C. (1968). *J. Cell Biol.* **37,** 729–746.

Wilgram, G. F. and Kennedy, E. P. (1963).*J. Biol. Chem.* **238,** 2615–2619.

Winkler, H. (1977). *Neuroscience* **2,** 657–683.

Wirtz, K. W. A. and Zilversmit, D. B. (1969). *Biochim. Biophys. Acta* **193,** 105–116.

Wirtz, K. W. A., Kamp, H. H., and Van Deenen, L. L. M. (1972). *Biochim. Biophys. Acta.* **274,** 606–617.

16

FOR COMPARISON: THE CASE IN ANIMAL CELLS

The amount of work done on both exocytosis and endocytosis in animal cells is immense. I do not think that I am exaggerating, therefore, when I estimate that, in terms of papers published, the literature on this subject in plant cells represents only 5% of this. A monograph on plant cell membranes cannot ignore this fact, but an attempt should be made to refrain from presenting a biased image of the situation in plant cells. The small amount of space available allows the case in animal cells to be presented in only an exemplary and cursory manner. For those wishing to delve deeper the following recent review articles are recommended; Brown et al. (1983), Helenius et al. (1983), Hopkins (1983), Morré (1981) Morré et al. (1979), Pastan and Willingham (1983), Steinman et al. (1983), and Tartakoff (1980).

16.1 MEMBRANES INVOLVED IN MOVEMENT TOWARD THE PM

16.1.1. Example: Synthesis and Transport of PM Glycoproteins

There are a number of PM glycoproteins for which the route from site of synthesis has been clearly mapped out. These include the glycoproteins of virus envelopes (Simons and Warren 1983) as well as cell surface antigens (Morré, 1981). Virus envelopes are chemically very similar to the PM of animal cells (Leonard and Compans, 1974). They possess glycoproteins which project as spikes from the external surface of the membrane and, depending on the type of virus, a protein at the inner surface as well which connects the nucleocapsid to the spike glycoproteins. In virus-infected cells the genetic machinery of the host is taken over to produce these various components which are assembled at the cell surface to produce an infectious virion by reserve pinocytosis.

The two best studied examples of this process are vesicular stomatitis virus (VSV) and Semliki Forest virus (SFV). The spike glycoproteins in both cases are synthesized with a short leader sequence at the NH_2 terminus and are inserted into the lipid bilayer of the ER where they are glycosylated with "high mannose oligosaccharides" (Rothman and Lodish, 1977). They are then transported to the GA as *trans*-membrane proteins where the "high mannose" oligosaccharides are processed into the "complex" type in a similar way to secretory proteins (Bergmann et al., 1981; Green et al., 1981), and finally reach the PM after about 60 min. In VSV-infected cells only one spike glycoprotein ("G protein") is synthesized (Lodish and Rothman, 1978), but in SFV cells these are two ("E_1 and E_2 proteins") which are synthesized sequentially from a single 26S polycistronic mRNA. Cleavage of the two polypeptides at a site just before the second, internally located leader sequence presumably occurs before their insertion into the ER membrane (Garoff et al., 1982).

The intracellular transport of these PM glycoproteins is thus analogous to that of secretory proteins. In contrast to the latter, which presumably lie separated from the membrane in the lumen of the ER, the "G protein" of VSV has a very short (29 AAs; about 3% of the total) COOH terminus which remains on the cytoplasmic side of the ER (Rose et al., 1980). If the structure of this domain is manipulated genetically, the intracellular transport is slowed down or completely arrested (Rose and Bergmann, 1983). This is a control feature which is not apparent in secretory protein transport.

16.1.2. Example: Regulated Secretion

There are a number of animal glands in which exocytosis is hormonally regulated. This results in the storage of secretory materials in secretory vesicles until the triggering of their discharge. Such is the case in the pancreas and parotid gland as well as in the chromaffin cells of the adrenal medulla (Tartakoff and Vassalli, 1978). The secretory vesicles are also called condensing vacuoles, since secretory material accumulates there over a period of hours (Jamieson and Palade, 1971). During this time no apparent increase in vesicle diameter is seen (Jamieson and Palade, 1967a).

Whereas there is no doubt that the secretory proteins are synthesized at the ER (Palade, 1975), there is uncertainty as to their route to the condensing vacuoles. Autoradiography of pulse-chased pancreas (see Figures 11 and 12 in Jamieson and Palade, 1967b) showed that silver grains were not present over the stacked cisternae of the GA, supporting the previous contention of Jamieson and Palade (1967a) that "secretory proteins are transported from the cisternae of the rER to condensing vacuoles via the small vesicles of the Golgi complex." The nature of the latter vesicles which lie at the periphery of the Golgi cisternal stack is not clear.

If they only came from the rER bringing secretory proteins to the condensing vacuoles, the membrane of the latter would increase in amount. Meldolesi (1974) suggested therefore that vesicles shuttled back excess membrane to the ER but failed to take into account the glycosylating role of the GA in his scheme. Novikoff and Novikoff (1977), among others, have proposed that the transfer of product from ER to GA occurs via tubular connections at the periphery of the Golgi cisternae. From this peripheral, "filling station" vesicles are budded off which travel to the condensing vacuole. Presumably, excess condensing vacuole membrane returns to the GA, since new condensing vacuoles appear to be formed there.

Although the origin and fate of the condensing vacuole membrane may not be clear, its structure is certainly not the same as the ER, GA, or PM, although there are some proteins present which are in common with both GA and PM (see Morré et al., 1979, p. 89, for summary). On the other hand, the chromograffin granule membrane, for instance, has much higher molar ratios of cholesterol and lysolecithin than the GA (Blaschko et al., 1967). The special nature of the condensing vacuole or secretory granule is also reflected in freeze-fracture studies in that very few intramembranous particles are present (De Camilli et al., 1976).

16.2. MEMBRANES INVOLVED IN MOVEMENT FROM THE PM

16.2.1. Example: Secretion-Coupled Endocytosis

These are several cell types in which secretion is immediately followed by an intense endocytotic activity. In one of these, the neuron, there is evidence that the endocytotic vesicles are converted directly into exocytotic vesicles without necessarily fusing with other organelles. Studies on frog neuronmuscular junctions (Heuser and Reese, 1973, 1981) and turtle retina (Schaeffer and Raviola, 1978) have shown that, within 1 min of stimulation of synaptic vesicle release, externally added electron-dense tracers [e.g., horse radish peroxidase (HRP)] are at the terminal PM. Within the next 15–30 min the synaptic vesicles became labeled and finally, after restimulating, the tracer was exocytosed.

The pathway of endocytosis in cells with regulated secretion has also been followed both with content markers (e.g., HRP and blue dextran) and with membrane markers [e.g., cationic ferritin (CF)]. The results obtained depend on the tracer and cell type used. Thus, in the pancreatic exocrine cell, dextran is taken up and finds its way to the GA, while HRP was found in the lysosomes (Herzog and Reggio, 1980). In thyroid follide cells, the majority of endocytosed CF is subsequently found in lysosomes, but a small amount was seen in the GA (Herzog and Miller, 1979). GERL has also been implicated as the ultimate recipient of endocytosed membrane in similar studies on neuroblastoma cells (Gonatas et al., 1977).

16.2.2. Example: Receptor-Mediated Endocytosis

As we have already seen (Chapter 8), cell surface receptors are internalized via coated pits. Now it is time to consider the fate of the endocytosed membrane with its receptors. It is clear from studies in which the number of binding sites at the cell surface for low-density lipoprotein (LDL—Goldstein et al., 1976) and macroglobulin protease (Kaplan, 1980) has been compared in normal and cycloheximide-treated cells that protein synthesis is not necessary for maintaining this population. Slow turnover rates of the order of many hours are seen in both cases, indicating that somehow the receptors are recycled in an undegraded form back to the PM.

The fate of the ligand is clear: in nearly all cases it reaches the lysosome where it is broken down by the acid hydrolases contained therein. What therefore prevents the receptor from sharing the same fate as the ligand? A clue to solving this problem is seen in the case of the iron transporter protein transferrin. This ligand remains bound to its receptor after internalization, releasing only its iron before it reappears at the cell surface (Dautry-Varsat et al., 1983). This suggests that separation of ligand and receptor must occur in a compartment other than the lysosome, a compartment having two exit possibilities—one to the lysosome for the ligand and the other leading back to the PM for the receptors—and having a low pH environment to enable dissociation of the ligand from the receptor.

Such a receptor-rich, nonlysosomal compartment has recently been isolated (Kahn et al., 1982; Marsh et al. 1983) and is termed, *endosome*. Moreover, the separation of ligand from the receptor in it has been beautifully demonstrated by immunocytochemical means (Geuze et al., 1983).

Morphologically, the endosome is a complex structure consisting of a ramifying network of cisternal, tubular, and "vacuolar" elements, extending from the cell surface to the juxtanuclear area (Wall et al., 1980). Some of the vacuoles are seen to contain a large number of small vesicles and are, therefore, probably equivalent to the previously described multivesicular bodies. Since in one and the same cell (even in the same coated pit—Carpentier et al., 1982) many different types of receptors may be internalized, some of which are degraded [e.g., epidermal growth factor (Das and Cox, 1978)] and many which are recycled, the sorting problem in the endosome must be a considerable one.

REFERENCES

Bergmann, I. E., Tokuyasu, K. T., and Singer, S. I. (1981). *Proc. Natl. Acad. Sci. USA* **78,** 1746–1750.

Blaschko, H., Firemark, H., Smith, A. D., and Winkler, H. (1967). *Biochem. J.* **104,** 545–549.

Brown, M. S., Anderson, R. G. W., and Goldstein, L. (1983) *Cell* **32,** 663–667.

Carpentier, T. L., Gorden, P., Anderson, R. G. W., Goldstein, T. L., Brown, M. S., Cohen, S., and Orci, L. (1982) *J. Cell Biol.* **95,** 73–77.

Das, M. and Fox, C. F. (1978). *Proc. Natl. Acad. Sci. USA* **75,** 2644–2648.

Dautry-Varsat, A., Geichanover, A., and Lodish, H. F. (1983). *Proc. Natl. Acad. Sci. USA* **80,** 2258–2262.

De Camilli, P., Paluchetti D. and Meldolesi, (1976) Z. *J. Cell Biol.* **70,** 59–74.

Garoff, H., Kondor-Koch, C., and Riedel, H. (1982). *Current Topics Microbiol.* **99,** 1–50.

Geuze, H. T., Stot, T. W., Strous, G. I. A. M., Lodish, H. F., and Schwartz, A. L. (1983). *Cell* **32,** 277–287.

Goldstein, J. L., Basa, S. K., Brunschede, G. Y., and Browa, M. S. (1976). *Cell* **7,** 85–95.

Gonatas, N. K., Kim, S. U., Streber, A., and Avrameas, S. (1977). *J. Cell Biol.* **73,** 1–13.

Green, J., Griffiths, G., Loward, D., Quinn, P., and Warren, G. (1981). *J. Mol. Biol.* **152,** 663–698.

Helenius, A., Mellman, I., Wall, D., and Hubbard, A. (1983). *TIBS* **8,** 245–250.

Herzog, V. and Miller, F. (1979). *Eur. J. Cell Biol.* **19,** 203–215.

Herzog, V. and Reggio, H. (1980). *Eur. J. Cell Biol.* **21,** 141–150.

Heuser, J. E. and Reese, T. (1973). *J. Cell Biol.* **57,**315–344.

Heuser, J. E. and Reese, T. S. (1981). *J. Cell Biol.* **88,** 564–580.

Hopkins, C. (1983). *Nature* **304,** 684–685.

Jamieson, J. D. and Palade, G. E., (1967a). *J. Cell Biol.* **34,** 577–596.

Jamieson, J. D. and Palade, (1967b). *J. Cell Biol.* **34,** 597–615.

Jamieson, J. D. and Palade, G. E. (1971). *J. Cell Biol.* **48,** 503–522.

Kahn, M. N., Posner, B., Kahn, R. J., and Bergeron, T. T. M. (1982). *J. Biol. Chem.* **257,** 5969–5976.

Kaplan, J. (1980). *Cell* **19,** 197–205.

Leonard, T. and Compans, W. (1974). *Biochim. Biophys. Acta* **344,** 51–94.

Lodish, H. F. and Rothman, J. E. (1978). *Sci. Amer.* **240,** 38–53.

Marsh, M., Bolzan, E., and Helenius, A. (1983). *Cell* **32,** 931–940.

Meldolesi, J. (1974). *Adv. Cytopharmacol.* **2,** 71–84.

Morré, D. J. (1981). In *International Cell Biology 1980–1981* (H. G. Schweiger, ed.), pp. 622–632, Springer Verlag, Berlin.

Morré, D. J., Kartenbeck, J., and Franke, W. W. (1979). *Biochem. Biophys. Acta* **559,** 72–152.

Novikoff, A. B. and Novikoff, P. M. (1977). *Histochem. J.* **9,** 525–551.

Palade, G. E. (1975). *Science* **189,** 347–358.

Pastan, I. and Willingham, M. C. (1983). *TIBS* **8,** 250–254.

Rose, T. K. and Bergmann, T. E. (1983). *Cell* **34,** 513–524.

Rose, T. K., Welch, W. T., Sefton, B. M., Esch, F. S., and Ling, N. C. (1980). *Proc. Natl. Acad. Sci. USA* **77,** 3884–3888.

Rothman, T. E. and Lodish, H. F. (1977). *Nature* **269,** 775–780.

Schaeffer, S. F. and Raviola, E. (1978). *J. Cell Biol.* **79,** 802–825.

Simons, K. and Warren, G. *Adv. Prot. Chem.* (1984) (in press).

Steinman, R. M., Mellman, I. S., Mullen, W. A., and Cohn, Z. A. (1983). *J. Cell Biol.* **96,** 1–27.

Tartakoff, A. (1980). *Int. Rev. Experm. Patholog.* **22,** 227–251.

Tartakoff, A. and Vassalli, P. (1978). *J. Cell Biol.* **79,** 694–707.

Wall, D. A., Wilson, G., and Hubbard, A. L. (1980). *Cell* **21,** 79–93.

17

THE CASE IN PLANT CELLS

Since studies with specific membrane proteins, comparable to those on animal cells (see Chapter 16), have not been undertaken, there remains as yet no direct evidence for membrane flow from the ER to the PM via the GA in plant cells. We assume, however, that it must take place, based on the passage of secretory materials through these organelles as determined by fractionation studies. This assumption is strengthened by ultrastructural work documenting the extreme polarity of the plant dictyosome, which is suggestive of a process for membrane moving through it (see Chapter 5). Nevertheless, one must realize that biochemically we are still in the Dark Ages when it comes to understanding the biogenesis of plant cell endo- and plasma membranes. In this respect, it seems correct to assume that plant cells will not possess basically different mechanisms from animal cells in this latter; instead it is probably more a question of their proportional usage.

17.1. SECRETION KINETICS AND MEMBRANE TURNOVER

17.1.1. Normal Cells

Table 17.1 gives a comparison of residence or displacement times for radioactively labeled secretory polymers in the membrane systems of plant and animal cells. In plant systems particularly, where radioactive sugar precursors have been employed, there are two possible sources of error:

1. direct incorporation into secretory vesicles or into GA vesicles which become detached upon homogenization (see Morré et al., 1979) and
2. the presence of GA-derived secretory vesicles in ER preparations (see Robinson, 1977).

With the exception of those animal types with a regulated discharge of secretory materials, the residence times tend to lie below 10 min for each fraction. The higher values for the GA in the cases of pancreatic exocrine and of goblet cells may reflect a contamination of these fractions with

TABLE 17.1. KINETICS OF GRANULOCRINE SECRETION BASED ON CONTENT MARKERS

				Displacement Time (min)		
References	Material	Method	Isotope	ER	GA	Secretory (GA) vesicles
A. Higher plants						
Chrispeels (1970) Gardiner and Chrispeels (1975)	Carrot root discs	Fractionation	[^{14}C]Proline	4	4	4–8
Bowles and Northcote (1974)	Maize root tips	Fractionation	[^{14}C]Glucose	7	2–3	2–3
Robinson *et al.* (1976)	Pea stem segments	Fractionation	[^{14}C]Glucose	—	3	3–4
B. Animals (a selection)						
Jamieson and Palade (1967)	Guinea pig: pancreas exocrine	Autoradiography	[^{14}C]Leucine	7	20	60[a]
Neutra and Leblond (1966a)	Rat: goblet cells	Autoradiography	[^{3}H]Glucose	—	15–35	60
Bergeron *et al.* (1978)	Rat: liver	Fractionation	[^{14}C/^{3}H]Leucine	2–4	6–7	3–8
Kimber (1981)	Schistocera: egg shell	Autoradiography	[^{3}H]Leucine	2–4	6–7	3–8

[a] This is an example of regulated discharge (i.e., secretory vesicle accumulation).

secretory vesicles which in both cell types are present in very large numbers in the apical part of the cell.

One notes that the total residence times for movement of secretory material from ER to PM lies well under 20 min, particularly for plant cells, which is much faster than those values given for transport of PM glycoproteins from the ER (see Chapter 16). This could be taken as evidence for vesicles shuttling between the organelles in question, but the discrepancy could just as well be explained by assuming that not all PM proteins synthesized in the ER are transported at equal rates.

Values for membrane displacement in the GA of some plant and animal systems are given in Table 17.2. Some of these values are based on morphometric observations either with or without the application of some sort of secretory inhibitor (see also below) or are estimations based on pulse-chase studies. Irrespective of this, or of the cell type employed, and being aware that the fundamental assumption in these calculations is that cisternae are shunted through the dictyosome from the *cis*- to *trans*-face in order to replenish material lost from the latter, it is surprising that the renewal times for dictyosomes are similar: of the order of 1 to several minutes. Depending on the number of cisternae in a dictyosome, values for its complete turnover of between 10 and 30 min are typical.

It has become customary to assume that the GA in plant cells is the immediate donor of membrane for the PM. However, Morré (1980) has cast doubts about this, saying that "in many cells which exhibit rapid surface growth or turnover, the Golgi apparatus is non-secretory and shows little sign of vesicle formation. A classic example of this is that of rapidly elongating plant cells." This is a curious statement since Morré himself (Morré and Van der Woude, 1976) has clearly shown that in rapidly elongating pollen tubes of *Lilium longiflorum* (6 $\mu m \cdot min^{-1}$) more than 1000 vesicles are released from the GA per minute, which account for an increase in PM of 300 $\mu m^2 \cdot min^{-1}$. Presumably, Morré has equated the existence of hypertrophied dictyosome vesicles with exocytotic activity which might be true for pollen tubes and root cap cells (see Chapter 5) but is not necessarily so for elongating cells. A large number of smaller vesicles released more frequently from the dictyosome can just as easily account for the necessary demands of membrane at the PM. This fact has been appreciated by Quaite et al. (1983) who have shown that for *Avena* coleoptile cells the average number of dictyosome-derived vesicles (di-

TABLE 17.2. MEMBRANE DISPLACEMENT IN THE GOLGI APPARATUS

References	Material	Renewal Times (min)	
		Individual cisternae	Entire dictyosomes
Schnepf and Koch (1966)	*Vacuolaria virescens*—water secretion	1	8–10
Neutra and Leblond (1966b)	Rat colon—mucin secretion	2–4[a]	Not given
Brown (1969)	*Pleurochrysis scherffelii*—scale secretion	1–2	20–30
Eisinger and Ray (1972)	*Pisum sativum*—cell wall secretion	1–2[a]	5–10
Heinrich (1973)	*Monarda fistulosa*—"water" secretion	1	8
Bowles and Northcote (1974)	Zea mays—slime secretion	0.3[a]	Not given
	cell wall secretion	2.5[a]	Not given
Williams (1974)	*Hymenomonas carterae*—scale secretion	6	30–40
Morré et al. (1979)	Rat liver—protein secretion	4–5[a]	15–20
Picton and Steer (1981)	*Tradescantia virginiana*—cell wall secretion	3.7	15–18.5
Kristen and Lockhausen (1982)	*Aptenia cordifolia*—slime secretion	7.3	66
Morré et al. (1983)	*Daucus carota*—cell wall secretion	2–4	Not given

From Robinson and Kristen (1982).
[a] Estimations not based on morphometry.

ameter ~ 0.09 μm) per cell is increased more than twofold in response to auxin-induced elongation.

17.1.2. Inhibitor Experiments

Because the higher plant GA is structurally more delineated from other membranes than it is in animal cells, it is easier to recognize and document changes in its structure as a result of perturbations in secretion through inhibitor application. Such changes, although not yet biochemically characterized, are quite significant for our understanding of biogenetic relationships between the endomembranes and PM. We may distinguish between two types of inhibitor: those effecting changes in GA structure and those affecting secretory vesicle transport.

Over 20 years ago Schnepf (1961) proposed an inverse relationship between secretory activity and cisternal number per dictyosome. He suggested that whereas the replenishment of dictyosomal membrane remains more or less constant, the loss of membrane through secretory vesicle release is not, so that during periods of intense secretory activity the dictyosomes become smaller. In contrast, when secretion is reduced, either through natural causes (e.g., low temperature or as a result of inhibitor treatment), the dictyosome enlarges. This may be expressed as an increase in the number of cisternae per dictyosome (see Figure 17.1a) or as a lateral "growth" of the cisternae (see Figure 17.1b). Polarity in the stack is, particularly in the latter case, difficult to recognize, but the growth seems to occur more at the *cis* cisternae than at the *trans* cisternae causing the dictyosome to curl up. The "cup-shaped" dictyosomes which result have been recorded on many occasions (see Table IV in Robinson and Kristen, 1982) and can be made to revert back to normal dictyosomes simply by removing the inhibitor in question.

Morré et al. (1983) have recently examined the short-term effects of the sodium-selective ionophore monensin on plant dictyosomal structure. This substance is known to block both intracellular PM glycoprotein transport (Johnson and Schlesinger, 1980) as well as secretory protein transport (Ledger et al., 1980) in animal cells. In both plant (Mollenhauer et al., 1982; Robinson, 1981a) and animal (Tartakoff and Vassalli, 1978) cells treatment with monensin inevitably results in a vacuolation of the *trans*-face cisternae and their separation from the stack. However, a brief exposure to monensin (minutes) allows the detection of an extra one to two

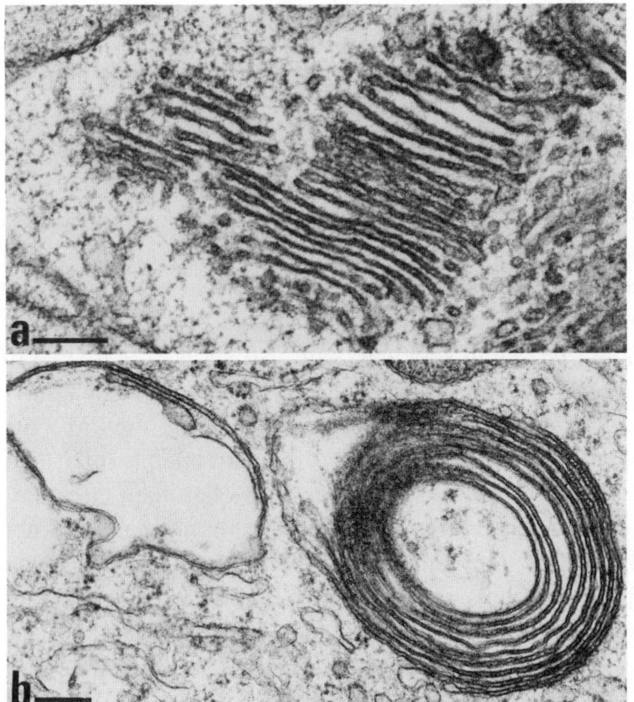

Figure 17.1. Inhibitor-induced changes in dictyosomal structure. (a) Increase in the number of cisternae in a stack obtained as a result of cold treatment of *Vicia faba* leaf mesophyll protoplasts. (b) Increase in the length of cisternae resulting in a cup-shape which is here sectioned at 90° to the "cup-axis". Dictyosome from a cold-treated root cap cell of *Zea mays* L. (for comparison see Figure 5.1c,d). Bar = 0.2 μm. Unpublished micrographs of Joachim and Robinson.

cisternae per dictyosomal stack before those at the *trans*-face become less recognizable.

Results such as these demonstrate that the inflow of membrane into the GA can be uncoupled from the outflow. One must hasten to add, however, that such effects are not restricted to the GA: the ER also increases in amount as shown both electron optically and biochemically (Robinson, 1981b, 1984). Since vesiculation processes are usually regarded as being energy dependent (Jamieson and Palade, 1968, 1971, Robinson and Ray, 1977; Tartakoff and Vassalli, 1978), and there is no reason to assume that the ER and GA might be different in this regard, it is difficult to understand how both ER and GA accumulate membrane. Perhaps this is an indication for chemical turnover in these membranes with proteolytic and lipolytic reactions being more affected than their synthetic counterparts.

The transport of secretory vesicles from the GA to the PM may be inhibited by the application of cytochalasin B or D. Usually the dictyosome is unaffected by this treatment (Kristen and Lockhausen, 1982; Mollenhauer and Morré, 1976; see, however, Volkmann and Czaja, 1981) and the secretory vesicles accumulate in the near vicinity. Robinson et al. (1976) have confirmed this with isolated secretory vesicle fractions from etiolated pea stem segments but also showed that secretion of cell wall polymers still took place; that is, a retention of secretory vesicles occurs rather than an inhibition of the fusion with the PM. Although the cytochalasins are well known as microfilament inhibitors (Brown and Spudich, 1981), it is by no means clear how they exert their effect on secretory vesicle migration, since microfilaments are not normally distinctive in the vicinity of the GA. Nevertheless, the rate at which vesicles accumulate as a result of cytochalasin treatment has been used to estimate the rate of vesicle production (Picton and Steer, 1981; Kristen and Lockhausen, 1982).

17.2. MEMBRANE FLOW, EQUILIBRIUM AND RECYCLING

In plant cells as in animal cells the amount of membrane reaching the PM in the form of secretory vesicles is quite considerable. Estimates of up to $10\%\cdot min^{-1}$ of the existing PM are not uncommon. This has often been equated with total production of GA vesicles (Kristen and Lockhausen, 1982) but, in my opinion, this assumption is only valid in cells which do not have vacuoles. The majority of higher plant cells have a tonoplast which, because it enlarges during growth, also requires membrane. There must then be a dichotomy of pathways, but not necessarily of the same degree for GA (or GA plus ER?) vesicles.

Not only is the higher plant cell different from the animal cell in this respect, the relative contributions of the ER to the GA is dissimilar. In Chapter 5 I have already pointed to the lack of so-called transition vesicles at the *cis*-face cisterna of the higher plant dictyosome. This feature is presumably coupled to the comparatively small amounts of protein which

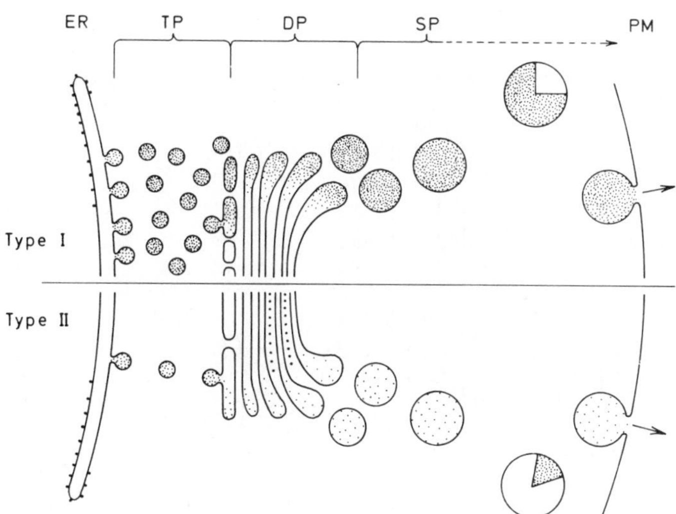

Figure 17.2. ER–GA relationships expressed in terms of the nature of the secretory product. The relative amounts of protein in the secretory product are indicated as stippled sectors in the "composition circles." Two extremes are visualized: Type I with a high proportion of protein and therefore a distinct vesicle traffic between ER and GA and Type II with mainly polysaccharide secretion and therefore a much reduced ER–GA vesicle transfer. TP—transition vesicle phase; DP—dictyosomal phase; SP—secretory vesicle phase. From Robinson and Kristen (1982).

are secreted by higher plant cells (see Figure 17.2). What therefore is the source and form of the membrane required to replenish the higher plant dictyosome? A direct recycling from the PM (see Figure 17.3) is a good candidate but evidence for it has been hard to provide. Morré and Mollenhauer (1983), for example, have attempted to follow the pathway of internalized PM by applying the PTAC stain (see Chapter 7) but were without success. The uptake of surface bound cationic ferritin via coated pits in bean leaf protoplasts has, however, been recently demonstrated (Joachim and Robinson, 1984). The presence of this label in vesicles attached to the trans face of the GA is suggestive of a direct recycling from the PM.

17.3. MEMBRANE BIOGENESIS IN GERMINATING SEED TISSUES

To this point my discourse on the synthesis of endo- and plasma membranes has been restricted to their maintenance in cells where they already

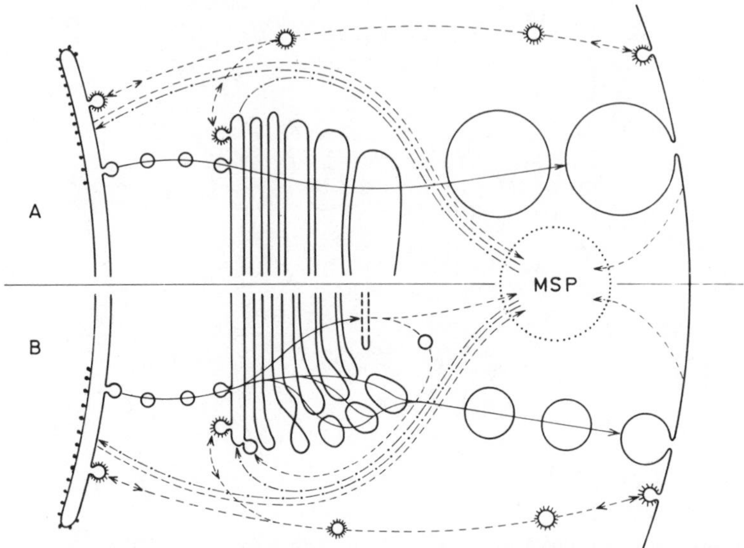

Figure 17.3. Pathways and possibilities of secretion-associated membrane flow and recycling in higher plant cells. In A a dictyosome of the type characteristic for some slime-secreting cells is depicted. Here the entire maturing face cisterna is released as a secretory vesicle. In B, the more usual case, the central portion of the *trans*-face cisterna degenerates and may be directly recycled back into the *cis*-face cisternae either in vesicle form or in "membrane subunit form." Recycling from the plasma membrane is presented in the form of endocytotic (coated) vesicles as well as through the release of membrane subunits which may constitute membrane subunit pools (MSP). ER and GA can be equally considered recipients for recycled membrane in these forms. From Robinson and Kristen (1982).

exist rather than the biogenesis of these membranes. However, interesting information can also be obtained from studies on germinating seeds and on wounded storage tissues. Since the latter is more often experimentally triggered, it will be covered in the next chapter.

In cotyledonary or endosperm tissue of freshly germinating seeds it is often very difficult to detect the GA; and the small amounts of ER present are often attached to LBs or PBs (see Chapter 11). During the first 24 hr of germination there is, together with the gradual disappearance of storage materials, a synthesis of endomembranes as documented in a number of ultrastructure papers (see, e.g., Bain and Mercer, 1966; Colborne et al., 1976; Harris and Chrispeels, 1980; Öpik, 1966). As to the origin of this membrane, particularly the ER, Mollenhauer et al. (1978) have presented evidence for the participation of so-called lamellar bodies as intermediates in the biogenesis of ER in maize embryo cells and cotyledon cells of *Pisum sativum and Phaseolus vulgaris*. These structures appear to emanate from the LBs and PBs at first in the form of myelinlike configurations. At later stages in germination direct attachments between these structures and rER were demonstrated. These observations require confirmation from other workers.

Biochemical evidence for a proliferation of ER during the early stages of germination is also available. An increase in phospholipids associated with ER fractions isolated from *Ricinus* endosperm has been recorded by Lord (1978). Similarly, the activities of enzymes involved in lecithin biosynthesis have been shown to rise during this period (Bowden and Lord, 1975; Vick and Beevers, 1978). Since this increase is also measurable in seeds germinated at 4 °C (Gilkes et al., 1979), it must be assumed that the enzymes were already present in the seeds before dessication. The activity of the ER marker enzyme NADH–CCR has also been shown to increase in activity several fold (Gilkes and Chrispeels, 1980; Jones, 1980). In germinating barley aleurone cells there is also a shift in the equilibrium density distribution of CCR activity: it moves from 1.13–1.14 to 1.11–1.12 $g{\cdot}cm^{-3}$ as germination proceeds (Jones, 1980). Information on the biogenesis of the GA in germinating seed tissue is not yet available.

REFERENCES

Bain, J. M. and Mercer, F. V. (1966). *Austr. J. Biol. Sci.* **19,** 69–84.

Bergeron, J. J. M., Borts, D., and Cruz, J. (1978). *J. Cell Biol.* **76,** 87–97.

Bowden, L. and Lord, J. M. (1975). *FEBS Lett.* **49,** 369–371.

Bowles, D. J. and Northcote, D. H. (1974). *Biochem. J.* **142,** 139–144.

Brown, R. M. (1969). *J. Cell Biol.* **41,** 109–123.

Brown, S. S. and Spudich, A. (1981). *J. Cell Biol.* **88,** 487–491.

Chrispeels, M. J. (1970). *Biochem. Biophys. Res. Comm.* **39,** 732–737.

Colborne, A. J., Morris, G., and Laidmann, L. (1976). *J. Exp. Bot.* **27,** 759–768.

Eisinger, W. and Ray, P. M. (1972). *Plant Physiol.* **49,** Suppl. 2.

Gardiner, M. and Chrispeels, M. J. (1975). *Plant Physiol.* **55,** 536–541.

Gilkes, N. R. and Chrispeels, M. J. (1980). *Plant Physiol.* **65,** 600–604.

Gilkes, N. R., Herman, E. M., and Chrispeels, M. J. (1979). *Plant Physiol.* **64,** 38–42.

Harris, N. and Chrispeels, M. J. (1980). *Planta* **148,** 293–303.

Heinrich, G. (1973). *Protoplasma* **77,** 271–278.

Jamieson, J. D. and Palade, G. E. (1967a). *J. Cell Biol.* **34,** 577–596.

Jamieson, J. D. and Palade, G. E. (1967b). *J. Cell Biol.* **34,** 597–615.

Jamieson, J. D. and Palade, G. E. (1968). *J. Cell Biol.* **39,** 589–603.

Jamieson, J. D. and Palade, G. E. (1971). *J. Cell Biol.* **48,** 503–522.

Joachim, S. A. and Robinson, D. G. (1984). *Eur. J. Cell Biol.*

Johnson, D. C. and Schlesinger, M. J. (1980). *Virology* **102,** 407–424.

Jones, R. L. (1980). *Planta* **150,** 70–81.

Kimber, S. J. (1981). *J. Cell Sci.* **50,** 225–243.

Kristen, U. and Lockhausen, J. (1982). *Eur. J. Cell Biol.* **29,** 262–267.

Ledger, R., Uchida, N. and Tanzer, M. L. (1980). *J. Cell Biol.* **87,** 663–671.

Lord, J. M. (1978). *J. Exp. Bot.* **29,** 13–23.

Mollenhauer, H. H. and Morré, D. J. (1976). *Protoplasma* **87,** 39–48.

Mollenhauer, H. H., Morré, D. J., and Jelsema, C. L. (1978). *Bot. Gaz.* **139,** 1–10.

Mollenhauer, H. H. and Morré, D. J., and Norman, J. O. (1982). *Protoplasma* **112,** 117–126.

Morré, D. J. (1980). In *Cell Compartmentation and Metabolic Channeling* (L. Nover, F. Lynen, and K. Mothers, ed.) pp. 47–61. Elsevier-North Holland Biomedical Press, Amsterdam.

Morré, D. J. and Van Der Woude, W. J. (1974). In *Macromolecules Regulating Growth and Development* (E. D. Hay, T. J. King, and J. Papaconstantinou, eds.), pp. 81–111. Academic Press, New York.

Morré, D. J. and Mollenhauer, H. H. (1983). *Eur. J. Cell Biol.* **29,** 126–132.

Morré, D. J., Boss, W. F., Grimes, H., and Mollenhauer, H. H. (1983). *Eur. J. Cell Biol.* **30,** 25–32.

Morré, D. J., Kartenbeck, J., and Franke, W. W. (1979). *Biochem. Biophys. Acta* **559,** 72–152.

Neutra, M. and Leblond, C. P. (1966a). *J. Cell Biol.* **30,** 119–136.

Neutra, M. and Leblond, C. P. (1966b). *J. Cell Biol.* **30,** 137–150.

Öpik, H. (1966). *J. Exp. Bot.* **17,** 427–439.

Picton, J. M. and Steer, M. W. (1981). *J. Cell Sci.* **49,** 261–272.

Quaite, E., Parker, R. E., and Steer, M. W. (1983). *Plant Cell and Environment* **6,** 429–432.

Robinson, D. G. (1977). *Adv. Botanical Res.* **5,** 89–151.

Robinson, D. G. (1981a). *Eur. J. Cell Biol.* **23,** 267–272.

Robinson, D. G. (1981b). In *Cell Walls '81* (D. G. Robinson and H. Quader, eds.), pp. 47–56. Wiss. Verlagsgesellsch., Stuttgart.

Robinson, D. G. In *Membranes and Compartmentation in the Regulation of Plant Function,* (A. Boudet, ed.), Vol. 23, pp. xxx–xxx. *Ann. Rev. Proc. Phytochem. Soc. Eur.* (in press).

Robinson, D. G. and Kristen, U. (1982). *Intl. Rev. Cytol.* **77,** 89–127.

Robinson, D. G. and Ray, P. M. (1977). *Cytobiol.* **15,** 65–77.

Robinson, D. G., Eisinger, W. R., and Ray, P. M. (1976). *Ber. Dtsch. Bot. Ges.* **89,** 147–161.

Schnepf, E. (1961). *Z. Naturforschg.* **16b,** 605–610.
Schnepf, E. and Koch, W. (1966). *Arch. Mikrobiol.* **54,** 229–236.
Tartakoff, A. and Vassalli, P. (1978). *J. Cell Biol.* **79,** 694–707.
Vick, B. and Beevers, H. (1978). *Plant Physiol.* **62,** 173–178.
Volkmann, D. and Czaja, A. W. P. (1981). *Exp. Cell Res.* **135,** 229–236.
Williams, J. A. (1980). *Am. J. Physiol.* **238,** 269–279.

18

A SPECIAL CASE: WOUNDED TISSUES

When plant storage tissues are cut into thin slices and incubated aerobically, a large number of effects are set into motion. Collectively these are described as "aging," which, unfortunately, is somewhat of a misnomer since a true senescence does not take place. The tissues are not dying but merely adapting to new conditions caused by wounding the dormant tissue.

Physiologically the "aging" is characterized by two phenomena: a pronounced rise in respiration, which was first observed almost a century ago (Boehm, 1887), and the development of an increased salt uptake capacity, which has also been known for many decades (Steward et al., 1932). At the cytological level almost all of the membranous organelles are affected, in general leading to their increased presence. The potentials of this system for studying membrane biogenesis are great, a fact which is becoming increasingly apparent.

18.1 THE CYTOLOGICAL RESPONSE

Freshly cut slices of storage tissue are notoriously difficult to prepare satisfactorily for thin sections in the electron microscope (Fowke and Setterfield, 1968; Barckhausen, 1978). The cells are parenchymatous, typically highly vacuolate and have only a thin layer of cytoplasm which usually does not traverse the vacuole. In the white potato, due to the relative lack of ribosomes, the cytoplasm is normally not richly stained. On the other hand, in Jerusalem artichoke numerous ribosomes are present. The most conspicuous membranous organelles are the nucleus, mitochondria, and plastids, which are present either as leucoplasts (amyloplasts) or chromoplasts. These tend to come together in clumps in the cell corners. Dictyosomes are rarely found and microbodies are only occasionally present. It has been claimed for red beetroot (Van Steveninck and Jackman, 1967) that slicing induces a rapid destruction of a lamellar ER system into small ER vesicles. In contrast, studies on potato tuber and Jerusalem artichoke (Barckhausen, 1978; Fowke and Setterfield, 1969) have not recorded such a change, the freshly cut and fixed slices evincing only rudimentary ER. Probably it is a question of whether or not the fixation is quicker than the various hydrolytic processes which are started immediately upon wounding.

Depending on the type of storage tissue involved, the aging process may be accompanied by cell divisions. This is the case in the white potato which has probably been used for biochemical studies more than any other tissue. Here the divisions tend to be restricted to cells immediately below the cut surface ("phellogen") with the outermost cells undergoing a suberinization and subsequently degenerating. Cell divisions can also be induced in other types by the addition of hormones to the aeration medium.

As a result of aging, a period which can vary from 20 to 200 hr, almost every organelle undergoes some sort of change. These changes are depicted diagrammatically in Figure 18.1. In general, more cytoplasm is present, strands of which now traverse the vacuole, and the various endomembranes are present in greater number. Clearly, since the tissues are nonphotosynthetic, the energy for these syntheses must come from the reserve substances and, as a rule, starch in amyloplasts disappears. The plastids do not, however, disappear but undergo a sort of differentiation. In the white potato, for example, large vesicles termed *intraplastid bodies* are seen to develop as outgrowths from prolammellar bodies and occupy almost the total volume of the plastid (Barckhausen, 1978).

The nucleolus enlarges with the granular portion (*pars granulosa*) now dominating. A so-called vacuolation also takes place giving rise to ring-shaped nucleoli in section (Rose, Setterfield, and Fowke, 1971). Interestingly, these "vacuoles" appear to contract and expand over periods of 1–3 hr. In the later stages of aging the fibrillar, chromatin, portion of the nucleolus disperses and is often seen as small lacunae within the *pars granulosa*. The increase in the granular portion of the nucleolus is an indication of increased ribosomal synthesis and has been confirmed by autoradiography (Setterfield, 1963; Rose et al., 1971; Byrne and Setterfield, 1977). Initially (i.e., for the first 20–40 min after wounding), labeling from radioactive uridine is localized over nonnucleolar chromatin, but this changes to an intense nucleolar location after about 1 hr and remains so for many subsequent hours.

Although there appears to be a difference in the aging response of white potato slices as against those of Jerusalem artichoke with respect to an increase in the number of ribosomes (compare Barckhausen, 1978, with Setterfield et al., 1978), there is agreement that aging results in an increase (from 10 to 65%) in the proportion of polysomes and (from 5 to 25%) in the proportion of membrane-bound ribosomes (Leaver and Key, 1967; Kahl, 1971; Sparkuhl et al., 1976; Sparkuhl and Setterfield, 1977; Cherry,

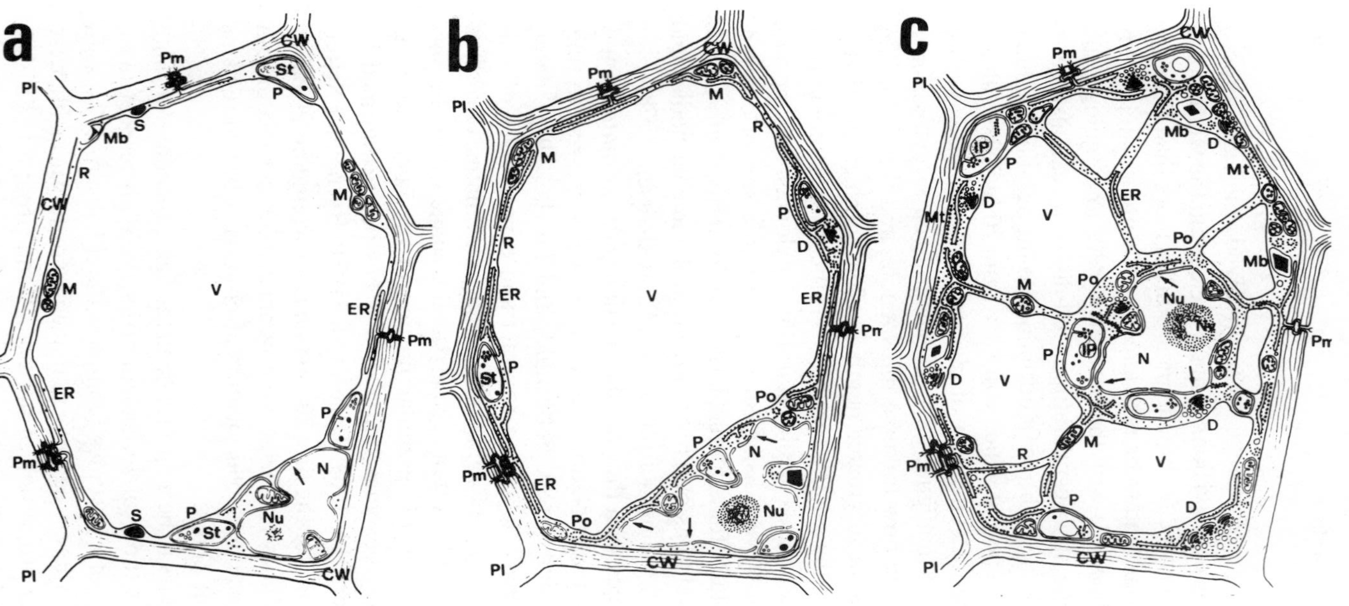

Figure 18.1. (a) Parenchymatous cell of intact storage tuber in diagrammatic form. The cell is highly vacuolate and has relatively few organelles. (b) The same but about 24 hr after wounding. An increase in cytoplasm and more organelles are to be seen. (c) A later stage in the differentiation of wounded storage tissue cells. The vacuole is now traversed by cytoplasmic strands and a considerable proliferation in number and types of membranous organelle has taken place. CW = cell wall; ER = endoplasmic reticulum; IP = intraplastid body; M = mitochondria; Mb = microbody; Mt = microtubules; N = nucleus; Na = nucleolus; NV = nucleolar vacuole; P = plastid; Pl = plasma membrane; Pm = plasmodesmata; R = ribosomes; S = spherosome; St = starch. Arrow indicates nuclear pore. (From Barckhausen, 1978).

1968). Clearly associated with the latter is an increase in the presence of ER. Usually this is rER, but in white potato a considerable amount of tubular gER is formed. This develops over a 48-hr period after wounding and then gradually declines over the next 6 days. It is presumably associated with the synthesis and transport of suberin precursors whose deposition also reaches a maximum in periods subsequent to 48 hr. Anomalous is also the beetroot, where it has been shown that protein crystals gradually develop in the lumen of rER when the washing period exceeds 2–3 days (Jackman and Van Steveninck, 1967).

Whether or not dicytosomes increase in number as a result of aging appears to depend on the tissue type and on the presence of hormones. In white potato and carrot, for instance, the addition of hormones to the aeration medium is not necessary, but this is not so for Jerusalem artichoke (Fowke and Setterfield, 1968). Here a response is only obtained upon the addition of auxin with or without kinetin. The biogenesis of dictyosomes is undoubtedly associated with an increase in cell wall production, and this fact has been utilized in studies on the synthesis and secretion of cell wall glycoproteins (Chrispeels et al., 1974).

Although there are biochemical indicators that microbodies, or at least their component enzymes, are synthesized as a result of aging, there is no definite *in situ* cytological evidence for this. Ultrastructural changes in mitochondria before and after aging are difficult to ascertain. Their greater number, however, is undisputed (Lee and Chasson, 1966; Verleur, 1969; Asahi et al., 1966).

18.2. THE BIOCHEMICAL RESPONSE

18.2.1. Initial Events: The Wounding Itself

It has been known for some time (Hanes and Baker, 1931) that one of the immediate effects of slicing storage tissues is a 5- to 10-fold increase in respiration. This effect is not simply one of increased O_2 availability, since measurements of O_2 tension in potato tubers, for instance, show values of at least 15% (Burton, 1950), which still lie in excess of that of the KmO_2 for cytochrome oxidase. Furthermore, increases in external O_2 do not cause a rise in the endogenous tuber respiration (Mapson and Burton, 1962).

Usually the endogenous respiration of the parent plant is CN resistant. This indicates the existence of an alternate electron transport pathway to molecular O_2 which by-passes the cytochromes and is characterized by the lack of a phosphorylation site and its sensitivity to hydroxamic acids (Henry and Nyns, 1975; Solomos, 1977). Depending on the type of storage tissue involved, slicing may result in the immediate loss of this CN resistance. Theologis and Laties (1980) have surveyed the incidence of this feature of storage tissue wounding and have placed, for example, white potato, beetroot, and Jerusalem artichoke in the group of the sensitive plants and sweet potato, carrot, and banana in the insensitive group. Although CN resistance and its loss have been used as characteristics of wound-induced respiration, it is just as important to know what the actual, or proportional, participation of the alternate and cytochrome pathways is in the two tissue types. As it turns out, the cytochrome pathway is employed exclusively by all storage tissues. The alternate pathway appears to be used in some CN-resistant types but only after aging.

Work, particularly that carried out with slices of white potato, has shown that the initial loss of CN resistance is associated with a change in respiratory substrate from glycolytic and TCA cycle intermediates, which are typical of the intact tuber (Barker, 1963), to lipid. This is to be inferred from the sudden ineffectiveness of malonate, a TCA cycle inhibitor (Laties, 1959, 1964) and from a comparison of C^{12}/C^{13} ratios in the major classes of compounds in intact and sliced tubers (Jacobsen et al., 1970; Smith and Jacobsen, 1976). Since free lipids are found in only small quantities in white potato tubers (Galliard, 1968), the source of the lipid for fresh slice respiration has been sought for in membrane lipids. As we have already seen (Jackman and Van Steveninck, 1967), there are direct electron microscopic observations supporting an extensive membrane degradation immediately upon wounding. Indirect support for this is to be seen in the difficulties involved in isolating intact active mitochondria from freshly cut tissue slices (Van Steveninck and Jackman, 1976; Castelfranco et al., 1971), and from the fact that some typical microsomal enzyme activities (e.g., NADH–cytochrome *c* reductase—Rungie and Wiskich, 1972; cytochrome P_{450}-mediated hydroxylation of cinnamic acid—Rich and Lamb, 1977) are not detectable in microsomal fractions isolated from freshly cut slices.

The immediate degradation of all classes of membrane lipids (phospholipids, glycolipids, and neutral lipids) with a rise in the amount of free fatty acids upon slicing has been documented by Theologis and Laties (1981) for those tissues whose CN respiration is sensitive to wounding. In addition, some of the fatty acids released, in particular the long-chain polyunsaturated α-linolenic and linoleic acids, are further subjected to peroxidation as indicated by the production of ethane (Konze and Elstner, 1978). Those tissues whose CN respiration is insensitive to wounding do not show membrane lipid degradation.

The three types of enzymes necessary for this lipid breakdown, lipid acyl hydrolases, phospholipases, lipoxygenases, and fatty acid hydroperoxide degradative enzymes, have all been found in potato tissue (Gallaird, 1970, 1978). The question of their intracellular localization and the triggering of their activities through wounding is most important, particularly with respect to the group whose CN resistance is not lost upon wounding. Since the extensive loss of membrane lipids caused by slicing is, of course, also operative as a result of homogenization and even occurs appreciably at 0°C (Galliard, 1970), it is almost impossible to use normal white potato tubers for studies on the intracellular localization of these enzymes. Wardale and Galliard (1975, 1977) have attempted to overcome this problem by choosing systems which have relatively low lipolytic activities e.g., potato shoots, brassica florets, and pea roots. The lipid acyl hydrolase activities in these tissues have been localized, together with other acid hydrolases in a "lysosomal" or vacuole fraction. To conclude that these enzymes are localized in the large central vacuole of the whote potato tuber but are absent in the vacuoles of those plants which do not loose their CN resistance upon wounding is not unreasonable. However, the story cannot be so simple since aged slices of white potato tuber possess lipid acyl hydrolase activity at the same level as in freshly cut slices but a massive membrane lipid degradation does not ensue upon rewounding them (Theologis and Laties, 1981). Laties (Theologies and Laties, 1980; Laties, 1978) has drawn attention to two further possibilities. One is that an acyl hydrolase inhibitor may be present in those tissues which do not lose their CN resistance and has indeed extracted a heat-labile protein from carrot roots which does in fact inhibit lipolysis in white potato homogenates. The other takes into account the Ca^{2+} stimulation of phospholipase activity (Björnstad, 1966; Hasson and Laties, 1976). Laties has postulated that slicing or homogenization could well lead to such an activation through contact of enzyme with cell wall Ca^{2+}. This would, however, implicate a difference in cell wall chemistry between the two types of storage tissue for which there is at present no evidence.

Not only is the burst of lipolytic activity responsible for changes in energy metabolism in CN-respiration-sensitive plants by attacking the mitochondria, the products themselves act deleteriously. Thus, free fatty acids give rise to an inhibition of oxidative phosphorylation (Wojtczack et al., 1969; Earnshaw et al., 1970), of glycolytic enzymes (Weber et al., 1966; Radamoss et al., 1976) and of glucose transport (Decker and Tanner, 1975). In addition to the large amounts of free fatty acids β-hydroxy α-ketoglutaric acid is synthesized as a result of the wounding in white potato slices and this inhibits the TCA cycle (Laties, 1964). Because the TCA cycle is inoperative, free fatty acids cannot be broken down by β-oxidation since the acetyl CoA produced by this pathway cannot be further utilized. Instead some of the free fatty acids are subjected to an α-oxidation which has been confirmed both by studies with labeled substrates (Laties et al., 1972) as well as through the inhibition of CO_2 evolution by the use of imidazole, a selective inhibitor of fatty acid α-oxidation (Martin and Stumpf, 1959). Other free fatty acids appear to be dealt with by ketone body formation (Theologies and Laties, 1981).

From the foregoing it can be appreciated that much work has been carried out on the respiratory metabolic aspects of the events which are triggered by wounding. In contrast, very little attention has been given to short-term changes in nucleic acids or proteins, particularly in those types where wounding induces a lypolytic degradation of cell membranes. It has been reported for Jerusalem artichoke that the amount of ribosomal RNA drops by nearly 30% in the first 2 hr after wounding (Sparkuhl et al., 1976), but information on a possible proteolysis of membrane proteins is not at hand. This is a pity since it is important to establish whether the various membrane proteins released into the cytosol as a result of lipolytic action are reusable for the subsequent membrane resynthesis which takes place during aging. Although it is known that wounding is the cause of a variety of effects on cytosolic proteins [e.g., some of the glycolytic enzymes are synthesized, some are degraded or inactivated and others remain unaffected (Kahl, 1978)], we can only cite the instability of NADPH–CCR when dissociated from the ER membrane in animal tissues (Kuriyansa et al., 1969; Kuriyansa and Omura, 1971) as an example of what might occur to membrane proteins after lipolytic membrane degradation.

18.2.2. Subsequent Events: The "Aging"

Aging in aerated water results in a further two- to threefold increase in respiration for tissues which both lose their CN resistance upon wounding and those which do not. In the former group the CN resistance is gradually regained, and a return to carbohydrate as the respiratory substrate, as indicated by malonate sensitivity, occurs (Laties, 1978). Despite this change the basic respiratory ability is the same for freshly cut and aged slices. This has been demonstrated by Theologis and Laties (1978a,b) who showed that uncouplers of phosphorylation, such as CCCP (carbonyl cyanid *m*-chlorophenyl hydrazone), can produce an increase in fresh slice respiration of the same order as that obtained through aging alone. Thus, as these authors have stressed (Theologis and Laties, 1980), the increase in respiration due to aging does not necessarily involve the synthesis of electron transport chain (cytochrome) components but rather entails substrate mobilization. However, there are conflicting reports as to actual changes in mitochondrial CCO activity as a result of aging. Thus, on the one hand, little or no increase and no effect of chloramphenicol (some of the subunits of CCO are synthesized on mitochondrial ribosomes) have been recorded (Hacket et al., 1960; Waring and Laties, 1977a). One thing is certain, however: irrespective of whether or not *de novo* synthesis of mitochondria occurs as a result of aging, it is doubtful whether such an increase in mitochondrial number (at best about 2-fold, Asahi et al., 1966) can account for the 20- to 30-fold difference in respiration between the dormant tuber and the aged slices which are prepared from it.

The respiratory response induced through aging is dependent upon nucleic acid synthesis and protein synthesis, as judged by its inhibition through treatment with actinomycin D and puromycin or cycloheximide, respectively (Glick and Hackett, 1963; Sampson and Laties, 1968), but only when these substances are applied during the first 8–10 hr of aging. During this period the synthesis of mRNA (as early as 20 min after wounding—Byrne and Setterfield, 1977), tRNA (after an initial breakdown a new synthesis—Sparkuhl et al., 1976) have all been shown to occur. There is also evidence for a considerable increase in DNA-dependent RNA polymerase activity as a result of wounding and aging (Kahl and Wielgat, 1976). Furthermore, the proportion of ribosomes involved in protein synthesis rises from 12 to 13% in dormant tuber tissue to 30% within 1 hr after wounding (Leaver and Key, 1967; Sparkuhl et al., 1976).

More important than the question as to which proteins are being synthesized during age-induced respiration is: for which of these proteins is new mRNA being transcribed? There are certainly a number of glycolytic enzymes which increase in amount in storage tissues due to *de novo* synthesis (Kahl, 1974), but the activity of a selection of these together with a number of TCA cycle and pentose phosphate cycle enzymes are unaffected by actinomycin D, irrespective of the time of application of the inhibitor (Waring and Laties, 1977a). In contrast, the latter authors have shown that, in aging white potato slices, several key enzymes of membrane phospholipid synthesis (phosphorylcholine-glyceride transferase, phosphorylcholine-cytidyl transferase, and phosphatidyl phosphatase) are inhibited by actinomycin D in a time-restricted way. That membrane lipid synthesis occurs as a result of aging is to be expected and is indeed well documented (see Mazliak and Kader, 1978, for a review); that it might be the cause of the age-induced respiration is not immediately apparent. Waring and Laties (1977b) have shown that both cerulenin, a specific inhibitor of fatty acid synthesis and dimethylaminoethanol, an analog of choline which alters membrane fluidity, prevent the development of the respiratory response. Since the lipid environment of membrane-bound enzymes is an acknowledged factor in the activity of such enzymes (Mazliak, 1978), it is attractive to see the cause of age-induced respiration in an incorporation of newly synthesized membrane lipid into the mitochondria.

As already mentioned in Chapter 15 the endomembranes but not the mitochondria, possess the enzymes necessary for the synthesis of phospholipid. Thus, a transfer of these lipids from their site(s) of synthesis to their intussusception in the mitochondria is a most necessary prerequisite of age-induced respiration. It should not therefore come as a surprise that the white potato is a preferred object for the isolation of plant phospholipid exchange proteins (Kader, 1975). The transfer of phospholipids from microsomal membranes to the mitochondria during aging presupposes that the former exists. Clearly, in the case of those storage organs where extensive membrane breakdown occurs on slicing (e.g., white potato), the resynthesis of the endomembrane organelles must occur before lipid transfer to the mitochondria can take place. Thus, increases in protein and phospholipid content together with antimycin-A-insensitive NADH–cytochrome *c* reductase activity have been measured for white potato (Ben Abdelkader, 1969, beetroot and turnip slices (Rungie and Wiskich,

1972), as well as for sweet potato (Tanaka et al., 1974) where wound-induced lipolysis does not occur. Interestingly, in the latter case a peak is reached after about 2 days of aging and is then followed by a decline. The decline in membrane-bound reductase activities is associated with a concomitant rise in the detectable activity of these enzymes in the cytosol fraction.

Aging also results in the development of microbodies. Based on their enzyme compliment, these are best termed *nonspecialized microbodies* (Huang and Beevers, 1971, 1973). In the case of the sweet potato increases in catalase activity occur mostly on the second day after slicing (Kanazawa et al., 1967), which thus lags behind the development of ER and mitochondria. The effects of aging on the GA in cells of storage slices has only been investigated with carrot tissue, a representative of the group which lacks wound-induced lipolytic activity. As a marker for the activity of the GA is the enzyme UDP–arabinosyl transferase, which forms short arabinosyl side chains attached via the hydroxyl group of hydroxyproline residues in extensin (Karr, 1972; see also Section 10.2). According to Chrispeels et al. (1974), the specific activity of this enzyme in a mixed membrane preparation (sedimenting between 25,000 and 45,000*g*) increases about fourfold over a 24-hr aging period. Although the values for fresh slices are not given, these authors show that the actual level of glycosylation *in vivo* of hydroxyproline-containing glycoproteins increases from 60% in 6-hr aged slices to almost 90% in 24-hr aged slices. A similar 50% increase in the specific activity of peptidyl hydroxylase is also recorded for this time period. Thus, the increase in the synthesis of cell wall glycoproteins is paralleled by corresponding increases in the activities of two of the enzymes responsible for posttranslational modifications of the protein.

Whereas it is not difficult to assume that ER-associated enzymes are synthesized by ER-bound ribosomes, does this also apply to the enzyme and other proteins of the GA? Or do some of these come directly from cytosolic ribosomes? At this time we do not know, but it seems appropriate to repeat these questions here (see also Chapter 15) because the system of aged slices of storage tissues appears to be a potentially excellent one. Particularly when slices are made from those tissues showing a wound-induced release of lipolytic activity, one has a unique opportunity to study the biogenesis of ER and GA membranes and to determine whether the synthesis of the latter is totally dependent upon the synthetic abilities of the former.

REFERENCES

Asahi, T., Honda, Y., and Uritani, I. (1966). *Arch. Biochem. Biophys.* **113,** 498–499.

Barckhausen, R. (1978). In *Biochemistry of Wounded Plant Tissues* (G. Kahl, ed.), pp. 1–42. De Gruyter, Berlin.

Barker, J. (1963). *Proc. Roy. Soc. B.* **158,** 143–155.

Ben Abdelkader, A. (1969). *Compt. Rend. Acad. Sci. (Paris) Ser. D.* **268,** 2406–2409.

Björnstad, P. (1966). *J. Lipid. Res.* **7,** 612–620.

Boehm, J. (1887). *Bot. Zeitung* **45,** 671–691.

Burton, W. G. (1950). *New Phytol.* **49,** 121–134.

Byrne, H. and Setterfield, G. (1977). *Planta* **136,** 203–210.

Castelfranco, P. A., Tang, W.-J., and Bolar, M. L. (1971). *Plant Physiol.* **48,** 795–800.

Cherry, J. (1968). In *Biochemistry and Physiology of Plant Growth Substances* (F. Wightman and G. Setterfield, eds.), pp. 417–431. Runge Press, Ottawa.

Chrispeels, M. J., Sadava, D., and Cho, Y. P. (1974). *J. Exp. Bot.* **25,** 1157–1166.

Decker, B. and Tanner, W. (1975). *Fed. Eur. Biochem. Soc. Lett.* **60,** 346–348.

Earnshaw, M. J., Truelove, B., and Butler, R. D. (1970). *Plant Physiol.* **45,** 318–321.

Fowke, L. and Setterfield, G. (1968). In *Biochemistry and Physiology of Plant Growth Substances* (F. Wightman and G. Setterfield, eds.), pp. 584–602. Runge Press, Ottawa.

Galliard, T. (1968). *Phytochem.* **7,** 1907–1914.

Galliard, T. (1970). *Phytochem.* **9,** 1725–1734.

Galliard, T. (1978). In *Biochemistry of Wounded Plant Tissues* (G. Kahl, ed.), pp. 155–201. De Gruyter, Berlin.

Glick, R. E. and Hackett, D. P. (1963). *P.N.A.S. (USA)* **50,** 243–247.

Hackett, D. P., Haas, D. W., Griffiths, S. K., and Niederpruem, D. J. (1960). *Plant Physiol.* **35,** 8–19.

Hanes, C. S., Barker, J. (1931). *Proc. Roy. Soc. (Lond). B* **108,** 95–118.

Hasson, E. P. and Laties, G. G. (1976). *Plant Physiol.* **57,** 142–147.

Henry, M. F. and Nyns, E. J. (1975). *Sub-Cell. Biochem.* **4,** 1–65.

Huang, A. H. C. and Beevers, H. (1971). *Plant Physiol.* **48,** 637–641.

Huang, A. H. C. and Beevers, H. (1973). *J. Cell Biol.* **58,** 242–248.

Jackman, M. E. and van Steveninck, R. F. M. (1967). *Austr. J. Biol. Sci.* **20,** 1063–1068.

Jacobsen, B. S., Smith, B., Epstein, S., and Laties, G. G. (1970). *J. Gen. Physiol.* **55,** 1–17.

Kader, J. C. (1975). *Biochim. Biophys. Acta* **380,** 31–44.

Kahl, G. (1971). *Z. Naturforsch.* **26B,** 1058–1064.

Kahl, G. (1974). *Bot. Rev.* **40,** 263–314.

Kahl, G. (1978). In *Biochemistry of Wounded Plant Tissues* (G. Kahl, ed.), pp. 262–283. De Gruyter, Berlin.

Kahl, G. and Wielgat, B. (1976). *Physiol. Veg.* **14,** 725–738.

Karv, A. L. (1972). *Plant Physiol.* **50,** 275–282.

Kanazana, Y., Asahi, T., and Uritani, T. (1967). *Plant and Cell Physiol.* **8,** 249–262.

Konze, J. R. and Elstner, E. F. (1978). *Biochim. Biophys. Acta* **528,** 213–221.

Laties, G. G. (1959). *Arch. Biochem. Biophys.* **79,** 364–377.

Laties, G. G. (1964). *Plant Physiol.* **39,** 654–663.

Laties, G. G.(1978). In *Biochemistry of Wounded Plant Tissues* (G. Kahl, ed.), pp. 421–466. De Gruyter, Berlin.

Laties, G. G., Hoelle, C., and Jacobsen, B. S. (1972). *Phytochem.* **11,** 3404–3411.

Leaver, C. J. and Key, J. L. (1967). *P.N.A.S. (USA)* **57,** 1338–1344.

Lee, S. G. and Chasson, R. M. (1966). *Physiol. Plant.* **19,** 199–206.

Mapson, L. W. and Burton, W. G. (1962). *Biochem. J.* **82,** 19–25.

Martin, R. O. and Stumpf, P. K. (1959). *J. Biol. Chem.* **234,** 2548–2554.

Mazliak, P. (1978). In *Lipids and Lipid polymers in Higher Plants* (M. Tevini and H. K., Lichfenthaler, ed.), pp. 48–74. Springer Verlag, Berlin.

Mazliak, P. and Kader, J.-C. (1978). In *Biochemistry of Wounded Plant Tissues* (G. Kahl, ed.), pp. 123–153. De Gruyter, Berlin.

Radamoss, C. S., Uyeda, K., and Johnston, J. M. (1976). *J. Biol. Chem.* **251,** 98–107.

Rich, P. R. and Lamb, C. J. (1977). *Eur. J. Biochem.* **72,** 353–360.

Rose, R. J., Setterfield, G., and Fowke, L. C. (1971). *Planta* **101,** 210–230.

Rungie, J. M. and Wiskich, J. T. (1972). *Austr. J. Biol. Sci.* **25,** 103–113.

Sampson, M. J. and Laties, G. G. (1968). *Plant Physiol.* **43,** 1011–1016.

Setterfield, G. (1963). *Symp. Soc. Exp. Biol.* **17,** 98–126.

Setterfield, G., Sparkuhl, J., and Byrne, H. (1978). In *Biochemistry of Wounded Plant Tissues* (G. Kahl, ed.), pp. 571–594. De Gruyter, Berlin.

Smith, B. N. and Jacobsen, B. S. (1976). *Plant and Cell Physiol.* **17,** 1089–1092.

Solomos, T. (1977). *Ann. Rev. Plant Physiol.* **28,** 279–297.

Sparkuhl, J. and Setterfield, G. (1977). *Planta* **135,** 267–273.

Sparkuhl, J., Gare, R. L., and Setterfield, G. (1976). *Planta* **129,** 97–104.

Steward, F. C., Wright, R., and Berry, W. E. (1932). *Protoplasma* **16,** 576–611.

Tanaka, Y., Kojima, M., and Urifani, I. (1974). *Plant and Cell Physiol.* **15,** 843–854.

Theologis, A. and Laties, G. G. (1978a). *Plant Physiol.* **62,** 232–237.

Theologis, A. and Laties, G. G. (1978b). *Plant Physiol.* **62,** 243–248.

Theologis, A. and Laties, G. G. (1980). *Plant Physiol.* **66,** 890–896.

Theologis, A. and Laties, G. G. (1981) *Plant Physiol.* **68,** 53–58.

van Steveninck, R. F. M. and Jackman, M. E. (1967). *Austr. J. Biol. Sci.* **20,** 749–760.

Verleur, J. D. (1969). *Z. Pflanzenphysiol.* **61,** 299–309.

Wardale, D. A. and Galliard, T. (1975) *Phytochem.* **14,** 2323–2329.

Wardale, D. A. and Galliard, T. (1977). *Phytochem.* **16,** 333–338.

Waring, A. J. and Laties, G. G. (1977a). *Plant Physiol.* **60,** 5–10.

Waring, A. J. and Laties, G. G. (1977b). *Plant Physiol.* **60,** 11–16.

Weber, G., Convery, H. J. H., Lea, M. A., and Stamm, N. B. (1966). *Science* **154,** 1357–1366.

Wojtczak, L., Bogucka, K., Sarzala, M. G., and Zaluska, H. (1969). *FEBS Symp.* **17,** 79–92.

INDEX